VOLUME SEVENTY TWO

ADVANCES IN BOTANICAL RESEARCH

The Molecular Genetics of Floral Transition and Flower Development

ADVANCES IN BOTANICAL RESEARCH

Series Editors

VOLUME SEVENTY TWO

ADVANCES IN BOTANICAL RESEARCH

The Molecular Genetics of Floral Transition and Flower Development

Volume Editor

FABIO FORNARA

Department of Biosciences,
University of Milan,
Italy

ELSEVIER

AMSTERDAM • BOSTON • HEIDELBERG • LONDON
NEW YORK • OXFORD • PARIS • SAN DIEGO
SAN FRANCISCO • SINGAPORE • SYDNEY • TOKYO
Academic Press is an imprint of Elsevier

Academic Press is an imprint of Elsevier
32 Jamestown Road, London NW1 7BY, UK
Radarweg 29, PO Box 211, 1000 AE Amsterdam, The Netherlands
The Boulevard, Langford Lane, Kidlington, Oxford, OX5 1GB, UK
225 Wyman Street, Waltham, MA 02451, USA
525 B Street, Suite 1800, San Diego, CA 92101-4495, USA

First edition 2014

ISBN: 978-0-12-417162-6
ISSN: 0065-2296

For information on all Academic Press publications visit our website at store.elsevier.com

CONTENTS

CONTRIBUTORS

Suvi K. Broholm
Department of Agricultural Sciences, University of Helsinki, Helsinki, Finland

Chiara Campoli
Max Planck Institute for Plant Breeding Research, Cologne, Germany

Paula Elomaa
Department of Agricultural Sciences, University of Helsinki, Helsinki, Finland

Vinicius Costa Galvão
Center for Integrative Genomics, Faculty of Biology and Medicine, University of Lausanne, Lausanne, Switzerland

Greg S. Golembeski
Department of Biology, University of Washington, Seattle, WA, USA

Emmanuelle Graciet
Smurfit Institute of Genetics, Trinity College Dublin, Dublin, Ireland; Department of Biology, National University of Ireland, Maynooth, Ireland

Hiro-Yuki Hirano
Department of Biological Sciences, Graduate School of Science, The University of Tokyo, Tokyo, Japan

Young Hun Song
Department of Biology, University of Washington, Seattle, WA, USA

Takato Imaizumi
Department of Biology, University of Washington, Seattle, WA, USA

Takeshi Izawa
Functional Plant Research Unit, National Institute of Agrobiological Sciences, Tsukuba, Ibaraki, Japan

Hannah A. Kinmonth-Schultz
Department of Biology, University of Washington, Seattle, WA, USA

Junko Kyozuka
Graduate School of Agriculture and Life Sciences, University of Tokyo, Yayoi, Bunkyo, Tokyo, Japan

Diarmuid S. O'Maoileidigh
Smurfit Institute of Genetics, Trinity College Dublin, Dublin, Ireland

Asami Osugi
Functional Plant Research Unit, National Institute of Agrobiological Sciences, Tsukuba, Ibaraki, Japan

Sungrye Park
Department of Molecular Biosciences, Plant Biology Graduate Program and Institute for Cellular and Molecular Biology, The University of Texas, Austin, TX, USA

Youngjae Pyo
Department of Molecular Biosciences, Plant Biology Graduate Program and Institute for Cellular and Molecular Biology, The University of Texas, Austin, TX, USA

Markus Schmid
Max Planck Institute for Developmental Biology, Tuebingen, Germany

Sibum Sung
Department of Molecular Biosciences, Plant Biology Graduate Program and Institute for Cellular and Molecular Biology, The University of Texas, Austin, TX, USA

Wakana Tanaka
Department of Biological Sciences, Graduate School of Science, The University of Tokyo, Tokyo, Japan

Teemu H. Teeri
Department of Agricultural Sciences, University of Helsinki, Helsinki, Finland

Beth Thompson
Biology Department, East Carolina University, Greenville, NC, USA

Taiyo Toriba
Department of Biological Sciences, Graduate School of Science, The University of Tokyo, Tokyo, Japan

Maria von Korff
Max Planck Institute for Plant Breeding Research, Cologne, Germany; Institute of Plant Genetics, Heinrich Heine University, Düsseldorf, Germany; Cluster of Excellence on Plant Sciences "From Complex Traits towards Synthetic Modules", Düsseldorf, Germany

Frank Wellmer
Smurfit Institute of Genetics, Trinity College Dublin, Dublin, Ireland

Yanpeng Xi
Department of Molecular Biosciences, Plant Biology Graduate Program and Institute for Cellular and Molecular Biology, The University of Texas, Austin, TX, USA

PREFACE

During their life cycle, plants undergo developmental transitions that profoundly change growth patterns. Regulation of the activity of meristems (groups of undifferentiated cells giving rise to all plant organs) is crucial to determine the correct progression through transitions and establish plant architecture. Different plant species have evolved complex regulatory networks to control meristems' fate and activity.

Upon perception of favourable environmental conditions and endogenous signals, plants initiate flowering and vegetative meristems, producing leaves and shoots, become inflorescence meristems. This transition is referred as the vegetative-to-reproductive or floral transition and commits the plant to flower. The timing of this transition is critical, because inflorescences are delicate organs eventually producing seeds, and plants need to flower when external conditions are optimal for offspring survival.

The first part of this book (Chapters One to Five) is dedicated to the molecular mechanisms that plants have evolved and adopted to measure environmental and endogenous parameters such as day length, temperature and hormonal levels, and how such information promotes or inhibits flowering by affecting expression of regulatory genes. Chapter 1 (Photoperiodic flowering regulation in *Arabidopsis thaliana* by Golembeski et al.), reviews the photoperiodic flowering pathway in *Arabidopsis*, the most studied plant model system, that has been instrumental to isolate the key players regulating flowering and to formulate current models for seasonal time measurement. Central to the photoperiodic flowering pathway is *FLOWERING LOCUS T* (*FT*), recently identified as the *florigen* and shown to be conserved across diverse plant lineages. In Chapter Two, Pyo et al. describe how specific ecotypes of *Arabidopsis* require exposure to cold in order to flower, a process known as vernalisation (Regulation of flowering by vernalisation in *Arabidopsis*). Stable repression of a transcription factor, *FLOWERING LOCUS C* (*FLC*), is essential to establish competence to respond to photoperiodic induction. The genetic and epigenetic regulation of *FLC* is very complex and requires remodelling of chromatin at the *FLC* locus. How this is achieved by distinct types of regulatory molecules is thoroughly described. Not only seasonal cues, but also the levels of internal signalling molecules, such as hormones and sugars, affect flowering. The contribution by V. Costa Galvão and M. Schmid (Chapter Three, Regulation of flowering by endogenous signals) provides an overview of the role of plant hormones on flowering. In this chapter the role of sugars as key nodes in regulatory networks is discussed, providing an exciting perspective of the connection between metabolism and gene regulation.

Arabidopsis is an extremely useful model to address the basic mechanisms of flowering in plants. However, not all plant species adopted the same developmental strategies to flower. In Chapter Four (Critical gates in day-length recognition to control the photoperiodic flowering), A. Osugi and T. Izawa describe how rice responds to changes in day length, flowering as days become shorter. The use of rice as model system has allowed the identification of novel regulatory mechanisms controlling photoperiodic flowering responses, clearly indicating that some molecular components specifically evolved in the monocot lineage and are not shared by dicot species. Further developing on monocot diversity, C. Campoli and M. vKorff present an overview of the pathways controlling flowering in temperate cereals, including wheat and barley (Chapter Five, Genetic control of reproductive development in temperate cereals). As opposed to rice, where no vernalisation pathway has evolved because of its tropical origins, flowering of temperate cereals is accelerated by exposure to low temperatures. Natural genetic variation at loci controlling flowering responses to photoperiod and low temperatures has been exploited by breeders to produce varieties better adapted to diverse cultivation areas.

Once committed to flower, inflorescence meristems produce branches and ultimately floral meristems that give rise to floral organs. Specification of distinct structures on the inflorescence main axis generates diverse architectures that constitute the focus of Chapters Six to Ten. D.S. O'Maoileidigh and colleagues set the stage in Chapter Six (Genetic control of *Arabidopsis* flower development), by describing how flowers are formed in *Arabidopsis* and how molecular cloning of regulatory genes from this species laid the foundation of models of flower development, largely applicable to many plant species. In Chapter Seven, J. Kyozuka describes the development of grass inflorescences (Grass inflorescence: basic structure and diversity), whose remarkable and distinctive characteristic is that they form spikelets, which are short and modified flowering branches. Rice flower development is the focus of Chapter Eight (Flower development in rice) authored by W. Tanaka et al. The ABC model of flower development, i.e. the basic molecular plan that instructs cells to form a flower, is largely conserved in rice. However, not all floral structures are shared between monocots and dicots, implying the evolution of regulatory mechanisms to establish the identity of novel organ types. In Chapter Nine (Genetic and hormonal regulation of maize inflorescence development), B. Thompson expands the discussion on grass inflorescence development, focusing on maize. Maize is a monoecious species, in which male and female flowers are produced on distinct inflorescence types formed on the same plant, providing a beautiful example of how some species have established regulatory mechanisms for sex specification. This chapter drives the reader through maize flower development, ultimately focusing on how hormonal pathways affect establishment of male or female identity. Flower shapes and colours are countless and it would be impossible to describe them all in one single book. However, the concluding chapter (Chapter Ten, Molecular Control of Inflorescence Development in Asteraceae) by Broholm and colleagues, addresses flower development in *Asteraceae*, a family characterized by producing a showy inflorescence called *capitulum* that is formed by the specific arrangement of different flower types. The beauty of such structures allows us to have a glimpse of Nature's endless work in shaping plant forms and to appreciate the sophisticated mechanisms that generate it.

Fabio Fornara

CHAPTER ONE

Photoperiodic Flowering Regulation in *Arabidopsis thaliana*

Greg S. Golembeski, Hannah A. Kinmonth-Schultz, Young Hun Song and Takato Imaizumi[1]
Department of Biology, University of Washington, Seattle, WA, USA
[1]Corresponding author: e-mail address: takato@u.washington.edu

Contents

Abstract

Photoperiod, or the duration of light in a given day, is an important cue that flowering plants utilise to effectively assess seasonal information and coordinate their reproductive development in synchrony with the external environment. The use of the model plant, *Arabidopsis thaliana*, has greatly improved our understanding of the molecular mechanisms that determine how plants process and utilise photoperiodic information to coordinate a flowering response. This mechanism is typified by the transcriptional activation of *FLOWERING LOCUS T (FT)* by the transcription factor CONSTANS (CO) under inductive long-day conditions in *Arabidopsis*. FT protein then moves from the leaves to the shoot apex, where floral meristem development can be initiated. As a point of integration from a variety of environmental factors in the context of a larger system of regulatory pathways that affect flowering, the importance of photoreceptors

Advances in Botanical Research, Volume 72
ISSN 0065-2296
http://dx.doi.org/10.1016/B978-0-12-417162-6.00001-8

and the circadian clock on CO regulation throughout the day is a key feature of the photoperiodic flowering pathway. In addition to these established mechanisms, the recent discovery of a photosynthate derivative trehalose-6-phosphate as an activator of *FT* in leaves has interesting implications for the involvement of photosynthesis in the photoperiodic flowering response.

1.1 INTRODUCTION

Seasonal variation in climate has selected for the ability of organisms to predict future environmental conditions and use this information to complete necessary adjustments to thrive. The tilt of the earth's axis relative to the sun throughout the solar year can lead to radical changes in weather patterns and temperature, especially in non-equatorial regions (Thomas & Vince-Prue, 1996). Survival often depends on the development of strategies to cope with sub-optimal conditions and the use of optimal ones as fully as possible. Precise timing of key events in the span of a life cycle is necessary for organisms faced with a seasonally shifting environment. The timing of the reproductive cycle is a good example of this phenomenon, as in a substandard environment, premature flowering can have severe implications for relative fitness. For plants dependent on pollinators for reproduction, flowering also must to be timed with the seasonal availability of other organisms (Hegland, Nielsen, Lázaro, Bjerknes, & Totland, 2009). As an irreversible process in most species, the timing of the reproductive transition in plants is especially critical (Kobayashi & Weigel, 2007).

The topic of how plants are able to recognise what constitutes optimal conditions for flowering has been an active area of research for almost a century. The United States Department of Agriculture researchers Wightman Garner and Henry Allard were the first to empirically describe that the duration of light in a 24-h period is a key cue for the induction of flowering in many plant species (Garner & Allard, 1920). Originally interested in explaining why soybeans planted sequentially over the summer decreased in days to flower as they were planted later in the season, they sought to find the casual variable behind the phenomenon. Over the course of 2 years from 1918 to 1920, they experimentally manipulated exposure of plants to light and dark cycles by moving plants from a common outdoor plot into darkened sheds. Through the careful control of light and dark duration to simulate different seasonal light conditions, they were able to determine critical durations of light or darkness that are required for induction of flowering in over 12 plant species and many different cultivars. To describe this general principle of an exhibited response triggered by a change in day length, they coined the term 'photoperiodism' (Garner & Allard, 1920). This revolutionary idea changed the thinking about seasonal responses by suggesting that the mechanism for sensing seasonal changes could be tied specifically to the sensing of duration of light in a given day. In addition, they found that plants could be classified into three different groups by their flowering response. Some plants flower as day length increases in late spring (long-day plants), some flower as day length wanes as autumn begins (short-day plants) and some plants flower at certain times regardless of the photoperiod (day-neutral plants) (Garner & Allard, 1920).

The determination of day length as a critical regulator of flowering time left several questions with regard to the physiology of the flowering response. Where is day length sensed in the plant and how is the signal for floral induction carried throughout the organism? Elegant grafting experiments performed first by the Russian physiologist Mikhail Chailakhyan determined that a mobile signal from leaf scions exposed to inductive photoperiods could induce flowering in non-induced graft stocks (Chailakhyan, 1937; Chailakhyan, 1968). Experimental evidence suggested that the transmissible signal could be universal or nearly universal among flowering plants. For instance, grafts in which leaves from induced short-day *Kalanchoë blossfeldiana* and long-day *Sedum spectabile* plants were able to induce flowering when grafted reciprocally with each other, suggesting that the flowering signal was common between long-day and short-day plants (Wellensiek, 1967; Zeevaart, 2006). Grafts between different species were also often found to lead to flowering induction (Zeevaart, 1976, 2006). These and other observations led Denis Carr and Lloyd Evans to eventually propose a model for two-step floral induction (Carr, 1967; Evans, 1971). The first stimulus would be involved in the sensing of photoperiod and the incorporation of other endogenous and environmental factors, and subsequently induce the secondary stimulus that was potentially universal and transmitted from the leaf.

The search for the chemical basis of florigen remained elusive and gradually fell out of favour until contributions from *Arabidopsis* facilitated the discovery of FLOWERING LOCUS T (FT) protein as a key candidate. The discovery of FT as a mobile signal in *Arabidopsis* along with recognition that its function is conserved in a range of distantly related plant species (Corbesier et al., 2007), has cemented the role of FT as a universal florigen (Abe et al., 2005; Kobayashi & Weigel, 2007; Kojima et al., 2002; Tamaki, Matsuo, Wong, Yokoi, & Shimamoto, 2007; Wigge et al., 2005). Increasingly, as our understanding of the photoperiodic sensing mechanism has expanded, we have found that similar regulatory networks govern flowering plant species other than *Arabidopsis*, and that the mechanism of photoperiodic flowering induction is highly conserved (Song, Ito, & Imaizumi, 2010).

In this chapter, we will review developments in understanding the molecular mechanism of the photoperiodic flowering response through the model organism *Arabidopsis thaliana*, and discuss recent discoveries highlighting the modulation of the photoperiodic sensing mechanism used to accommodate both external environmental factors such as light quality through the action of photoreceptor proteins as well as internal physiological status through the sensing of photosynthetic accumulation.

1.2 PHOTOPERIODIC FLOWERING AND THE EXTERNAL COINCIDENCE MODEL

The key question that emerged with the discovery of photoperiodic flowering responses was the mechanism for how photoperiod was sensed. Since the early eighteenth century with the experiments of the astronomer De Mairan, plants have been known to have oscillatory leaf movements that occur in 24-h cycles even in the absence of light, as if a light stimulus was present (De Mairan, 1729). These rhythms, which show a period of around 24h (hence circadian), are indicative of an inherited entrainment to the rotation of the earth that persists even after many generations of exposure to alternative day lengths in the laboratory (Bünning, 1960). This internal 'clock' has extreme selective value through the regulation of internal biochemical processes of the cell and the organism throughout the day, which we can now appreciate given the advances in molecular biology in the last few decades (Baudry & Kay, 2008). The connection between the internal clock and photoperiodic responses, however, was not immediately clear. First proposed by Erwin Bünning in 1936, and later refined by Colin Pittendrigh, the 'external coincidence' model proposed that photoperiodic phenomena could be explained by the interaction of light stimuli and the clock (Bünning, 1936; Pittendrigh & Minis, 1964). The clock would set the pace of the 24-h rhythm, and define a period of photosensitivity to which light exposure would induce a photoperiodic response (Pittendrigh, 1972). In non-inductive photoperiods, the presence of darkness during the photosensitive period of the circadian cycle would result in no elicited reaction. In contrast, the encroachment of light into the photosensitive period during longer inductive photoperiods would cause a physiological response (Figure 1.1).

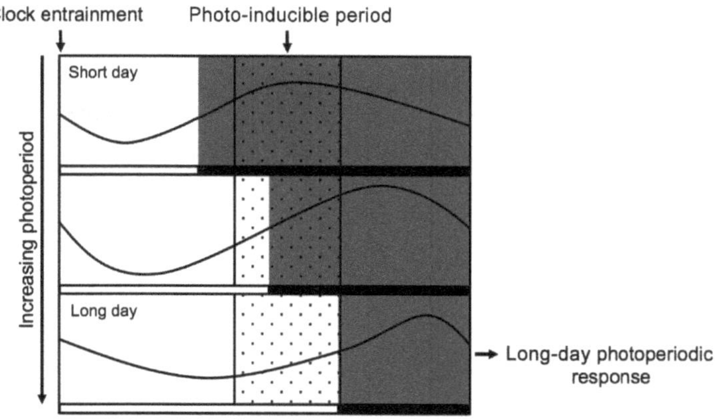

Figure 1.1 The external coincidence model for photoperiodic phenomena. The following example represents a photoperiodic response that occurs in the afternoon of long days, as in photoperiodic flowering in *Arabidopsis*. The circadian clock generates a rhythm that determines a specific period of the day in which a light signal can induce the response. This period is similar regardless of day length. In short-day conditions the photo-inducible period does not coincide with a light signal, so no response occurs. As days lengthen with the coming of spring and summer, light begins to encroach on the photo-inducible period, eliciting the photoperiodic response. Light serves a dual purpose: to reset the clock at dawn and dusk and to be present or absent during the photo-inducible phase, to promote or halt the response.

For more than 30 years, it remained controversial that the endogenous circadian clock regulated the photoperiodic flowering response. Key experiments that unequivocally linked flowering to the clock were performed by Murray Coulter and Karl Hamner on the short-day plant *Glycine max* in 1964, by giving light pulses at different time points after transfer of plants into continuous darkness. One of the prevailing counter-hypotheses of the time posited that night duration was the primary cue for the photoperiodic response, and that this was mediated by the turnover kinetics of the photoreceptor phytochrome. According to this hypothesis, for short-day plants, in which photoperiods below a certain threshold are inductive, directing light pulses at different times of night should affect the photoperiodic flowering response equally as long as a certain night length was prevented. It was found, instead, that light pulses during the night (referred to as night breaks) affected the flowering response in a rhythmic fashion (Carpenter & Hamner, 1964; Coulter & Hamner, 1964). Additional experiments performed by Halaban in 1968 in the short-day plant *Coleus frederici* showed that the phases in which flowering was inhibited by night break pulses always correlated with leaf movement position rather than the duration of night (Halaban, 1968a, 1968b). This was true for plants placed under several different photoperiods. These early findings helped to cement the clock as a crucial component in determining photoperiodic flowering responses.

1.2.1 Genetics of Photoperiodic Flowering in *Arabidopsis*

Most *Arabidopsis* accessions that were initially collected for use in the laboratory belong to the summer annual class of wild *Arabidopsis*, mainly due to the ease of flowering without vernalisation treatment and compact stature. Interestingly, some of the earliest mutations described in *Arabidopsis* are part of the regulatory framework that determines the photoperiodic flowering response, as mutations in these genes often convert compact summer annual accessions into phenotypes with long vegetative phases of growth. Mutagenic screens performed by Gyorgy Rédei in 1962 isolated *GIGANTEA (GI)* and *CONSTANS (CO)* as supervital mutants, far earlier than the forward genetic screens that would later more clearly define the regulatory networks that govern the flowering response (Rédei, 1962).

The advent of molecular markers in *Arabidopsis* in the late 1980s by Maarten Koorneef and colleagues enabled the systematic categorisation of genes involved in the regulation of flowering time and mapping of their associated loci. Initial genetics of late flowering mutants of *Arabidopsis* found that *CO, GI* and *FT* were likely components of the same regulatory pathway (Koornneef, Hanhart, & van der Veen, 1991).

1.2.2 CO–FT Module in *Arabidopsis*

The *co* and *gi* mutant phenotype initially interested researchers studying the genetic basis of flowering time because these mutants exhibited a 'day-neutral' phenotype (Park et al., 1999; Putterill, Robson, Lee, Simon, & Coupland, 1995). Under inductive long-day conditions, they flowered much later than wild type plants, but flowered at about the same time as wild type under non-inductive short-day conditions. Additional phenotypic analyses led to the conclusion that *CO* is a limiting factor for flowering under short-day conditions and that *CO* can promote flowering in a dose-dependent manner under inductive photoperiods (Putterill et al., 1995). Transgenic analysis of plants expressing *CO* under a dexamethasone inducible construct found that plants could be induced to flower regardless of the external photoperiod when *CO* is highly expressed (Simon, Igeno, & Coupland, 1996). Generation of mutants involved in the regulation of the circadian clock and light signalling also commonly affected the photoperiodic flowering response. Mutations in *LATE ELONGATED HYPOCOTYL* (*LHY*),

CIRCADIAN CLOCK ASSOCIATED 1 (CCA1), EARLY FLOWERING 3 (ELF3), TIMING OF CAB EXPRESSION 1 (TOC1), FLAVIN-BINDING, KELCH REPEAT, F-BOX 1 (FKF1), PSEUDO-RESPONSE REGULATOR 5 (PRR5), PRR7, PRR9, and *CRYPTOCHROME 2 (CRY2)* all displayed aberrant flowering phenotypes, which suggested that the clock on a molecular level was key to the proper induction of a photoperiodic response (El-Din El-Assal, Alonso-Blanco, Peeters, Raz, & Koornneef, 2001; Hicks et al., 1996; Ito et al., 2008; Nelson, Lasswell, Rogg, Cohen, & Bartel, 2000; Park et al., 1999; Sato, Nakamichi, Yamashino, & Mizuno, 2002; Schaffer et al., 1998; Somers, Schultz, Milnamow, & Kay, 2000). *CO* mRNA abundance was found to show a pronounced circadian oscillation under long-day conditions, and was found to continue to occur after plants entrained to long-day conditions were transferred to continuous light (Yanovsky & Kay, 2002). This suggested that the circadian clock regulated *CO* transcription. Additionally, the *CO* transcriptional pattern was significantly affected by mutations in clock components such as *toc1-1,* resulting in early flowering in short days. *toc1-1* mutants have a shortened circadian period of about 21 h. When light/dark cycles were artificially shortened to 21 h to compensate for the short-period defect in *toc1-1*, however, proper *CO* expression and function was restored. *CO* transcripts continue to oscillate under short-day conditions, but CO protein was initially shown to be highly unstable and actively degraded in the dark (Valverde et al., 2004; Yanovsky & Kay, 2002). This discrepancy between transcript abundance and protein stability explains how the constriction of active CO protein to the afternoon of long days enables a photoperiodic response, and fits nicely with our understanding of the external coincidence model in reference to photoperiodic phenomena (Figure 1.1). Coupled with experimental evidence that CO was a transcriptional activator of *FT* (Kardailsky et al., 1999; Kobayashi, Kaya, Goto, Iwabuchi, & Araki, 1999; Onouchi, Igeno, Perilleux, Graves, & Coupland, 2000; Samach et al., 2000) and that *FT* was directly involved in signalling the activation of floral meristem differentiation, a CO–FT module in which clock- and light-regulated CO would perceive photoperiodic information and signal for the induction of downstream flowering responses through the activation of *FT* transcription began to take shape.

Thus, in-line with earlier experimental data from the 1960s and 1970s, molecular evidence suggested that the circadian clock could regulate the photoperiodic response through the transcriptional and post-translational regulation of CO, and that this could lead to flowering under inductive conditions.

Since this discovery, the regulation of photoperiodic flowering pathway has become increasingly complex, and many factors have been shown to regulate CO and FT through a variety of mechanisms (Andrés & Coupland, 2012).

1.3 CURRENT MOLECULAR MECHANISM OF PHOTOPERIODIC FLOWERING IN *ARABIDOPSIS*

In *Arabidopsis*, long days promote flowering through the function of FT protein (Andrés & Coupland, 2012; Wigge, 2011). The protein, a mobile florigen, is synthesised in phloem companion cells of leaves and translocated to the shoot apical meristem, where the floral primordia are formed (Corbesier et al., 2007). The timing of flowering is strongly correlated with the relative amount of *FT*; the high levels of *FT* transcript under longer photoperiods induce more rapid flowering compared with the low levels typical of shorter photoperiods (Kobayashi et al., 1999). The transcriptional activator CO directly induces the expression of *FT* gene in a day-length-dependent manner (Samach et al., 2000; Tiwari et al., 2010). *CO* gene expression is controlled by the circadian clock (Suárez-López et al., 2001), and CO protein abundance is modulated by light signalling, allowing CO protein to be stabilised in the afternoon of long days (Song, Smith, To, Millar, & Imaizumi, 2012b; Valverde et al., 2004). Together, these processes explain how day length is measured and how the floral transition is mediated under inductive photoperiod.

1.3.1 Regulation of *CO* Transcription

To accurately control the timing of seasonal flowering in *Arabidopsis*, the circadian clock-regulated *CO* expression is a crucial mechanism to precisely measure the difference in day length. *CO* transcription is controlled by many circadian clock proteins, such as CCA1, LHY and PRRs (Imaizumi, 2010). These clock proteins directly or indirectly regulate the gene expression of *CYCLING DOF FACTORs* (*CDFs*), transcriptional repressors of *CO* (Song et al., 2010). *CDF1* directly binds to the *CO* promoter and represses its transcription in the morning redundantly with other CDF proteins, CDF2, CDF3 and CDF5 (Fornara et al., 2009; Imaizumi, Schultz, Harmon, Ho, & Kay, 2005; Sawa, Nusinow, Kay, & Imaizumi, 2007). The expression level of *CDF1* gene is positively regulated by CCA1 and LHY proteins (Nakamichi et al., 2007), which are most abundant at dawn (Schaffer et al., 1998; Wang & Tobin, 1998). Consequently, the expression level of *CDF1* transcript remains high during the morning (Imaizumi et al., 2005).

In the afternoon, the abundance of *CDF* transcripts is reduced through the function of four PRR family members, TOC1, PRR5, PRR7 and PRR9. These PRR proteins physically associate with the *CCA1* and *LHY* loci and repress *CCA1* and *LHY* gene expression (Huang et al., 2012; Nakamichi et al., 2010). TOC1, PRR5, PRR7 and PRR9 proteins also negatively regulate the expression of *CDF1* gene (Ito et al., 2008; Nakamichi et al., 2007). In addition, PRR5 and PRR7 directly bind to the *CDF2* and *CDF5* loci to repress their transcription (Liu, Carlsson, Takeuchi, Newton, & Farré, 2013; Nakamichi et al., 2012). *CDF2*, *CDF3* and *CDF5* transcripts are also high in the morning, similar to *CDF1* (Fornara et al., 2009). Clock regulation of *CDF* expression, which keeps *CO* expression low in the morning, lays the groundwork for determining the photosensitive period later in the afternoon of long days, preventing early flowering under shorter photoperiods.

In long days, the repression of *CO* gene expression by CDF proteins is released through the function of FKF1–GI complex in the afternoon (Sawa et al., 2007). FKF1 protein is a blue light photoreceptor (Imaizumi, Tran, Swartz, Briggs, & Kay, 2003; Sawa et al., 2007) and possesses E3 ubiquitin ligase activity that mediates proteasome-dependent degradation of target proteins (Imaizumi et al., 2005). Once the expression patterns of FKF1 and GI proteins coincide with light in the afternoon, FKF1 absorbs blue light and is activated. Then, the blue light-activated FKF1 forms a protein complex with GI. The FKF1–GI complex recognises *CO* repressors, the CDF proteins, and removes those repressors by ubiquitin-dependent degradation on the *CO* promoter (Sawa et al., 2007). FKF1 homologues, ZEITLUPE (ZTL) and LOV KELCH PROTEIN 2 (LKP2) proteins, both of which interact with FKF1 and GI proteins, are also involved in the destabilisation of CDF2 protein (Fornara et al., 2009). Removal of CDF proteins through the function of FKF1 protein constricts the action of CDF repressors to the morning of long days and facilitates the expression of the *CO* gene during the late afternoon, while light is still present (Figure 1.2). Maintaining this window of *CO* expression to the late afternoon allows for the subsequent peak of activation of *FT* at dusk during long days, enabling the photoperiodic flowering response.

In contrast to long days, the expression of FKF1 and GI proteins is out of phase in short-day conditions. Little functional complex between the proteins exists in the daytime under these conditions, which results in the accumulation of *CO* transcripts only during the dark period, which subsequently causes an extremely low level of *FT* expression throughout the day. Transcriptional regulation of *CO* gene expression thus is critical for sensing

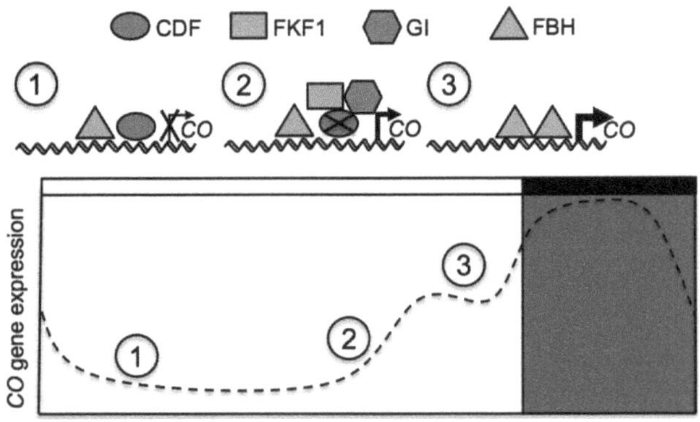

Figure 1.2 *CONSTANS (CO)* oscillatory transcription is dependent on multiple factors throughout the day. Under inductive long-day conditions, the peak of *CO* expression is constrained to the afternoon before dusk. In the morning, CYCLING DOF FACTOR (CDF) family transcription factors bind to the *CO* promoter to repress its transcription. Beginning in the afternoon, FLAVIN-BINDING, KELCH REPEAT, F-BOX 1 (FKF1) and GIGANTEA (GI) form a protein complex that ubiquitinates CDFs through FKF1 and targets them for proteasomal degradation, releasing the *CO* promoter from repression. FLOWERING BHLH (FBH) transcriptional activators are then recruited to the *CO* genomic locus, resulting in increased transcription of *CO* before dusk. Constraining *CO* mRNA expression to the late afternoon and stabilisation of resultant CO protein result in *FLOWERING LOCUS T* expression at dusk and promotion of flowering in long days. (See the colour plate.)

day length and differentiating between inductive and non-inductive photoperiods to coordinate the flowering response (Sawa et al., 2007).

Once *CO* repression by CDF proteins is relieved, four basic helix–loop–helix (bHLH) transcription factors, FLOWERING BHLH 1 (FBH1), FBH2, FBH3 and FBH4, activate *CO* expression (Ito et al., 2012). These FBH proteins directly bind to the E-box elements in the *CO* promoter and redundantly induce *CO* expression during the late afternoon and the dark under both long- and short-day conditions (Figure 1.2). It is proposed that FBH-mediated *CO* activation is conserved in other plant species, because overexpression of *FBH* homologue genes of rice and poplar highly upregulates *CO* transcripts in *Arabidopsis* (Ito et al., 2012). To date, our knowledge of transcriptional repression of *CO* is much more developed than its activation (Song, Ito, & Imaizumi, 2013), and more work needs to be done to determine additional factors involved as well as time-dependent impacts of *CO* activators on *CO* transcription.

1.3.2 Post-translational Regulation of CO Protein

Along with the transcriptional regulation of the *CO* gene, the post-translational regulation of CO protein is crucial for the day-length-dependent *FT*

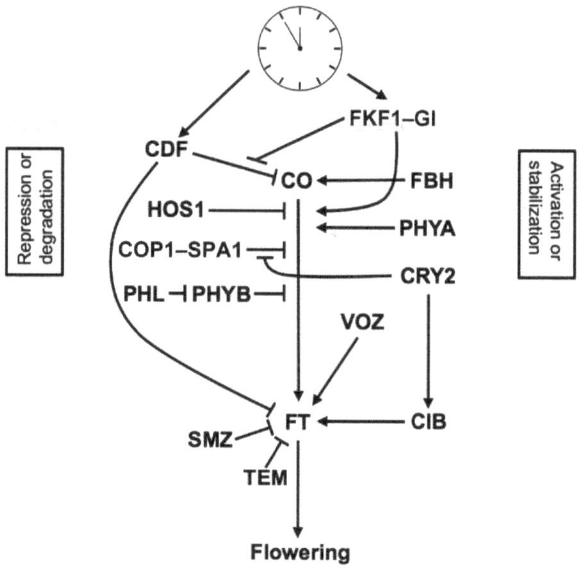

Figure 1.3 Flowering under inductive long days requires *FLOWERING LOCUS T (FT)* expression in the late afternoon. *CONSTANS (CO)* transcription, CO protein stability and *FT* transcription are critical to the photoperiodic flowering response. Blue light promotes flowering through FLAVIN-BINDING, KELCH REPEAT, F-BOX 1 (FKF1)-dependent degradation of CYCLING DOF FACTORs (CDFs) and stabilisation of CO protein, direct activation of *FT* through CRYPTOCHROME-INTERACTING BASIC HELIX–LOOP–HELIX (*CIB*) transcription factors, and stabilisation of the CONSTITUTIVE PHOTOMORPHOGENIC 1 (COP1)–SUPPRESSOR OF PHYA-105 1 (SPA1) complex by CRYPTOCHROME 2 (CRY2), which normally destabilises CO protein in the dark. PHYTOCHROME B (PHYB) inhibits flowering through destabilisation of CO protein under red light. Several factors reduce the inhibitory effect of PHYB on flowering. PHYTOCHROME-DEPENDENT LATE FLOWERING (PHL) may interfere with PHYB-dependent destabilisation of CO by sheltering CO protein. VASCULAR PLANT ONE ZINC-FINGER (VOZ) transcription factors activate *FT* expression, and their activity is regulated by PHYB. Far-red light promotes flowering through increased stability of CO protein by PHYA. Several other factors influence *FT* transcription directly. SCHLAFMÜTZE (SMZ), TEMPRANILLO 1 (TEM1) and related transcription factors are able to directly repress *FT* transcription. The promotion or inhibition of each respective component can affect the flowering output, and thus serves to integrate multiple environmental signals such as day length, light quality and temperature.

activation (Figure 1.3). In both long- and short-day conditions, the highest accumulation of *CO* mRNA occurs in the dark (Suárez-López et al., 2001). However, the expression of *FT* peaks at dusk in long days (Suárez-López et al., 2001). Various light signalling and proteasome-dependent protein degradation mechanisms have been shown to control CO protein stability and allow the protein to accumulate only in the late afternoon of long days, which accounts for the day-length-dependent expression of *FT* (Jang et al., 2008; Lazaro, Valverde, Pineiro, & Jarillo, 2012; Liu, Zhang, et al., 2008; Song, Smith, et al., 2012; Valverde et al., 2004). Red light delays flowering through the destabilisation of CO protein, and far-red and blue light promote flowering through the stabilisation of the protein (Valverde et al., 2004). PHYTOCHROME A (PHYA) and PHYB mediate far-red and red light responses, respectively, and CRY2 and FKF1 mediate blue light responses. Two RING finger E3 ubiquitin ligases, CONSTITUTIVE PHOTOMORPHOGENIC 1 (COP1) and HIGH EXPRESSION OF OSMOTICALLY RESPONSIVE GENES 1 (HOS1), negatively regulate CO protein stability (Jang et al., 2008; Lazaro et al., 2012; Liu, Zhang, et al., 2008).

CO protein is stable under far-red light and unstable under red light in wild type *Arabidopsis* plants. In addition, the amount of the protein is reduced in a *phyA* mutant background throughout the daytime and, by contrast, increased in a *phyB* mutant background especially in the morning (Valverde et al., 2004). In natural conditions, the ratio of red to far-red light is high during the daytime and relatively low at dusk. Reflecting this ratio, the levels of CO protein are reduced in the morning and increased in the late afternoon. This seems to indicate that PHYA and PHYB antagonistically modulate the stability of the protein.

Recent evidence has suggested that the PHYB dependent regulation of CO protein stability is quite complex, and may require several factors that both positively and negatively affect CO. Mutations in *PHYTOCHROME-DEPENDENT LATE FLOWERING (PHL)* cause a late flowering phenotype under long days, similar to other photoperiodic flowering pathway components (Endo, Tanigawa, Murakami, Araki, & Nagatani, 2013). Double mutant combinations with *phyB* abolish the late flowering phenotype, suggesting that *PHL* affects the ability of PHYB to repress flowering. *PHL* does not appear to regulate *CO* transcription, but CO protein and PHL interact (Endo et al., 2013). The PHL protein is thus a likely factor involved in sheltering CO from PHYB-dependent degradation (Endo et al., 2013). Similarly, it has been found that the VASCULAR PLANT ONE ZINC-FINGER1 (VOZ1) and VOZ2, two NAC (NAM, ATAF1/2 and CUC2) domain transcription factors, interact with PHYB, and positively regulate flowering under long days. Like PHYB, VOZ1 and VOZ2 are expressed in the cytoplasm and are translocated into the nucleus (Yasui et al., 2012). Their expression is also vascular specific, together with other photoperiodic flowering components (Yasui et al., 2012). The discovery of these factors adds a new layer of complexity with regard to PHYB regulation of photoperiodic flowering, but the exact mechanisms for how PHYB destabilises CO protein remain to be determined. How these PHYB-dependent positive regulators of flowering fit into the larger framework of antagonistic PHYB and PHYA signalling with regard to CO will have important implications for light quality dynamics and their impact on the photoperiodic response.

The HOS1 E3 ubiquitin ligase mediates degradation of CO protein in the morning by directly interacting with CO (Lazaro et al., 2012). Another E3 ubiquitin ligase COP1 forms a protein complex with SUPPRESSOR OF PHYA-105 1 (SPA1). The COP1–SPA1 complex binds to CO protein and degrades it during the night. Other SPA proteins, SPA2, SPA3 and SPA4, physically interact with CO protein and redundantly regulate its destabilisation (Jang et al., 2008; Laubinger et al., 2006; Saijo et al., 2003). CRY2 is also involved in CO stabilisation by forming a protein complex with SPA1 (Zuo, Liu, Liu, Liu, & Lin, 2011). The binding of photoactivated CRY2 to SPA1 enhances the interaction between CRY2 and COP1 in response to blue light, resulting in the suppression of COP1–SPA1 activity and in turn the accumulation of CO during the daytime (Zuo et al., 2011). This function of CRY2 partially explains how blue light accelerates flowering through the stabilisation of CO protein and induction of *FT* transcripts.

While we have a good idea of which factors contribute to CO protein stabilisation and destabilisation, the relationship between these factors throughout the day and how they compete or interact dynamically for CO protein needs to be further clarified. As has been discussed, three photoreceptors, PHYA, PHYB and CRY2 and two E3 ubiquitin ligases, HOS1 and COP1, regulate CO protein stability. However, functions of the photoreceptors cannot fully account for the question about how CO protein is stabilised only at the end of the day under long-day conditions because these photoreceptors are constitutively expressed throughout the day (Mockler et al., 2003). The function of another blue light photoreceptor, FKF1, provides a clue to answer the question. FKF1 protein physically interacts with CO protein in a blue light-enhanced manner, and the FKF1–CO interaction increases CO stability at a specific time of day, in the afternoon, under long-day conditions (Song, Smith, et al., 2012). Together with the similar expression profile of these proteins (Imaizumi et al., 2003; Valverde et al., 2004), the blue light-enhanced FKF1–CO interaction supports the notion that FKF1 determines both the timing of CO stabilisation, and the timing of *CO* expression during the light phase under long-day conditions that is crucial for *FT* induction. Thus a model emerges in which both gene expression and the protein accumulation of CO are regulated by FKF1 function. As the core clock components CCA1 and LHY regulate the timing of *FKF1* (Imaizumi et al., 2003), the circadian regulation of the FKF1 photoreceptor function is likely the molecular basis of the photosensitive phase proposed in the external coincidence model in *Arabidopsis*.

1.3.3 Transcriptional Regulation of *FT* Gene

The photoperiodic flowering pathway serves as a conduit for a large variety of environmental parameters that convert external information and integrate it into regulation of *FT* expression. These environmental signals merge to control *FT* expression through several transcription factors (Figure 1.3) (Song et al., 2010; Song et al., 2013). Several classes of transcriptional repressors control *FT* gene expression. SCHLAFMÜTZE (SMZ) gene encodes an APETALA2 (AP2)-related transcription factor that binds downstream of the *FT* locus and represses *FT* transcription (Mathieu, Yant, Mürdter, Küttner, & Schmid, 2009), and the expression of the gene is negatively regulated by GI function mediated through a microRNA pathway (Jung et al., 2007). GI protein positively regulates microRNA172 (miR172) accumulation under long days. miR172 targets *SMZ* transcripts reducing their abundance (Mathieu et al., 2009). TEMPRANILLO 1 (TEM1) protein directly associates with the 5′-UTR (untranslated region) of *FT* gene and represses the gene expression throughout the day under long-day conditions. TEM1 is involved in the regulation of *FT* expression redundantly with TEM2 (Castillejo & Pelaz, 2008). In addition, GI protein interacts with TEM1 and TEM2 in the nucleus in tobacco cells and probably changes the activities of TEM proteins (Sawa & Kay, 2011).

Interestingly, the *CO* transcriptional regulator CDF1 also associates with the *FT* promoter near the transcriptional start site and represses *FT* transcription in the morning (Song, Smith, et al., 2012). Other CDF proteins (CDF2, CDF3 and CDF5) also likely regulate *FT* gene expression. The repression of *FT* transcription by CDF1 is released by the function of FKF1–GI complex on the *FT* promoter in the afternoon (Song, Smith, et al., 2012), concomitantly with the removal of CDF1 repression on the *CO* promoter. Together with CO protein stabilisation, these observations suggest that FKF1 protein controls *FT* induction through a multiple-feed forward motif, which allows strong activation of flowering signals in long-day conditions.

In the activation of *FT* transcription, two classes of transcription factors play major roles. As previously discussed, CO, a member of the B-box transcription factor family, acts as a strong direct activator of *FT* expression (Putterill et al., 1995; Robson et al., 2001; Tiwari et al., 2010). CO protein contains two functional motifs; two B-box domains at the N-terminus and the CCT (CO, CO-like and TOC1) domain at the C-terminus (Robson et al., 2001). The CO protein associates with the *FT* promoter and activates *FT* gene expression through two modes of action (Song, Lee, Lee, Imaizumi, & Hong, 2012; Song, Smith, et al., 2012; Tiwari et al., 2010; Wenkel et al., 2006); (1) by directly binding to the CO responsive element via the CCT motif (Tiwari et al., 2010), and (2) by recruitment of the CCAAT box-binding proteins including selected subunits of Nuclear Factor-Y and ASYMMETRIC LEAVES 1 that both physically interact with CO protein (Song, Lee, et al., 2012; Wenkel et al., 2006). *FT* induction is largely CO-dependent, as the relative abundance of *FT* transcripts greatly increases when *CO* expression is constitutive, regardless of day length (Valverde et al., 2004). Members of a transcription factor family characterised by a bHLH domain, including CRYPTOCHROME-INTERACTING BASIC HELIX–LOOP–HELIX 1 (CIB1), CIB2, CIB4 and CIB5, are involved in *FT* induction (Liu, Yu, et al., 2008; Liu, Li, Li, Liu, & Lin, 2013). CIB1 protein forms a complex with CRY2 protein in a blue light-dependent manner and acts as a *FT* activator by directly binding to the *FT* promoter (Liu, Yu, et al., 2008). The blue light-dependent CIB1 accumulation is positively regulated by ZTL and LKP2, but not by FKF1 (Liu, Wang, et al., 2013). All other CIB proteins also interact with CRY2 in vitro but only CIB2 and CIB5 form complexes with CRY2 in vivo (Liu, Li, et al., 2013). CIB proteins redundantly regulate *FT* transcription. CIB1 protein forms heterodimeric complexes with other CIBs, and the heterodimerisation increases the DNA-binding affinity of CIB1 protein to the specific *cis*-element in the *FT* promoter (Liu, Li, et al., 2013). As described above, blue light signalling plays a pivotal role in the regulation of *FT* induction through degradation of *FT* repressors and stabilisation of *FT* activators in *Arabidopsis*.

1.3.4 Movement of FT Protein

Where FT is synthesised differs from where it functions; therefore, understanding how FT moves is also of great interest. FT protein, once synthesised in phloem companion cells in the leaves, is loaded into the phloem and migrates towards its eventual destination at the shoot apex. Initial debate upon the discovery of FT as a primary component of the florigen occurred over whether the mobile signal was *FT* mRNA or FT protein. Multiple studies have since confirmed that the movement of FT protein explains the florigenic signal (Corbesier et al., 2007; Jaeger & Wigge, 2007; Mathieu, Warthmann, Küttner, & Schmid, 2007; Yoo, Hong, Jung, & Ahn, 2013). Grafting experiments in *Cucurbita moschata* in particular have proved a useful system for the study of FT movement. Reverse-transcription polymerase chain reactions (RT-PCR) and mass spectrometry analysis on phloem sap detected no *FT* transcript but observed FT protein (Lin et al., 2007). Cross-species grafting experiments using *C. moschata* and *Cucurbita maxima* also showed that FT peptides belonging to the induced scion were detected in the phloem sap, but not *FT* mRNA (Yoo, Chen, et al., 2013). Additional work in this system has given a picture in which FT movement is regulated in different ways as it moves. Mutations in *FT* that prevent movement into the shoot apex have been shown to have the capacity to move through the companion cell to sieve-tube element barrier. This is supported by evidence that protein size affects the ability of tagged FT to enter the phloem and that specific regions of FT protein are important for movement out of the phloem and into the shoot apex (Yoo, Chen, et al., 2013). This suggests a combination of FT movement by diffusion from the companion cell and into the phloem stream as well as a more active transport mechanism through plasmodesmata to move FT protein into the cells of the shoot apex (Yoo, Chen, et al., 2013). Several candidate proteins involved in interaction with or facilitated movement of FT have been identified, but their roles need to be further clarified and a more nuanced model for FT movement at each step needs to be elucidated (Liu et al., 2012; Yoo, Chen, et al., 2013).

Once FT reaches the shoot apex, a complex cascade of interactions occurs that leads to the activation of downstream developmental patterning genes, giving rise to floral meristem initiation. FT protein interacts with the bZIP (basic-leucine zipper) transcription factor FD and 14-3-3 to activate transcription of downstream floral targets such as *AP1* and *LEAFY* (Abe et al., 2005; Kardailsky et al., 1999; Taoka et al., 2011; Wigge, 2011). Modelling of the interactors at the shoot apex has shown that maintenance of steady state levels of FT and other interactors at the shoot apex are necessary to maintain and push the reprogramming of the vegetative meristem forward into the inflorescence meristem (Jaeger, Pullen, Lamzin, Morris, & Wigge, 2013). This mechanism is reminiscent of classical feed-forward genetic mechanisms found in *Drosophila* development (Thuringer & Bienz, 1993), and suggests that threshold levels of FT movement may be critical for the reproductive transition. It will be interesting to see experimentally the quantitative effects of FT protein on the floral transition. Classical grafting experiments have shown that cross-species grafts for floral induction can induce some partners but be insufficient for others, suggesting that threshold levels of FT may be different between species (Evans, 1971). Modelling interactions of FT and its downstream targets during the floral transition in other species may have interesting implications for the dynamics of the reproductive transition across evolutionary lines.

1.4 PHOTOSYNTHATES AS A COMPONENT OF THE PHOTOPERIODIC FLOWERING STIMULUS

Photosynthesis and photosynthetic assimilates have also been recognised in historical experiments to be involved in seasonal flowering, but determining the relationship between inductive photoperiods, the florigenic signal and the photosynthetic status of the plant could not easily be disentangled in the past and the connection is far from concrete in the present (see (Evans, 1971; Zeevaart, 1976) for review of historical work). Recent molecular genetics evidence suggests that photosynthetic components can act in leaves in a photoperiodic manner to contribute in tandem to the known photoperiodic signalling pathway. This new information sheds light on older experimental data demonstrating that photosynthetic status may alter the ability to respond to an optimum photoperiod in *Arabidopsis*.

1.4.1 Early Evidence for the Involvement of Photosynthesis in the Photoperiodic Flowering Response

Many experiments have been performed historically to determine the effect of changes in photosynthetic activity on the transition from vegetative to reproductive development. Although photoperiod remains a strict determinant of flowering in many species, the capacity of a plant to respond to an inductive photoperiodic signal can depend on other factors. Experiments that use the application of 3-(3,4-dichlorophenyl)-1,1-dimethylurea (DCMU), an inhibitor of photosynthesis, showed that flowering could be severely delayed in *Lolium temulentum*, a long-day grass (Evans & Wardlaw, 1966). However, DCMU seemed to have no effect on many short-day species, but not universally (Evans, 1971). Prolonged growth in elevated CO_2 coupled with inductive day lengths has been observed to accelerate flowering in several long-day species (Reekie, Hicklenton, &Reekie, 1994). In contrast to these results, however, experiments utilising albino *Arabidopsis* mutants grown on 1% glucose showed that flowering could still be induced, suggesting that carbon availability rather than photosynthesis influences the flowering response (Brown & Klein, 1971). Recently, DCMU treatment and removal of CO_2 have also been shown to influence the period of the circadian clock. Thus, it may be possible that part of the observed flowering time changes in response to photosynthetic inhibition could be through changes in the pace of the clock, in a manner similar to the delay or acceleration of flowering in several circadian clock mutant backgrounds (Haydon, Mielczarek, Robertson, Hubbard, & Webb, 2013; Yanovsky & Kay, 2002).

Although inhibition or increase in photosynthetic activity seemed to be involved in flowering induction, it was not clear where in the plant photosynthates were acting, nor was it clear whether they were acting through the same mechanism or separately from the floral stimulus. During the 1980s and 1990s, when the idea of a universal transmissible signal had fallen out of favor (Zeevaart, 2006), several studies demonstrated a marked increase in sucrose or glucose at the shoot apex of both long- and short-day species around the time of the flowering induction (Lejeune, Bernier, Requier, & Kinet, 1993; Milyaeva & Komarova, 1996; Mirolo, Bodson, & Bernier, 1985; Perilleux & Bernier, 1997). Yet, in *Sinapsis alba* during a single displaced short day (8 h of light at the end of a subjective 16-h day) and in *L. temulentum* after a single inductive long day, an appreciable mobilisation of carbohydrates to the shoot apex did not occur until after the floral stimulus left the leaf (Bodson, King, Evans, & Bernier, 1977; Perilleux & Bernier, 1997). This lead Bodson and colleagues to speculate whether photosynthates could lead to floral induction in the leaves rather than at the shoot apex.

Arabidopsis, like *S. alba*, can be induced to flower by exposure to a single long day or to a displaced short day. Mutations in *phosphoglycerate/bis phospho-glycerate mutase (pgm)* result in the inability to accumulate starch. *pgm* mutants show delayed floral induction and no increase in sucrose exported from the leaves in a single displaced short-day treatment compared to wild type plants or *pgm* exposed to one long day. This flowering repression of *pgm* in displaced short days, however, could be partially restored by application of sucrose at their apices (Corbesier, Lejeune, & Bernier, 1998). Laurent Corbesier and colleagues concluded that sufficient sucrose mobilisation from the leaves was needed for flowering induction, and that both a florigenic signal as well as a photosynthetic component was required for the proper photoperiodic flowering response.

1.4.2 Photosynthates Act in the Leaves to Promote Flowering

Recent work regarding trehalose-6-phosphate (T6P) has provided more detailed insight into the involvement of photosynthates during floral induction. The amount of T6P increases parallel to that of sucrose in the leaves

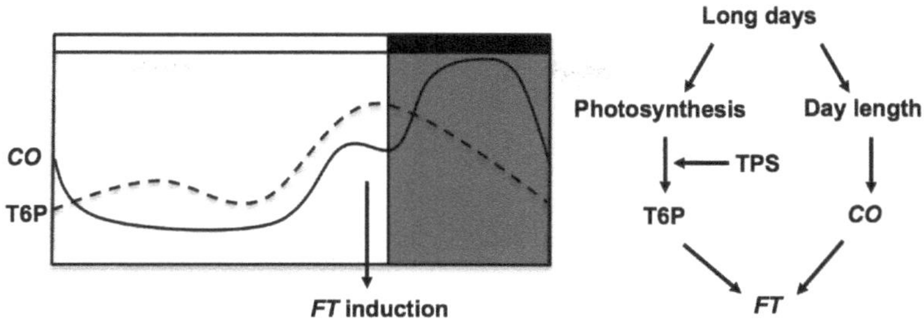

Figure 1.4 Trehalose-6-phospate (T6P) regulates *FLOWERING LOCUS T (FT)* expression under long days. T6P levels peak in the afternoon of long days in *Arabidopsis* and coincide with the afternoon peak of *FT* transcription. Loss of function of *trehalose-6-phosphate synthase 1 (tps1)*, the enzyme that produces T6P, results in a significant reduction in the dusk peak of *FT* expression. *CO* expression levels are unaltered by the *TPS1* mutation, suggesting that T6P regulation of *FT* in leaves occurs in a *CO*-independent manner. Inductive long days may thus produce a photoperiodic flowering response through *FT* via multiple regulatory pathways.

and its levels correlate to increasing starch synthesis (Ponnu, Wahl, & Schkid, 2011). It has been implicated as a signal for carbohydrate status in the plant; although new research has shown substantial starch accumulation cannot be induced by T6P alone (Martins et al., 2013). In *Arabidopsis*, the level of T6P increases at dusk similar to the transcription pattern of *FT* in long days (Imaizumi et al., 2003; Wahl et al., 2013). Loss of *trehalose-6-phosphate synthase 1* (*tps1*) markedly reduces the dusk peak of *FT* and delays flowering in long days. Together, this evidence suggests a link between photosynthetic assimilation, long-day induction of *FT*, and flowering. Expression of *CO* was only minimally altered in *tps1* mutants, suggesting that increase in *FT* transcripts is *CO* independent (Figure 1.4) (Wahl et al., 2013).

Experimental evidence from *S. alba*, which is closely related to *Arabidopsis*, has also shown that photosynthate production during phases of the day can influence the flowering response. High-intensity light provided by fluorescent lamps coupled with removal of CO_2 from the air failed to promote flowering when the treatment occurred during the first 8 h of a long-day cycle. However, flowering was strongly induced when the treatment occurred during the last 8 h of the daytime. To our knowledge, parallel work has not been done in *Arabidopsis*, however, removal of CO_2 throughout the entire day from *Arabidopsis* plants transferred from short days to long days resulted in a significant downregulation of *FT* transcription compared to controls grown under normal levels of CO_2 (King, Hisamatsu, Goldschmidt, & Blundell, 2008). If T6P indeed interacts with the photoperiodic pathway to induce flowering in leaves, a time-dependent sensitivity to photosynthate accumulation could explain how T6P promotes *FT* transcription only at dusk. It will be interesting to see if T6P levels are strongly reduced by removal of CO_2 early in the day; however, this remains to be tested.

King and colleagues demonstrated that high-intensity (270 μmol/m^2/s) fluorescent light, presumably increasing photosynthetic intake, led to relatively early flowering in short days compared to normal (100 μmol/m^2/s) light intensity in the *ft* mutants. This indicates that photosynthesis may be able to override the lack of *FT* signal under short-day conditions (King et al., 2008). Clarifying a possible mechanism, Wahl and colleagues found that unlike the *ft* mutant, a *TPS1* deficiency delayed flowering in *Arabidopsis* in short days as well as long days, indicating that T6P could interact with floral signals besides *FT* (Wahl et al., 2013). Further, loss of *TPS1* resulted in reduced expression of *SQUAMOSA PROMOTER BINDING PROTEIN-LIKE 3* (*SPL3*), *SPL4* and *SPL5* at the shoot apex. The SPL protein family is a known component of the age-dependent flowering pathway in *Arabidopsis*. Reduced *SPL* expression appeared to be accomplished partially through and partially independently of miR156, which delays the vegetative–reproductive phase transition. Mature miR156 was initially higher in *tps1* mutants compared to wild type, and although it declined to wild type levels over time, *SPL3*, *4* and *5* accumulated more slowly in *tps1* mutants (Wahl et al., 2013). Finally, *SUPPRESSOR OF OVEREXPRESSION OF CONSTANS 1* and *FRUITFUL* were not altered in the *tps1* mutants, although they have been implicated as inducing *FT* downstream of the SPL proteins in the leaves (Wahl et al., 2013). It seems that T6P acts to regulate *FT* in the leaves mainly independently of the age-dependent pathways, while it acts to induce flowering directly at the shoot apex in response to plant age (Samach et al., 2000; Teper-Bamnolker & Samach, 2005). Because of this, T6P probably occupies a role as a stimulus of flowering in both a photoperiodic and non-photoperiodic context based on tissue specificity.

The FT protein is still the primary component of the transmissible signal in the lengthening days of spring and summer in *Arabidopsis*. Now, it is becoming clear that photosynthetic by-products can lead to induction of *FT* at the leaf level, additively with the established photoperiod-sensing mechanism through *CO*. This synergy appears to be long-day specific. Exposure of plants to short-day conditions characterised by higher light intensities compared to long days has been reported (Yanovsky & Kay, 2002). Such method tends to normalise the amount of energy received by plants grown under different day lengths. Under such short-day conditions, *FT* induction does not occur (Yanovsky & Kay, 2002). Therefore, greater accumulation of photosynthates from high-intensity light in short days seems not to override the requirement of late-afternoon light for *FT* induction. Further, it appears that photosynthesis enhances the photoperiodic response, but cannot completely abrogate it, as *FT* did not decline to short-day levels when CO_2 was removed from the air (King et al., 2008). By what mechanism do photosynthesis and photosynthates interact with the photoperiodic pathway to induce *FT* and flowering? Because *FT* induction through T6P is likely *CO* independent, a heretofore-unknown factor or pathway must be involved in *FT* transcriptional regulation in response to photosynthetic accumulation. Clearly, more work to determine the effects of T6P on photoperiodic pathway components is needed.

Earlier studies into the mechanisms of photoperiodic flowering and photosynthetic involvement in the flowering response highlight interactions between age and photoperiod that we do not fully understand. Many early experiments that were able to induce flowering by a single inductive long day did so by first growing their plants in short days for several weeks (Corbesier et al., 1998; Evans & Wardlaw, 1966; King et al., 2008). It appears, therefore, that age or carbohydrate status increases the amount or reduces the threshold requirement of the floral stimulus, or both. The activity of T6P at the shoot apex, proposed as a fail-safe to ensure flowering will occur even in the absence of inductive conditions (Wahl et al., 2013), suggests one mechanism. The decline of miR156 in the leaves over time resulting in upregulation of *FT* suggests another (Srikanth & Schmid, 2011). Although T6P seems to act independently of FT at the shoot apex, it is possible these two pathways act in parallel to modify the plant's response. A better understanding of how age and the carbohydrate statuses of plants interact with photoperiodic induction either at the leaves or the shoot apex is critical to determine the threshold of FT necessary to promote flowering under different timescales and spatial contexts.

1.5 CONCLUSIONS

While our current understanding of the underlying mechanisms that confer a photoperiodic flowering response in *Arabidopsis* has now largely expanded, the number of factors that are involved in the process makes it a very intricate regulatory system. Circadian clock control of a variety of *CO* and *FT* regulators, light perception through photoreceptors as well as photosynthetic status through T6P can affect the photoperiodic response synergistically to promote flowering in *Arabidopsis* as days get longer in the springtime. Brought into the larger context of flowering time regulation, there are several players whose roles and domains may not be easily defined and whose outputs affect feedback within the system. One of the greater challenges in the future will be to understand how the plant is able to assimilate information regarding day length, light quality, temperature, precipitation, photosynthetic status, developmental age and other external and physiological characteristics, and to incorporate this information in a way that is meaningful towards timing the floral transition. To this end, systems-level approaches will be necessary to untangle the influence of so many factors on one output, and this will be critical if we are to understand how flowering time functions under natural conditions. At a surface level, we assume the large amount of redundancy, overlap and crosstalk within and among flowering pathway regulators must be necessary and of selective value in coordinating the flowering response; but is this the case? As our mechanistic knowledge of flowering improves, we should continue to look out among natural populations in *Arabidopsis* and other species to see whether what we presume is indeed the case. Can we see that these factors affect fitness as plants expand and contract across geographic ranges, climates and latitudes?

As detailed, work in *Arabidopsis* has established that the CO–FT module is critical for day-length sensing, and recent developments have confirmed the highly conserved nature of this mechanism for flowering and its co-option for other photoperiodic outputs across the angiosperm lineage (Böhlenius et al., 2006; Kloosterman et al., 2013; Song et al., 2010). While the limited information we have on other species points to this similarity, much work is needed to better characterise mechanisms of photoperiodic sensing in other plants. With the improved genomic and functional systems at our disposal, hopefully these will shed light on the commonalities and divergence of seasonal adaptation and how plants utilise that information to survive. Hopefully we can use that knowledge to better adapt the plants that we depend on to flourish in changing habitats.

ACKNOWLEDGEMENTS

This work was supported by a Pre-doctoral Developmental Biology Training Grant (5T32HD007183) from the National Institutes of Health to G.S.G., the National Science Foundation Graduate Research Fellowship Program to H.K.S., funding from the Next-Generation BioGreen 21 Program (SSAC, PJ009495) to Y.H.S., and the National Institutes of Health Grant (GM079712) to T.I.

REFERENCES

Abe, M., Kobayashi, Y., Yamamoto, S., Daimon, Y., Yamaguchi, A., Ikeda, Y., et al. (2005). FD, a bZIP protein mediating signals from the floral pathway integrator FT at the shoot apex. *Science, 309*, 1052–1056.

Andrés, F., & Coupland, G. (2012). The genetic basis of flowering responses to seasonal cues. *Nature Reviews Genetics, 13*, 627–639.

Baudry, A., & Kay, S. (2008). Clock control over plant gene expression. *Advances in Botanical Research, 48*, 69–105.

Bodson, M., King, R. W., Evans, L. T., & Bernier, G. (1977). The role of photosynthesis in flowering of the long-day plant *Sinapis alba*. *Functional Plant Biology, 4*, 467–478.

Böhlenius, H., Huang, T., Charbonnel-Campaa, L., Brunner, A. M., Jansson, S., Strauss, S. H., et al. (2006). *CO/FT* regulatory module controls timing of flowering and seasonal growth cessation in trees. *Science, 312*, 1040–1043.

Brown, J. A., & Klein, W. H. (1971). Photomorphogenesis in *Arabidopsis thaliana* (L.): threshold intensities and blue-far-red synergism in floral induction. *Plant Physiology, 47*, 393–399.

Bünning, E. (1936). Die endogene tagesrhythmik als grundlage der photoperiodischen reaktion. *Berichte der Deutschen Botanischen Gesellschaft, 54*, 590–607.

Bünning, E. (1960). Opening address – biological clocks. *Cold Spring Harbor Symposia on Quantitative Biology, 25*, 1–9.

Carpenter, B. H., & Hamner, K. C. (1964). The effect of dual perturbations on the rhythmic flowering response of *Biloxi* soybean. *Plant Physiology, 39*, 884–889.

Carr, D. J. (1967). Relationship between florigen and flower hormones. *Annals of the New York Academy of Sciences, 144*, 305–312.

Castillejo, C., & Pelaz, S. (2008). The balance between CONSTANS and TEMPRANILLO activities determines *FT* expression to trigger flowering. *Current Biology, 18*, 1338–1343.

Chailakhyan, M. K. (1937). Concerning the hormonal nature of plant development processes. *Doklady Akademii Nauk SSSR, 16*, 227–230.

Chailakhyan, M. K. (1968). Internal factors of plant flowering. *Annual Review of Plant Physiology, 19*, 1–36.

Corbesier, L., Lejeune, P., & Bernier, G. (1998). The role of carbohydrates in the induction of flowering in *Arabidopsis thaliana*: comparison between the wild type and a starchless mutant. *Planta, 206*, 131–137.

Corbesier, L., Vincent, C., Jang, S., Fornara, F., Fan, Q., Searle, I., et al. (2007). FT protein movement contributes to long-distance signaling in floral induction of *Arabidopsis*. *Science, 316*, 1030–1033.

Coulter, M. W., & Hamner, K. C. (1964). Photoperiodic flowering response of *Biloxi* soybean in 72-hour cycles. *Plant Physiology, 39*, 848–856.

De Mairan, J. J. (1729). Observation botanique. *Histoire de l'Academie Royale des Sciences*, 35–36.

El-Din El-Assal, S., Alonso-Blanco, C., Peeters, A. J., Raz, V., & Koornneef, M. (2001). A QTL for flowering time in *Arabidopsis* reveals a novel allele of *CRY2*. *Nature Genetics, 29*, 435–440.

Endo, M., Tanigawa, Y., Murakami, T., Araki, T., & Nagatani, A. (2013). PHYTOCHROME-DEPENDENT LATE-FLOWERING accelerates flowering through physical interactions with phytochrome B and CONSTANS. *Proceedings of the National Academy of Sciences of the United States of America, 110*, 18017–18022.

Evans, L. T. (1971). Flower induction and florigen concept. *Annual Review of Plant Physiology, 22*, 365–394.

Evans, L. T., & Wardlaw, I. F. (1966). Independent translocation of C_{14}-labelled assimilates and of the floral stimulus in *Lolium temulentum*. *Planta, 68*, 310–326.

Fornara, F., Panigrahi, K. C., Gissot, L., Sauerbrunn, N., Rühl, M., Jarillo, J. A., et al. (2009). *Arabidopsis* DOF transcription factors act redundantly to reduce *CONSTANS* expression and are essential for a photoperiodic flowering response. *Developmental Cell, 17*, 75–86.

Garner, W. W., & Allard, H. A. (1920). Effect of the relative length of day and night and other factors of the environment on growth and reproduction in plants. *Journal of Agricultural Research, 18*, 553–606.

Halaban, R. (1968a). Circadian rhythm of leaf movement of *Coleus blumei* × *C frederici* a short day plant: effects of light and temperature signals. *Plant Physiology, 43*, 1887–1893.

Halaban, R. (1968b). Flowering response of *Coleus* in relation to photoperiod and circadian rhythm of leaf movement. *Plant Physiology, 43*, 1894–1898.

Haydon, M. J., Mielczarek, O., Robertson, F. C., Hubbard, K. E., & Webb, A. A. (2013). Photosynthetic entrainment of the *Arabidopsis thaliana* circadian clock. *Nature, 502*, 689–692.

Hegland, S. J., Nielsen, A., Lázaro, A., Bjerknes, A.-L., & Totland, Ø. (2009). How does climate warming affect plant-pollinator interactions? *Ecology Letters, 12*, 184–195.

Hicks, K. A., Millar, A. J., Carre, I. A., Somers, D. E., Straume, M., Meeks-Wagner, D. R., et al. (1996). Conditional circadian dysfunction of the *Arabidopsis early-flowering 3* mutant. *Science, 274*, 790–792.

Huang, W., Perez-Garcia, P., Pokhilko, A., Millar, A. J., Antoshechkin, I., Riechmann, J. L., et al. (2012). Mapping the core of the *Arabidopsis* circadian clock defines the network structure of the oscillator. *Science, 336*, 75–79.

Imaizumi, T. (2010). *Arabidopsis* circadian clock and photoperiodism: time to think about location. *Current Opinion in Plant Biology, 13*, 83–89.

Imaizumi, T., Schultz, T. F., Harmon, F. G., Ho, L. A., & Kay, S. A. (2005). FKF1 F-box protein mediates cyclic degradation of a repressor of *CONSTANS* in *Arabidopsis*. *Science, 309*, 293–297.

Imaizumi, T., Tran, H. G., Swartz, T. E., Briggs, W. R., & Kay, S. A. (2003). FKF1 is essential for photoperiodic-specific light signalling in *Arabidopsis*. *Nature, 426*, 302–306.

Ito, S., Niwa, Y., Nakamichi, N., Kawamura, H., Yamashino, T., & Mizuno, T. (2008). Insight into missing genetic links between two evening-expressed pseudo-response regulator genes *TOC1* and *PRR5* in the circadian clock-controlled circuitry in *Arabidopsis thaliana*. *Plant and Cell Physiology, 49*, 201–213.

Ito, S., Song, Y. H., Josephson-Day, A. R., Miller, R. J., Breton, G., Olmstead, R. G., et al. (2012). FLOWERING BHLH transcriptional activators control expression of the photoperiodic flowering regulator *CONSTANS* in *Arabidopsis*. *Proceedings of the National Academy of Sciences of the United States of America, 109*, 3582–3587.

Jaeger, K. E., Pullen, N., Lamzin, S., Morris, R. J., & Wigge, P. A. (2013). Interlocking feedback loops govern the dynamic behavior of the floral transition in *Arabidopsis*. *The Plant Cell, 25*, 820–833.

Jaeger, K. E., & Wigge, P. A. (2007). FT Protein acts as a long-range signal in *Arabidopsis*. *Current Biology, 17*, 1050–1054.

Jang, S., Marchal, V., Panigrahi, K. C., Wenkel, S., Soppe, W., Deng, X. W., et al. (2008). *Arabidopsis* COP1 shapes the temporal pattern of CO accumulation conferring a photoperiodic flowering response. *The EMBO Journal, 27*, 1277–1288.

Jung, J. H., Seo, Y. H., Seo, P. J., Reyes, J. L., Yun, J., Chua, N. H., et al. (2007). The *GIGANTEA*-regulated microRNA172 mediates photoperiodic flowering independent of *CONSTANS* in *Arabidopsis*. *The Plant Cell, 19*, 2736–2748.

Kardailsky, I., Shukla, V. K., Ahn, J. H., Dagenais, N., Christensen, S. K., Nguyen, J. T., et al. (1999). Activation tagging of the floral inducer FT. *Science, 286*, 1962–1965.

King, R. W., Hisamatsu, T., Goldschmidt, E. E., & Blundell, C. (2008). The nature of floral signals in *Arabidopsis*. I. Photosynthesis and a far-red photoresponse independently regulate flowering by increasing expression of *FLOWERING LOCUS T (FT)*. *Journal of Experimental Botany, 59*, 3811–3820.

Kloosterman, B., Abelenda, J. A., Gomez Mdel, M., Oortwijn, M., de Boer, J. M., Kowitwanich, K., et al. (2013). Naturally occurring allele diversity allows potato cultivation in northern latitudes. *Nature, 495*, 246–250.

Kobayashi, Y., Kaya, H., Goto, K., Iwabuchi, M., & Araki, T. (1999). A pair of related genes with antagonistic roles in mediating flowering signals. *Science, 286*, 1960–1962.

Kobayashi, Y., & Weigel, D. (2007). Move on up, it's time for change-mobile signals controlling photoperiod-dependent flowering. *Genes and Development, 21*, 2371–2384.

Kojima, S., Takahashi, Y., Kobayashi, Y., Monna, L., Sasaki, T., Araki, T., et al. (2002). *Hd3a*, a rice ortholog of the *Arabidopsis FT* gene, promotes transition to flowering downstream of *Hd1* under short-day conditions. *Plant and Cell Physiology, 43*, 1096–1105.

Koornneef, M., Hanhart, C. J., & van der Veen, J. H. (1991). A genetic and physiological analysis of late flowering mutants in *Arabidopsis thaliana*. *Molecular and General Genetics, 229*, 57–66.

Laubinger, S., Marchal, V., Gentilhomme, J., Wenkel, S., Adrian, J., Jang, S., et al. (2006). *Arabidopsis* SPA proteins regulate photoperiodic flowering and interact with the floral inducer CONSTANS to regulate its stability. *Development, 133*, 3213–3222.

Lazaro, A., Valverde, F., Pineiro, M., & Jarillo, J. A. (2012). The *Arabidopsis* E3 ubiquitin ligase HOS1 negatively regulates CONSTANS abundance in the photoperiodic control of flowering. *The Plant Cell, 24*, 982–999.

Lejeune, P., Bernier, G., Requier, M., & Kinet, J. (1993). Sucrose increase during floral induction in the phloem sap collected at the apical part of the shoot of the long day plant *Sinapis alba L. Planta, 90*, 71–74.

Lin, M. K., Belanger, H., Lee, Y. J., Varkonyi-Gasic, E., Taoka, K., Miura, E., et al. (2007). FLOWERING LOCUS T protein may act as the long-distance florigenic signal in the Cucurbits. *The Plant Cell, 19*, 1488–1506.

Liu, H., Wang, Q., Liu, Y., Zhao, X., Imaizumi, T., Somers, D. E., et al. (2013). *Arabidopsis* CRY2 and ZTL mediate blue-light regulation of the transcription factor CIB1 by distinct mechanisms. *Proceedings of the National Academy of Sciences of the United States of America, 110*, 17582–17587.

Liu, H., Yu, X., Li, K., Klejnot, J., Yang, H., Lisiero, D., et al. (2008). Photoexcited CRY2 interacts with CIB1 to regulate transcription and floral initiation in *Arabidopsis*. *Science, 322*, 1535–1539.

Liu, L., Liu, C., Hou, X., Xi, W., Shen, L., Tao, Z., et al. (2012). FTIP1 is an essential regulator required for florigen transport. *PLoS Biology, 10*, e1001313.

Liu, L. J., Zhang, Y. C., Li, Q. H., Sang, Y., Mao, J., Lian, H. L., et al. (2008). COP1-mediated ubiquitination of CONSTANS is implicated in cryptochrome regulation of flowering in *Arabidopsis*. *The Plant Cell, 20*, 292–306.

Liu, T., Carlsson, J., Takeuchi, T., Newton, L., & Farré, E. M. (2013). Direct regulation of abiotic responses by the *Arabidopsis* circadian clock component PRR7. *The Plant Journal, 76*, 101–114.

Liu, Y., Li, X., Li, K., Liu, H., & Lin, C. (2013). Multiple bHLH proteins form heterodimers to mediate CRY2-dependent regulation of flowering-time in *Arabidopsis*. *PLoS Genetics, 9*, e1003861.

Martins, M. C., Hejazi, M., Fettke, J., Steup, M., Feil, R., Krause, U., et al. (2013). Feedback inhibition of starch degradation in *Arabidopsis* leaves mediated by trehalose 6-phosphate. *Plant Physiology, 163*, 1142–1163.

Mathieu, J., Warthmann, N., Küttner, F., & Schmid, M. (2007). Export of FT protein from phloem companion cells is sufficient for floral induction in *Arabidopsis*. *Current Biology, 17*, 1055–1060.

Mathieu, J., Yant, L. J., Mürdter, F., Küttner, F., & Schmid, M. (2009). Repression of flowering by the *miR172* target SMZ. *PLoS Biology*, 7, e1000148.

Milyaeva, E. L., & Komarova, E. N. (1996). Changes in the sugar content in stem apices of the short-day plant *Perilla nankinensis* at floral transition. *Russian Journal of Plant Physiology, 43*, 149–154.

Mirolo, C., Bodson, M., Bernier, G. (1985). Effects of flower induction on the import of C_{14} assimilates into the apical bud of *Xanthium*. *Archives Internationales de Physiologie et de Biochemie, 93,* 13.

Mockler, T., Yang, H., Yu, X., Parikh, D., Cheng, Y. C., Dolan, S., et al. (2003). Regulation of photoperiodic flowering by *Arabidopsis* photoreceptors. *Proceedings of the National Academy of Sciences of the United States of America, 100,* 2140–2145.

Nakamichi, N., Kiba, T., Henriques, R., Mizuno, T., Chua, N. H., & Sakakibara, H. (2010). PSEUDO-RESPONSE REGULATORS 9, 7, and 5 are transcriptional repressors in the *Arabidopsis* circadian clock. *The Plant Cell, 22,* 594–605.

Nakamichi, N., Kiba, T., Kamioka, M., Suzuki, T., Yamashino, T., Higashiyama, T., et al. (2012). Transcriptional repressor PRR5 directly regulates clock-output pathways. *Proceedings of the National Academy of Sciences of the United States of America, 109,* 17123–17128.

Nakamichi, N., Kita, M., Niimura, K., Ito, S., Yamashino, T., Mizoguchi, T., et al. (2007). *Arabidopsis* clock-associated pseudo-response regulators PRR9, PRR7 and PRR5 coordinately and positively regulate flowering time through the canonical *CONSTANS*-dependent photoperiodic pathway. *Plant and Cell Physiology, 48,* 822–832.

Nelson, D. C., Lasswell, J., Rogg, L. E., Cohen, M. A., & Bartel, B. (2000). *FKF1*, a clock-controlled gene that regulates the transition to flowering in *Arabidopsis*. *Cell, 101,* 331–340.

Onouchi, H., Igeno, M. I., Perilleux, C., Graves, K., & Coupland, G. (2000). Mutagenesis of plants overexpressing *CONSTANS* demonstrates novel interactions among *Arabidopsis* flowering-time genes. *The Plant Cell, 12,* 885–900.

Park, D. H., Somers, D. E., Kim, Y. S., Choy, Y. H., Lim, H. K., Soh, M. S., et al. (1999). Control of circadian rhythms and photoperiodic flowering by the *Arabidopsis GIGANTEA* gene. *Science, 285,* 1579–1582.

Perilleux, C., & Bernier, G. (1997). Leaf carbohydrate status in *Lolium temulentum* during the induction of flowering. *New Phytologist, 135,* 59–66.

Pittendrigh, C. S. (1972). Circadian surfaces and the diversity of possible roles of circadian organization in photoperiodic induction. *Proceedings of the National Academy of Sciences of the United States of America,* 2734–2737.

Pittendrigh, C. S., & Minis, D. H. (1964). The entrainment of circadian oscillations by light and their role as photoperiodic clocks. *The American Naturalist,* 261–294.

Ponnu, J., Wahl, V., & Schmid, M. (2011). Trehalose-6-Phosphate: connecting plant metabolism and development. *Frontiers in Plant Science, 2,* 70.

Putterill, J., Robson, F., Lee, K., Simon, R., & Coupland, G. (1995). The *CONSTANS* gene of *Arabidopsis* promotes flowering and encodes a protein showing similarities to zinc finger transcription factors. *Cell, 80,* 847–857.

Rédei, G. P. (1962). Supervital mutants of *Arabidopsis*. *Genetics, 47,* 443–460.

Reekie, J. Y. C., Hicklenton, P. R., & Reekie, E. G. (1994). Effects of elevated CO_2 on time of flowering in four short-day and four long-day species. *Canadian Journal of Botany, 72,* 533–538.

Robson, F., Costa, M. M., Hepworth, S. R., Vizir, I., Pineiro, M., Reeves, P. H., et al. (2001). Functional importance of conserved domains in the flowering-time gene *CONSTANS* demonstrated by analysis of mutant alleles and transgenic plants. *The Plant Journal, 28,* 619–631.

Saijo, Y., Sullivan, J. A., Wang, H., Yang, J., Shen, Y., Rubio, V., et al. (2003). The COP1-SPA1 interaction defines a critical step in phytochrome A-mediated regulation of HY5 activity. *Genes and Development, 17,* 2642–2647.

Samach, A., Onouchi, H., Gold, S. E., Ditta, G. S., Schwarz-Sommer, Z., Yanofsky, M. F., et al. (2000). Distinct roles of CONSTANS target genes in reproductive development of *Arabidopsis*. *Science, 288,* 1613–1616.

Sato, E., Nakamichi, N., Yamashino, T., & Mizuno, T. (2002). Aberrant expression of the *Arabidopsis* circadian-regulated *APRR5* gene belonging to the *APRR1/TOC1* quintet results in early flowering and hypersensitiveness to light in early photomorphogenesis. *Plant and Cell Physiology*, *43*, 1374–1385.

Sawa, M., & Kay, S. A. (2011). GIGANTEA directly activates *FLOWERING LOCUS T* in *Arabidopsis thaliana*. *Proceedings of the National Academy of Sciences of the United States of America*, *108*, 11698–11703.

Sawa, M., Nusinow, D. A., Kay, S. A., & Imaizumi, T. (2007). FKF1 and GIGANTEA complex formation is required for day-length measurement in *Arabidopsis*. *Science*, *318*, 261–265.

Schaffer, R., Ramsay, N., Samach, A., Corden, S., Putterill, J., Carre, I. A., et al. (1998). The *late elongated hypocotyl* mutation of *Arabidopsis* disrupts circadian rhythms and the photoperiodic control of flowering. *Cell*, *93*, 1219–1229.

Simon, R., Igeno, M. I., & Coupland, G. (1996). Activation of floral meristem identity genes in *Arabidopsis*. *Nature*, *384*, 59–62.

Somers, D. E., Schultz, T. F., Milnamow, M., & Kay, S. A. (2000). *ZEITLUPE* encodes a novel clock-associated PAS protein from *Arabidopsis*. *Cell*, *101*, 319–329.

Song, Y. H., Ito, S., & Imaizumi, T. (2010). Similarities in the circadian clock and photoperiodism in plants. *Current Opinion in Plant Biology*, *13*, 594–603.

Song, Y. H., Ito, S., & Imaizumi, T. (2013). Flowering time regulation: photoperiod- and temperature-sensing in leaves. *Trends in Plant Science*, *18*, 575–583.

Song, Y. H., Lee, I., Lee, S. Y., Imaizumi, T., & Hong, J. C. (2012). CONSTANS and ASYMMETRIC LEAVES 1 complex is involved in the induction of *FLOWERING LOCUS T* in photoperiodic flowering in *Arabidopsis*. *The Plant Journal*, *69*, 332–342.

Song, Y. H., Smith, R. W., To, B. J., Millar, A. J., & Imaizumi, T. (2012). FKF1 conveys timing information for CONSTANS stabilization in photoperiodic flowering. *Science*, *336*, 1045–1049.

Srikanth, A., & Schmid, M. (2011). Regulation of flowering time: all roads lead to Rome. *Cell and Molecular Life Sciences*, *68*, 2013–2037.

Suárez-López, P., Wheatley, K., Robson, F., Onouchi, H., Valverde, F., & Coupland, G. (2001). *CONSTANS* mediates between the circadian clock and the control of flowering in *Arabidopsis*. *Nature*, *410*, 1116–1120.

Tamaki, S., Matsuo, S., Wong, H. L., Yokoi, S., & Shimamoto, K. (2007). Hd3a protein is a mobile flowering signal in rice. *Science*, *316*, 1033–1036.

Taoka, K., Ohki, I., Tsuji, H., Furuita, K., Hayashi, K., Yanase, T., et al. (2011). 14-3-3 proteins act as intracellular receptors for rice Hd3a florigen. *Nature*, *476*, 332–335.

Teper-Bamnolker, P., & Samach, A. (2005). The flowering integrator FT regulates *SEPALLATA3* and *FRUITFULL* accumulation in *Arabidopsis* leaves. *The Plant Cell*, *17*, 2661–2675.

Thomas, B., & Vince-Prue, D. (1996). *Photoperiodism in plants*. Academic Press.

Thuringer, F., & Bienz, M. (1993). Indirect autoregulation of a homeotic *Drosophila* gene mediated by extracellular signaling. *Proceedings of the National Academy of Sciences of the United States of America*, *90*, 3899–3903.

Tiwari, S. B., Shen, Y., Chang, H. C., Hou, Y., Harris, A., Ma, S. F., et al. (2010). The flowering time regulator CONSTANS is recruited to the *FLOWERING LOCUS T* promoter via a unique *cis*-element. *New Phytologist*, *187*, 57–66.

Valverde, F., Mouradov, A., Soppe, W., Ravenscroft, D., Samach, A., & Coupland, G. (2004). Photoreceptor regulation of CONSTANS protein in photoperiodic flowering. *Science*, *303*, 1003–1006.

Wahl, V., Ponnu, J., Schlereth, A., Arrivault, S., Langenecker, T., Franke, A., et al. (2013). Regulation of flowering by trehalose-6-phosphate signaling in *Arabidopsis thaliana*. *Science*, *339*, 704–707.

Wang, Z.Y., & Tobin, E. M. (1998). Constitutive expression of the *CIRCADIAN CLOCK ASSOCIATED 1 (CCA1)* gene disrupts circadian rhythms and suppresses its own expression. *Cell, 93*, 1207–1217.

Wellensiek, S. J. (1967). Relations between flower inducing factors in *Silene Armeria* L.. *Zeitschrift Fur Pflanzenphysiologie, 56*, 33–39.

Wenkel, S., Turck, F., Singer, K., Gissot, L., Le Gourrierec, J., Samach, A., et al. (2006). CONSTANS and the CCAAT box binding complex share a functionally important domain and interact to regulate flowering of *Arabidopsis*. *The Plant Cell, 18*, 2971–2984.

Wigge, Philip A. (2011). FT, a mobile developmental signal in plants. *Current Biology, 21*, R374–R378.

Wigge, P. A., Kim, M. C., Jaeger, K. E., Busch, W., Schmid, M., Lohmann, J. U., et al. (2005). Integration of spatial and temporal information during floral induction in *Arabidopsis*. *Science, 309*, 1056–1059.

Yanovsky, M. J., & Kay, S. A. (2002). Molecular basis of seasonal time measurement in *Arabidopsis*. *Nature, 419*, 308–312.

Yasui, Y., Mukougawa, K., Uemoto, M., Yokofuji, A., Suzuri, R., Nishitani, A., et al. (2012). The phytochrome-interacting VASCULAR PLANT ONE-ZINC FINGER1 and VOZ2 redundantly regulate flowering in *Arabidopsis*. *The Plant Cell, 24*, 3248–3263.

Yoo, S. C., Chen, C., Rojas, M., Daimon, Y., Ham, B. K., Araki, T., et al. (2013). Phloem long-distance delivery of FLOWERING LOCUS T (FT) to the apex. *The Plant Journal, 75*, 456–468.

Yoo, S. J., Hong, S. M., Jung, H. S., & Ahn, J. H. (2013). The cotyledons produce sufficient FT protein to induce flowering: evidence from cotyledon micrografting in *Arabidopsis*. *Plant and Cell Physiology, 54*, 119–128.

Zeevaart, J. A. (1976). Physiology of flower formation. *Annual Review of Plant Physiology, 27*, 321–348.

Zeevaart, J. A. (2006). Florigen coming of age after 70 years. *The Plant Cell, 18*, 1783–1789.

Zuo, Z., Liu, H., Liu, B., Liu, X., & Lin, C. (2011). Blue light-dependent interaction of CRY2 with SPA1 regulates COP1 activity and floral initiation in *Arabidopsis*. *Current Biology, 21*, 841–847.

CHAPTER TWO

Regulation of Flowering by Vernalisation in *Arabidopsis*

Youngjae Pyo, Sungrye Park, Yanpeng Xi and Sibum Sung[1]
Department of Molecular Biosciences, Plant Biology Graduate Program and Institute for Cellular and Molecular Biology, The University of Texas, Austin, TX, USA
[1]Corresponding author: e-mail address: sbsung@austin.utexas.edu

Contents

Abstract

Plants have evolved several mechanisms to control flowering time in response to environmental and endogenous signals. In particular, changes in temperature and day length throughout the year provide plants with clues to sense seasonal changes. Many plants in temperate climates respond to a long-term cold temperature of winter to be competent to flower in the following spring, a process known as vernalisation. In *Arabidopsis*, *FLOWERING LOCUS C* (*FLC*) is a major floral repressor that inhibits floral integrator genes and is subject to the epigenetic repression by vernalisation. Therefore, the stable repression of *FLC* by vernalisation permits plants to flower when inductive day length is achieved in spring. The epigenetic repression of *FLC* by vernalisation includes multiple levels of gene regulation ranging from chromatin modifications to non-coding RNAs. Here, we describe the current understanding of the molecular basis of vernalisation in *Arabidopsis*.

Advances in Botanical Research, Volume 72
ISSN 0065-2296
http://dx.doi.org/10.1016/B978-0-12-417162-6.00002-X

2.1 INTRODUCTION

Flowering is a critical developmental transition from vegetative to reproductive growth in the life cycle of plants. This transition is not reversible. Thus, proper timing of flowering is crucial for successful reproduction. To ensure the reproductive success, the initiation of flowering must occur only under favourable seasons for fertilisation and seed maturation. Accordingly, plants have evolved sophisticated mechanisms to incorporate changes in environmental cues, such as day length and temperature, into their developmental decisions.

In the model plant *Arabidopsis*, the timing of flowering is under the control of five major pathways, namely the photoperiod pathway, the vernalisation pathway, the autonomous pathway, the ambient temperature pathway and the gibberellin (GA) pathway (Srikanth & Schmid, 2011) (Figure 2.1). The photoperiod pathway promotes flowering in response to day length. The vernalisation pathway promotes flowering in response to prolonged cold exposure. The ambient temperature pathway promotes flowering in response to warm temperature, but delays flowering in response to cool temperature. The autonomous pathway promotes flowering independently of environmental signals. The GA pathway promotes flowering mainly in response to endogenous developmental signals and is essential to initiate flowering under non-inductive short-day conditions. These flowering pathways are often interconnected by the expression of common flowering genes. For example, both the autonomous and vernalisation pathways repress the expression of *FLOWERING LOCUS C* (*FLC*), a major repressor of flowering (Michaels & Amasino, 1999, 2001; Sheldon, Rouse, Finnegan, Peacock, & Dennis, 2000). In turn, FLC represses several floral integrator genes, including *FLOWERING LOCUS T* (*FT*) and *SUPPRESSOR OF OVEREXPRESSION OF CONSTANS 1* (*SOC1*) (Hepworth, Valverde, Ravenscroft, Mouradov, & Coupland, 2002; Michaels, Himelblau, Kim, Schomburg, & Amasino, 2005; Searle et al., 2006). In contrast, the photoperiod pathway antagonistically promotes the expression of *FT* through the activation of *CONSTANS* (*CO*) (Corbesier et al., 2007; Kardailsky et al., 1999; Kobayashi, Kaya, Goto, Iwabuchi, & Araki, 1999; Samach et al., 2000). FT promotes *SOC1* expression (Michaels et al., 2005; Yoo et al., 2005), and the expression of *SOC1* is also directly activated by the GA pathway (Michaels et al., 2005; Yoo et al., 2005). GA also promotes flowering under long-day conditions through the activation of *FT* and *TWIN SISTER OF*

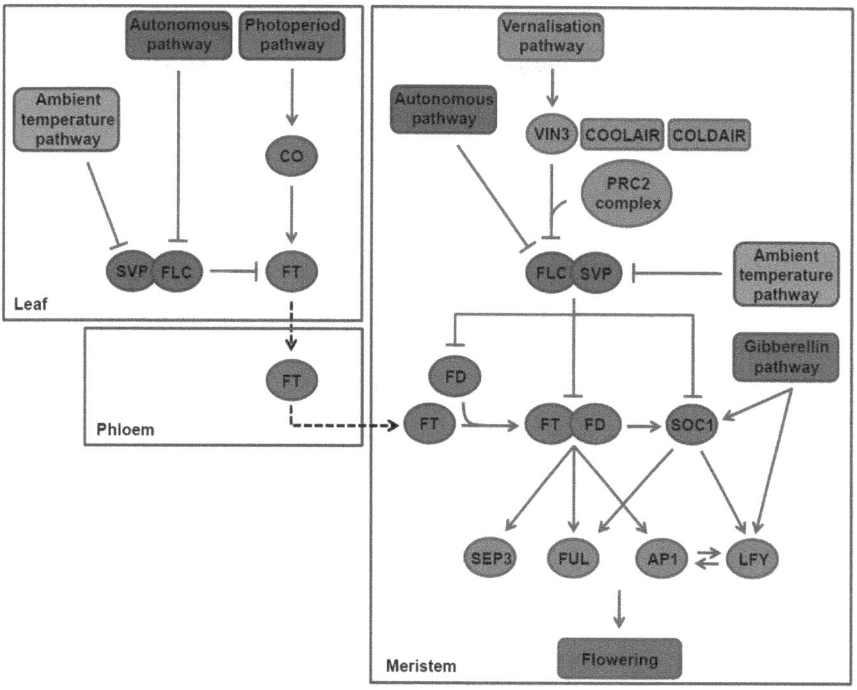

Figure 2.1 Flowering time pathways in *Arabidopsis*. Timing of flowering is controlled by the integration of various flowering pathways that incorporate environmental and developmental cues. There are five major flowering pathways in *Arabidopsis*. In the photoperiod pathway, CO activates the transcription of *FLOWERING LOCUS T (FT)* in response to inductive long days in the leaf. FT protein moves to the shoot apical meristem (SAM) via the phloem. In the SAM, FT protein interacts physically with FD protein. The FT–FD complex promotes the expression of *SUPPRESSOR OF OVEREXPRESSION OF CONSTANS 1 (SOC1)* and several other floral meristem identity genes, including *SEP3*, *FUL*, *AP1* and *LFY*. In the vernalisation pathway, *VERNALIZATION INSENSITIVE 3* and two long noncoding RNAs (COOLAIR and COLDAIR) are induced at various times during exposure to cold temperatures. *FLOWERING LOCUS C (FLC)* is negatively regulated by the autonomous pathway. Thus, both the vernalisation and autonomous pathways converge to repress *FLC*. FLC protein physically interacts with SVP protein and the FLC–SVP complex represses the expression of the floral integrator genes, such as FT, FD and SOC1 in the leaf and SAM. In the ambient temperature pathway, SVP protein is accumulated under cooler temperatures and represses the expression of floral integrator genes and delays flowering. The gibberellin (GA) pathway promotes flowering through the activation of *SOC1* and *LFY*. Arrows indicate the positive regulation and bars indicate the negative regulation. (See the colour plate.)

FT (*TSF*) in leaves (Galvao, Horrer, Kuttner, & Schmid, 2012; Porri, Torti, Romera-Branchat, & Coupland, 2012). Therefore, multiple flowering regulatory pathways converge to control the activation of floral integrator genes, such as *FT* and *SOC1* (Figure 2.1).

Many plants in temperate climates utilise the prolonged cold of winter as an important signal to sense seasonal changes, a phenomenon known as vernalisation. *FLC* is an essential component for the vernalisation requirement and response. A high level of *FLC* expression is necessary for the vernalisation requirement, and the repression of *FLC* by winter cold is a major event in the vernalisation response. Over the past few decades, considerable progress has been made in understanding the molecular mechanisms of vernalisation. They include distinct layers of regulation of gene expression, including histone modifications and regulation by non-coding RNAs. In this chapter, we describe the current understanding of the vernalisation pathway in *Arabidopsis*.

2.2 THE VERNALISATION RESPONSE

Flowering plants can be classified into three groups on the basis of their life cycles: annual, biennial and perennial plants (Amasino, 2004; Andres & Coupland, 2012). Annual plants complete their life cycle within a year, whereas biennial plants usually take 2 years to complete their life cycle. Perennial plants live for more than 2 years. Annual plants include two sub-groups: summer-annual and winter-annual plants. The life cycle of summer-annual plants spans from spring to fall, whereas that of winter-annual plants spans from the fall to the next spring. *Arabidopsis* is an annual species, with summer-annual and winter-annual strains. Winter-annual, biennial and some perennial plants that go through the winter season are able to sense the prolonged cold of winter as an environmental stimulus to achieve the competence to flower, known as vernalisation. Vernalisation is defined as "the acquisition or acceleration of the ability to flower by a chilling treatment" (Chouard, 1960). The term, vernalisation, comes from the Latin word, *vernus* (of the spring), reflecting that most vernalisation-required winter-annual and biennial plants flower in the spring. Even after vernalisation, plants do not flower immediately; rather they acquired the ability to flower under favourable inductive photoperiodic conditions. Therefore, vernalisation is an adaptive process ensuring that the reproductive development occurs only under favourable seasons of the year.

Classical physiological studies demonstrated that the shoot apex is the site of cold perception for the vernalisation response (Lang, 1965). For example, in a winter-annual pennycress (*Thlaspi arvense*), the shoot apex needs to be exposed to cold to flower early but flowering is not accelerated when roots or leaves are exposed to low temperatures (Metzger, 1988).

Therefore, other parts of the plant are not responding to chilling treatments. A subsequent study showed that actively dividing tissues are responsive to vernalisation treatments (Wellensiek, 1962), indicating that vernalisation takes place in cells undergoing active cell division and/or DNA replication.

The effect of vernalisation treatment persists throughout the mitotic cell division. This was demonstrated by physiological studies that took advantage of a biennial variety of henbane plants (*Hyoscyamus niger*) (Lang, 1965). The biennial variety of henbane is an obligate long-day plant that strictly requires inductive long-day conditions to initiate flowering. When the biennial henbane plants were kept in non-inductive short-day conditions after exposure to cold, they did not initiate flowering. However, when the same plants were moved to inductive long-day conditions, even after a long period of vegetative growth under non-inductive short day, they could flower. However, the vernalised state does not pass on to the next generation. Progeny of vernalised plants must go through the winter cold again to flower rapidly. Thus, once plants are exposed to vernalising cold, plants remember their vernalised state throughout mitotic cell divisions.

Plants need to be tolerant to low temperatures to survive during the prolonged period of winter cold. Low temperatures impose chilling and freezing stresses depending on the temperature. In general, chilling stress occurs at temperatures from 0 to 15 °C, while freezing stress occurs at temperatures below 0 °C (Chinnusamy, Zhu, & Zhu, 2007; Miura & Furumoto, 2013). Plants endemic to temperate climates are considered to be chilling tolerant to variable degrees, and their tolerance to freezing conditions can increase by exposure to low, non-freezing temperatures, a process known as cold acclimation (Thomashow, 1999). Plants adapted to tropical and subtropical regions are generally sensitive to chilling stress and do not have a mechanism to establish cold acclimation (Miura & Furumoto, 2013).

Apparently, there are similarities between cold acclimation and vernalisation. First, both cold acclimation and vernalisation are responses to low temperatures. Second, they occur under cold but above-freezing temperatures, suggesting that they both need metabolic activity that is inhibited by freezing temperatures (Sung & Amasino, 2004a). However, cold acclimation and vernalisation are distinct processes in several ways. First, cold acclimation is a short-term response that can be induced within a relatively short period of time while vernalisation is a long-term response that takes place over a period of several weeks. Second, cold acclimation is initiated through changes in membrane fluidity at low temperatures, altered calcium ion fluxes and protein phosphorylation (Chinnusamy et al., 2007). However, there is no evidence that such changes associated to cold acclimation affect the vernalisation response. Therefore, cold acclimation and vernalisation may use different signal transduction pathways. Finally, *C-REPEAT/DRE-BINDING FACTOR* (*CBF*) genes, which encode the transcription factors required for the cold acclimation, are rapidly induced by cold (Gilmour et al., 1998; Liu et al., 1998; Medina, Bargues, Terol, Perez-Alonso, & Salinas, 1999). However, *FLC* expression is not repressed but increased by over-expression of CBF1, CBF2 and CBF3 (Liu, Gilmour, Thomashow, & Van Nocker, 2002; Seo et al., 2009).

2.3 GENETIC ANALYSIS OF VERNALISATION

Arabidopsis is a facultative long-day plant. The term 'facultative' indicates that although *Arabidopsis* flowers more rapidly under long-day conditions, it will eventually flower even under short-day conditions. *Arabidopsis* is widely distributed throughout the northern hemisphere, including Europe, Asia, Africa and America, and locally adapted natural varieties exist (Al-Shehbaz & O'Kane, 2002; Weigel, 2012). Natural accessions of *Arabidopsis* can be classified as summer annuals and winter annuals based on their vernalisation requirement. Summer annuals flower rapidly without vernalisation when grown under long days, and they require only a few weeks to complete their life cycle. In contrast, winter annuals are late-flowering accessions that require vernalisation to accelerate flowering. Winter annuals can grow for months without flowering under long days. However, if they are exposed to prolonged cold, they flower rapidly under long days. Most commonly used laboratory strains, such as Columbia (Col) and Landsberg erecta (Ler), are summer annuals that flower rapidly. However, many naturally occurring accessions are winter annuals, which flower late unless vernalised.

Studies using the natural variation of *Arabidopsis* identified genes involved in the vernalisation requirement. Crosses between early-flowering and late-flowering accessions revealed that the winter-annual habit is conferred by a single dominant gene, named *FRIGIDA* (*FRI*) carried by many accessions (Clarke & Dean, 1994; Lee, Bleecker, & Amasino, 1993; Napp-Zinn, 1987). Many late-flowering accessions have a dominant allele of *FRI*, whereas early-flowering accessions, including Col and Ler, carry loss-of-function *FRI* alleles (Gazzani, Gendall, Lister, & Dean, 2003; Johanson et al., 2000). Another dominant gene, *FLC*, is also needed for the vernalisation requirement in natural accessions. *FLC* encodes an MADS-box DNA-binding protein that functions as a potent repressor of flowering (Michaels & Amasino, 1999;

Sheldon, Conn, Dennis, & Peacock, 2002). *FLC* is highly expressed in the shoot apex in winter-annual strains of *Arabidopsis*. In winter annuals, FRI is necessary to transcriptionally activate *FLC* to a level sufficient to repress flowering. *FRI* encodes a plant-specific nuclear protein with coiled-coil domain that acts to delay flowering solely through the activation of *FLC* (Johanson et al., 2000). FRI is a scaffold protein required to form the FRI-containing transcription activator complex that includes both general transcription factors and chromatin-modifying factors (Choi et al., 2011; Ding, Kim, & Michaels, 2013). FLC represses the expression of floral promoting factors, such as *FT* and *SOC1*, through its transcriptional repression activity (Hepworth et al., 2002; Searle et al., 2006). Thus, the presence of dominant alleles of *FRI* and *FLC* synergistically cause delayed flowering. The vernalisation response is achieved mainly through the repression of *FLC*, which allows *FT* and *SOC1* to be expressed to evoke the floral transition.

In *Arabidopsis*, several mutants that do not respond to vernalisation treatment have been isolated from winter-annual accessions or the autonomous pathway mutants. These genetic screens, have isolated *VERNALIZATION 1* (*VRN1*), *VRN2*, *VERNALIZATION INSENSITIVE 3* (*VIN3*), *VIN3-LIKE 1* (*VIL1*)/*VRN5* and *PROTEIN ARGININE METHYLTRANSFERASE 5* (*AtPRMT5*) (Chandler, Wilson, & Dean, 1996; Greb et al., 2007; Schmitz, Sung, & Amasino, 2008; Sung & Amasino, 2004b). Mutations in these genes commonly cause late flowering even after vernalisation. The impaired vernalisation response in these mutants is due to failure to maintain *FLC* in a transcriptionally repressed state. In winter-annual accessions of *Arabidopsis*, mRNA levels of *FLC* are reduced during cold, and the repressed state of *FLC* transcription is maintained after return to warm growth conditions. The level of *FLC* transcripts during the course of vernalisation provides a molecular explanation of the stability of vernalisation response. In *vrn1* and *vrn2* mutants, the *FLC* mRNA expression is repressed during cold, but the repressed state of *FLC* is not maintained when plants are returned to warm growth conditions (Gendall, Levy, Wilson, & Dean, 2001; Sung & Amasino, 2004b). *VRN2* encodes a nuclear-localised zinc finger protein with similarity to Polycomb group (PcG) proteins and is a homologue of Suppressor of Zeste 12 (Su(z)12) (Gendall et al., 2001). Su(z)12 is a component of Polycomb Repressive Complex 2 (PRC2). PRC2 is a methyltransferase complex that is responsible for the repression of gene expression by catalysing histone H3 Lys 27 trimethylation (H3K27me3) through its catalytic component Enhancer of Zeste (E(z)), a SET-domain containing methyltransferase (Cao et al., 2002; Kuzmichev et al., 2002;

Muller et al., 2002). In *vrn2* mutants, the levels of histone H3 Lys 9 (H3K9) and H3K27 methylation are not increased at *FLC* chromatin during the cold (Bastow et al., 2004; Sung & Amasino, 2004b). *VRN1* encodes a DNA-binding protein containing two plant-specific B3 domains (Levy, Mesnage, Mylne, Gendall, & Dean, 2002). In *vrn1* mutants, the level of H3K27 methylation is increased, but the level of H3K9 methylation is not increased at *FLC* chromatin during and after vernalisation (Bastow et al., 2004; Sung & Amasino, 2004b). These data showed that H3K9 and H3K27 methylation can occur independently and that *VRN1* is preferentially involved in H3K9 methylation.

VIN3 and *VIL1/VRN5* encode plant homeodomain (PHD) finger proteins with a fibronectin III (FNIII) domain and conserved C-terminal region (Greb et al., 2007; Sung & Amasino, 2004b; Sung, Schmitz, & Amasino, 2006). The PHD and FNIII domains are found throughout eukaryotes, and the C-terminal domain is unique in plants (Sung & Amasino, 2004b). PHD finger proteins are found as components of chromatin remodelling complexes (Bienz, 2006; Mellor, 2006). Mutations in either *VIN3* or *VIL1/VRN5* result in a vernalisation-insensitive phenotype. *VIN3* is expressed specifically during exposure to a prolonged period of cold, and its expression is quickly reduced when plants are returned to warm temperatures (Sung & Amasino, 2004b). *VIN3* expression is inversely correlated with *FLC* expression during cold. In *vin3* mutants, *FLC* expression is not repressed during vernalisation and vernalisation-mediated histone modifications including H3 acetylation, H3K9 and H3K27 methylation do not occur at *FLC* chromatin (Sung & Amasino, 2004b). Ectopic expression of *VIN3* alone is not sufficient to repress the transcription of *FLC*, suggesting that other vernalisation-specific components are necessary for the vernalisation response (Sung & Amasino, 2004b). *VIL1/VRN5* was identified independently through a yeast two-hybrid screen for VIN3-interacting proteins (Sung, Schmitz, et al., 2006) and by a genetic screen for mutants with a reduced vernalisation response (Greb et al., 2007). VIN3 and VIL1/VRN5 physically interact through a C-terminal domain (Greb et al., 2007; Sung, Schmitz, et al., 2006). Similar to *VIN3*, *VIL1/VRN5* is also required for the repressive histone modifications including methylation of both H3K9 and H3K27 at *FLC* chromatin (Greb et al., 2007; Sung, Schmitz, et al., 2006).

AtPRMT5 encodes a type II protein arginine *N*-methyltransferase, also known as SHK1-binding protein 1 (SKB1) (Deng et al., 2010; Hong et al., 2010; Pei et al., 2007; Sanchez et al., 2010; Schmitz et al., 2008; Wang et al., 2007;

Zhang et al., 2011). In wild type, vernalisation increases the level of symmetric dimethylation of arginine 3 of histone 4 (H4R3sme2) at *FLC* chromatin. In *atprmt5/skb1* mutants, the expression of *FLC* is increased, the level of H4R3sme2 is reduced at *FLC* chromatin and the vernalisation response is reduced. *AtPRMT5* is required for stably maintaining *FLC* repression after the cold exposure and H4R3sme2 appears to be required for the vernalisation-mediated enrichment of both H3K9 and H3K27 methylation (Schmitz et al., 2008). In addition to late flowering, *atprmt5/skb1* mutants also show other pleiotropic phenotypes, including growth retardation, and dark green and curly leaves, suggesting that *AtPRMT5/SKB1* plays important roles in various developmental processes (Pei et al., 2007; Sanchez et al., 2010; Wang et al., 2007).

2.4 VERNALISATION-MEDIATED CHANGES IN GENE EXPRESSION

2.4.1 The *FLC* Gene Family

In *Arabidopsis*, *FLC* is a major floral repressor that confers the vernalisation requirement and is repressed as a result of vernalisation. Interestingly, *flc* null mutants still show the vernalisation response, indicating that other floral repressors operate in the vernalisation response. FLC is an MADS-box DNA-binding protein and belongs to a small gene family that consists of *FLC* and other five *FLC*-like genes, including *FLOWERING LOCUS M* (*FLM*)/*MADS AFFECTING FLOWERING 1* (*MAF1*), *MAF2*, *MAF3*, *MAF4* and *MAF5*. Interestingly, *MAF2–5* are located in a tandem array. All members of the *FLC* gene family act as floral repressors (Gu et al., 2013; Kim & Sung, 2013; Ratcliffe, Kumimoto, Wong, & Riechmann, 2003; Ratcliffe, Nadzan, Reuber, & Riechmann, 2001; Scortecci, Michaels, & Amasino, 2001).

The *FLC* gene family is differentially expressed during the course of vernalisation (Kim & Sung, 2013) (Figure 2.2). Before cold, *FLC* is expressed at a high level that is sufficient to prevent flowering. Upon exposure to cold, *FLC* expression gradually decreases to a level that permits flowering. After cold, the repression of *FLC* is maintained. Before and during cold, *FLM/MAF1*, *MAF2* and *MAF3* are expressed at a high level that is sufficient to delay flowering. After cold, *FLM/MAF1*, *MAF2* and *MAF3* are transcriptionally repressed. Both *flm* and *maf2* mutants showed earlier flowering compared to the wild type regardless of cold exposure. In contrast, a *maf3* mutant showed an early flowering phenotype only after exposure to 40 days

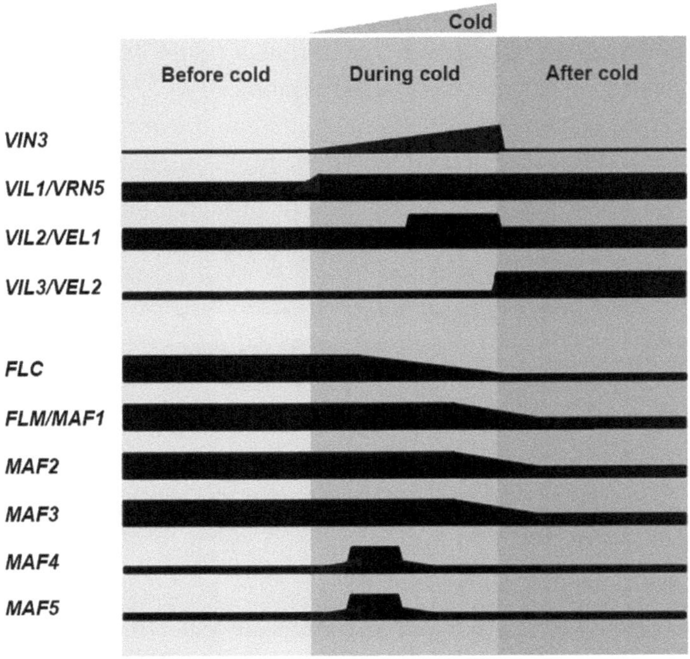

Figure 2.2 Vernalisation-mediated changes in expressions of *VERNALISATION INSENSITIVE 3 (VIN3)* and *FLOWERING LOCUS C (FLC)* gene families. All members of the VIN3 gene family are differentially expressed over the course of vernalisation. VIN3 is only expressed during cold. *VIN3-LIKE 1* (VIL1) and VIL2 are rather constitutively expressed with a small increase during cold. VIL3 is induced after cold. All members of the FLC gene family are also differentially expressed during the course of vernalisation. FLC, *FLOWERING LOCUS M*, *MADS AFFECTING FLOWERING 2* (MAF2) and MAF3 are repressed by vernalisation. The expression of MAF4 and MAF5 is transiently increased during early periods of cold exposure but eventually reduced to the basal level when cold persists. The VIN3 gene family is required for the regulation of expression of the FLC gene family by vernalisation. (See the colour plate.)

of cold. Unlike other members of the *FLC* gene family, *MAF4* and *MAF5* are expressed at low levels prior to the cold. However, *MAF4* and *MAF5* are induced early during cold exposure, with a peak at 20 days after beginning of the cold treatment. Increased levels of *MAF4* and *MAF5* transcription are eventually reduced to levels comparable to those observed before the cold exposure. In addition, both *maf4* and *maf5* mutants showed an early flowering phenotype when they were exposed to 20 days of cold but did not show a significant difference after 40 days of cold. These results indicate that *MAF4* and *MAF5* play a role in preventing a precocious vernalisation response.

2.4.2 The *VIN3* Gene Family

VIN3 belongs to a small gene family in *Arabidopsis*, including *VIL1/VRN5*, *VIL2/VEL1*, *VIL3/VEL2* and *VIL4/VEL3* (Greb et al., 2007; Sung, Schmitz, et al., 2006). However, *VIL4/VEL3* appears to be a pseudogene (Greb et al., 2007; Sung, Schmitz, et al., 2006). The PHD finger domain is widely known as a reader motif that recognises and binds to certain modified histones, recruiting transcription factors and nucleosome-associated complexes to chromatin (Saksouk et al., 2009). VIN3, VIL1 and VIL2 biochemically co-purify with core components of PRC2, including VRN2 and CURLY LEAF (CLF) (De Lucia, Crevillen, Jones, Greb, & Dean, 2008; Wood et al., 2006). This result suggests that they function together as part of the PRC2 in the vernalisation response.

Members of the *VIN3* gene family show differential expressions during the course of vernalisation (Kim & Sung, 2013) (Figure 2.2). *VIL1/VRN5* and *VIL2/VEL1* are expressed at a similar level before, during and after cold. *VIN3* expression is gradually induced during cold and is rapidly abolished after cold. *VIL3/VEL2* is barely expressed before and during cold but significantly increased after cold. This result indicates that the VIN3 family of proteins functions at different stages of vernalisation. All VIN3 family members directly associate with chromatin at the *FLC* and *MAFs* loci at particular times during the course of vernalisation. This coordinates proper repression of the *FLC* gene family by vernalisation (Kim & Sung, 2013; Sung & Amasino 2004b; Sung, Schmitz, et al., 2006). VIN3 protein is associated to chromatin of all *FLC* family members during exposure to prolonged periods of cold (Kim & Sung, 2013). In *vin3* mutants, all members of the *FLC* gene family are de-repressed during and after cold. Thus, VIN3 is absolutely required for the repression of all members of the *FLC* gene family by vernalisation. VIL1/VRN5 is enriched only at *FLC* and *FLM/MAF1* chromatin during the cold, and its abundance is increased and maintained even after the cold (Kim & Sung, 2013). In *vil1/vrn5* mutants, both *FLC* and *FLM* are initially repressed during cold but are de-repressed after cold. Mutations in *VIL1/VRN5* do not affect the expression of other members of the *FLC* gene family, and VIL1/VRN5 is not significantly enriched at *MAF2–MAF5* chromatin, demonstrating the specificity ofVIL1/VRN5 for *FLC* and *FLM/MAF1*. VIL2/VEL1 is especially enriched at *MAF4* and *MAF5* chromatin during the cold. In *vil2/vel1* mutants, *MAF4* and *MAF5* are not repressed by longer periods of cold exposure. Therefore, VIL2/VEL1 plays a role in establishing the repression of *MAF4* and *MAF5* during cold. VIL3/VEL2 accumulates at *MAF2–MAF5* chromatin but not at *FLC* and *FLM/MAF1* chromatin. In *VIL3/VEL2* RNAi lines, *MAF2–MAF5* are not repressed after cold. Thus, VIL3/VEL2 play a role in the maintenance of repressions of *MAF2–MAF5* after cold (Kim & Sung, 2013).

2.5 HISTONE MODIFICATIONS AND HISTONE-MODIFYING COMPLEXES

In eukaryotes, histone modifications are epigenetic mechanisms involved in the regulation of gene expression. Some histone modifications, such as histone acetylation and histone H3 Lys 4 trimethylation (H3K4me3), are marks for actively transcribed genes, whereas others such as histone H3 Lys9 dimethylation (H3K9me2) and histone H3K27me3 are marks for transcriptional repression. The H3K27me3 enrichment increases at the chromatin of *FLC* gene family members over the course of vernalisation. Depending on target specificities, the VIN3 family of proteins is necessary for the H3K27me3 enrichment at different subsets of *FLC* gene family chromatin (Kim & Sung, 2013).

2.5.1 Active Histone Marks at *FLC* Chromatin

Before cold, the high levels of *FLC* expression that create the vernalisation requirement are associated with active histone marks, such as H3 acetylation, H3K4me3 and histone H3 Lys 36 trimethylation (H3K36me3) (He, Doyle, & Amasino, 2004; Kim, Doyle, Sung, & Amasino, 2009; Ko et al., 2010; Pien et al., 2008; Tamada, Yun, Woo, & Amasino, 2009). Genes required for the activation of *FLC* expression encode components of histone-modifying complexes that are required for the deposition of active histone modifications at *FLC* chromatin (Figure 2.3 (A)).

H3K4 methylation is catalysed by a group of SET-domain containing histone methyltransferases including the yeast Set1 and the *Drosophila* Trithorax (Trx). The SET domain is important for histone methyltransferase activity, which can catalyse the methylation of several lysine residues in histone H3 and H4 (Lachner & Jenuwein, 2002). In yeast, Set1 is the only H3K4 methyltransferase, which is a subunit of the Complex Proteins Associated with Set1 (COMPASS) complex (Krogan et al., 2003). In mammals, there are five COMPASS-like complexes known as human SET1 and mixed-lineage leukemia 1 (MLL1), MLL2, MLL3 and MLL4 (Shilatifard, 2008).

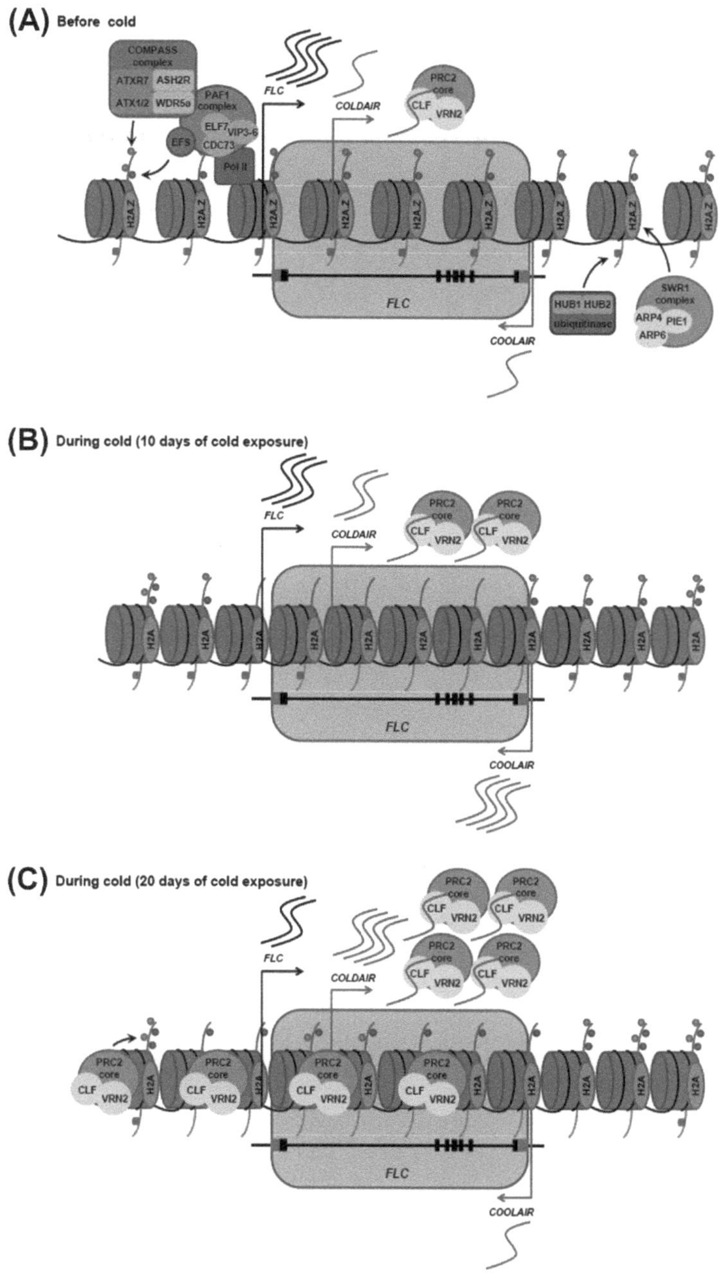

Figure 2.3 The vernalisation-mediated changes at *FLC* chromatin. (A) Before cold, *FLOWERING LOCUS C (*FLC) is highly expressed in winter-annual strains of *Arabidopsis*. H3 Lys 4 trimethylation (H3K4me3) and H3K36me3 are enriched at FLC chromatin. In addition, H2B mono-ubiquitination (H2Bub1) and H2A.Z are enriched at FLC chromatin when FLC is actively transcribed. Low levels of H3K27me3 and Polycomb Repressive Complex 2 (PRC2) enrichment at FLC chromatin are detected. COOLAIR and COLDAIR are expressed at low levels. (B) During cold, COOLAIR expression is rapidly increased and reaches a peak within 14 days of cold exposure. (C) The expression of COLDAIR increases with a peak expression at 20 days of cold exposure. COLDAIR interacts with PRC2 and functions to enrich PRC2 at FLC chromatin. Increased presence of PRC2 mediates the H3K27me3 enrichment at FLC chromatin.

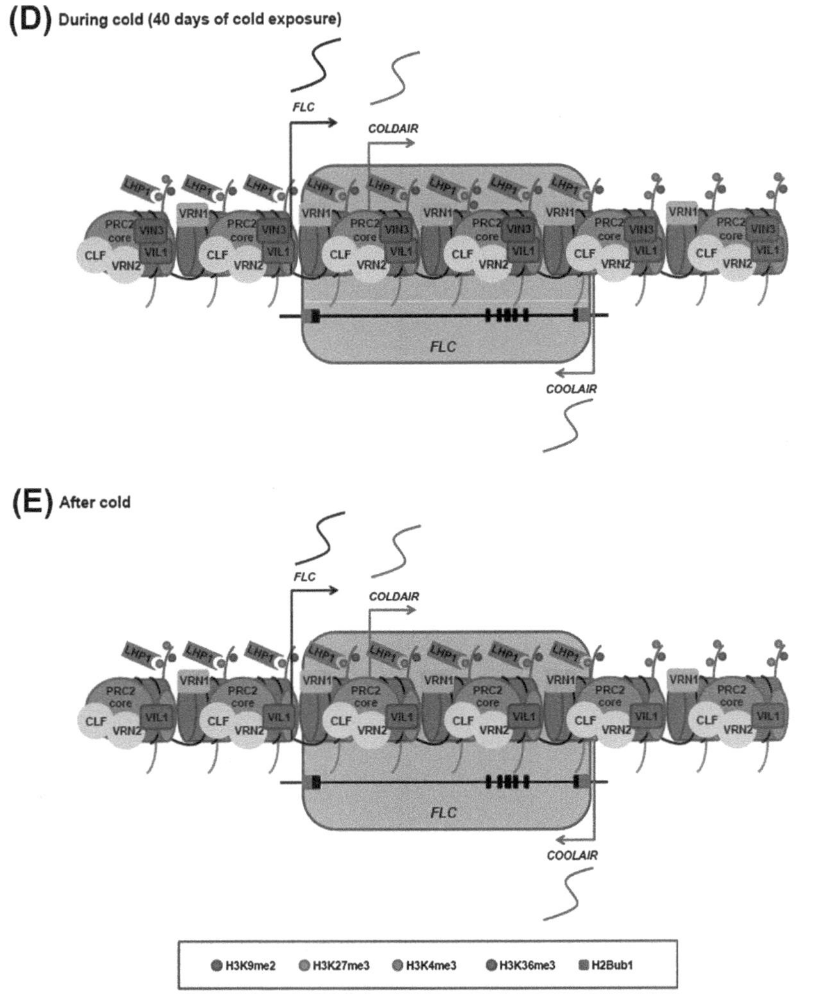

Figure 2.3 (D) *VERNALIZATION INSENSITIVE 3* (VIN3) and *VIN3-LIKE 1* (VIL1)/*VERNALIZATION 5* (VRN5) associate with PRC2 and enhance the H3K27 methylation activity of PRC2. H3K27me3 enrichment at FLC chromatin increases. LIKE-HETEROCHROMATIN PROTEIN 1 (LHP1) is associated with FLC chromatin. (E) After cold, VIN3 is no longer expressed and removed from the PRC2 complex. VIL1/VRN5 is still associated with the PRC2 complex at FLC chromatin. (See the colour plate.)

All of these complexes contain four core components including a H3K4 methyltransferase and structural components, known as WD repeat domain 5 (WDR5), Absent, small or homeotic discs 2 (Ash2) and Retinoblastoma-binding protein 5 (RbBP5), homologues of the yeast SWD3, BREFELDIN A SENSITIVITY2 (BRE2) and SWD1, respectively (Shilatifard, 2008;

Southall, Wong, Odho, Roe, & Wilson, 2009; Takahashi et al., 2011). Homologues of components of the yeast COMPASS complex and human COMPASS have been identified in *Arabidopsis*. *ARABIDOPSIS TRITHORAX1* (*ATX1*) and its close homologue *ATX2* encode proteins with similarity to the *Drosophila* Trx and mammalian MLL1 (Alvarez-Venegas et al., 2003). *ARABIDOPSIS TRITHORAX-RELATED7* (*ATXR7*) encodes a protein, which is closely related to yeast Set1 (Tamada et al., 2009). *WDR5a* and *ASH2 RELATIVE* (*ASH2R*) encode proteins that are closely related to WDR5 and ASH2, respectively (Jiang, Gu, & He, 2009; Jiang, Kong, Gu, Li, & He, 2011). These genes are required for H3K4 methylation and transcriptional activation of *FLC* and delayed flowering. Both single and double mutations of *ATX1* and *ATX2* cause abnormal floral organ identity and slightly early flowering. Early flowering is due to the reduced levels of *FLC* expression and the reduced H3K4me3 deposition at *FLC* chromatin (Pien et al., 2008). ATX1 forms a nuclear complex, which contains WDR5a and ASH2R (Jiang et al., 2009; Jiang et al., 2011). The mutation in *ATXR7* causes decreased levels of *FLC* expression along with the decreased enrichment of H3K4 methylation and the increased enrichment of H3K27 methylation at *FLC* chromatin (Tamada et al., 2009). *EARLY FLOWERING IN SHORT DAYS* (*EFS*)/*SET DOMAIN GROUP 8* (*SDG8*) encodes a Trx-like protein with a SET domain that mediates di- and trimethylation of both H3K4 and H3K36 enrichment at *FLC* chromatin prior to cold (Ko et al., 2010). The mutation in *EFS* causes the reduced levels of *FLC* expression and the reduced enrichment of H3K4me3 and H3K36me3 enrichment at *FLC* chromatin, and thus results in early flowering similar to mutations in RNA polymerase II-associated factor 1 (PAF1)-COMPASS components (Kim et al., 2005; Xu et al., 2008; Zhao, Yu, Meyer, Wu, & Shen, 2005).

To catalyse H3K4 methylation, the COMPASS complex acts together with the PAF1 complex (Krogan et al., 2003). The PAF1 complex is required for the association of COMPASS with RNA polymerase II machinery to regulate gene expression. In yeast, the minimum PAF1 complex consists of five subunits, Paf1, Ctr9, Cdc73, Rtf1 and Leo1. *Arabidopsis* homologues of components of yeast and human PAF1 complexes have been identified mainly from genetic screens that were designed to identify positive regulators of *FLC*. *EARLY FLOWERING 7* (*ELF7*) and *ELF8*/*VERNALIZATION INDEPENDENCE 6* (*VIP6*) encode proteins with similarities to PAF1 and CTR9, respectively (He et al., 2004). *VIP3* encodes a protein related to Ski8, which is a subunit of the human PAF1 complex (Zhang, Ransom, Ludwig, & van Nocker, 2003). *VIP4* and *VIP5* encode proteins similar to Leo1 and Rtf1, respectively (Oh, Zhang, Ludwig, & van Nocker, 2004; Zhang & van Nocker, 2002). Mutations in these genes commonly result in early flowering that is due to reduced levels of H3K4me3 enrichment at *FLC* chromatin and thus low levels of *FLC* expression.

In eukaryotes, histone H2B mono-ubiquitination (H2Bub1) is associated with active gene transcription. H2Bub1 is an important prerequisite for the proper deposition of H3K4me3 and the transcriptional activation of target genes (Wood et al., 2003). In yeast, H2Bub1 is mediated by a protein complex that contains BREFELDIN A SENSITIVITY1 (BRE1) and RADIATION SENSITIVE6 (RAD6) (Robzyk, Recht, & Osley, 2000). RAD6 has the E2 ubiquitin conjugating activity, and BRE1 has the E3 ubiquitin ligase activity. They function to mono-ubiquitinate histone H2B. In *Arabidopsis*, there are three RAD6 homologues, UBIQUITIN-CONJUGATING ENZYME 1 (UBC1), UBC2 and UBC3 and two BRE1 homologues, HISTONE MONOUBIQUITINATION 1 (HUB1) and HUB2 (Cao, Dai, Cui, & Ma, 2008; Gu, Jiang, Wang, Bachmair, & He, 2009). UBC1 and UBC2 are involved in flowering time control and act redundantly to mediate the enrichment of H2Bub1 at *FLC* chromatin, whereas UBC3 is dispensable for the regulation of *FLC* (Xu et al., 2009). Mutations in either *HUB1* or *HUB2* result in early flowering and the loss of H3K4me3 enrichment at the promoter region of *FLC*. UBIQUITIN-SPECIFIC PROTEASE 26 (UBP26) catalyses the de-ubiquitination of histone H2B and is required for H2Bub1 at *FLC* chromatin (Schmitz, Tamada, Doyle, Zhang, & Amasino, 2009). A mutation in *UBP26* results in rapid flowering that is due to the loss of *FLC* expression. Interestingly, *ubp26* mutants have reduced levels of H3K36me3 enrichment, without altering the level of H3K4me3 enrichment at *FLC* chromatin (Schmitz et al., 2009).

Histone variants also play roles in the regulation of gene expression. In yeast, the SWR1 complex catalyses the replacement of H2A histone by its variant, H2A.Z (Kobor et al., 2004). H2A.Z deposition by the SWR1 complex is required for transcriptional regulation, maintenance of heterochromatic barriers and genome stability in yeast (Kamakaka & Biggins, 2005). Homologues of the yeast SWR1 complex have been identified as positive regulators of *FLC* in *Arabidopsis* (Choi et al., 2005; Choi et al., 2007; Deal, Kandasamy, McKinney, & Meagher, 2005; Deal, Topp, McKinney, & Meagher, 2007; Lazaro, Gomez-Zambrano, Lopez-Gonzalez, Pineiro, & Jarillo, 2008; Martin-Trillo et al., 2006; Noh & Amasino, 2003). *PHOTOPERIOD INDEPENDENT EARLY FLOWERING 1* (*PIE1*) encodes a protein homologous to yeast SWR1 (Noh & Amasino, 2003). PIE1 is required for the activation of *FLC* and mediates H2A.Z deposition onto target chromatin including *FLC* (Deal et al., 2007). Mutations in *PIE1* result in the loss of H2A.Z from *FLC* chromatin leading to reduced levels of *FLC* expression and early flowering (Deal et al., 2007). In addition to the catalytic subunit SWR1, other subunits of the SWR complex have been identified in *Arabidopsis*. These include ACTIN-RELATED PROTEIN 4 (ARP4) (Kandasamy, Deal, McKinney, & Meagher, 2005), SUPPRESSOR OF FRIGIDA 3 (SUF3)/ACTIN-RELATED PROTEIN 6 (ARP6)/ EARLY IN SHORT DAYS 1 (ESD1) (Choi et al., 2005; Deal et al., 2005; Martin-Trillo et al., 2006) and SWC6 (Lazaro et al., 2008). Mutations in these genes also suppress *FLC* expression and thus lead to early flowering. Taken together, the deposition of H2A.Z into *FLC* chromatin by the SWR1 complex is necessary for the activation of *FLC* to repress flowering. Given the findings that a number of chromatin-remodelling complexes are necessary for the proper activation of *FLC*, chromatin-level regulation appears to be the main regulatory theme for *FLC*.

2.5.2 Repressive Histone Marks at *FLC* Chromatin

Multiple chromatin marks associated with transcriptional activation exist at *FLC* chromatin when *FLC* is highly expressed. The level of *FLC* expression is gradually reduced in response to a prolonged period of cold. This reduction in *FLC* expression is associated with changes in chromatin landscapes, including increased deposition of certain repressive histone modifications at *FLC* chromatin, such as histone deacetylation, H3K9me2, H3K27me3 and H4R3sme2 (Bastow et al., 2004; Greb et al., 2007; Schmitz et al., 2008; Sung & Amasino, 2004b; Sung, Schmitz, et al., 2006).

In higher eukaryotes, PRC2 catalyses methylation of H3K27, a repressive histone mark, to silence or repress gene expression. In *Drosophila*, the PRC2 complex consists of four core components: the catalytic SET-domain containing protein, Enhancer of Zeste (E(z)), the WD40 domain protein, Extra sex combs (ESC), the Zn-finger protein, Suppressor of Zeste, (Su(z)12), and the nucleosome-remodelling factor 55 (NURF-55) (Cao et al., 2002; Cao & Zhang, 2004; Muller et al., 2002; Schuettengruber, Chourrout, Vervoort, Leblanc, & Cavalli, 2007). *Arabidopsis* has three E(z) homologues (CURLY LEAF (CLF), SWINGER (SWN) and MEDEA (MEA)), three Su(z)12 homologues (FERTILIZATION INDEPENDENT SEED2 (FIS2), EMBRYONIC FLOWER2 (EMF2) and VERNALIZATION2 (VRN2)), one homologue of ESC (FERTILIZATION INDEPENDENT ENDOSPERM (FIE)) and five p55 homologues (MULTICOPY SUPPRESSOR

OF IRA (MSI) 1–5) (Hennig & Derkacheva, 2009). There are at least three PRC2-like complexes in *Arabidopsis*, known as FIS, VRN and EMF complexes (Hennig & Derkacheva, 2009; Hsieh, Hakim, Ohad, & Fischer, 2003). These complexes play roles in various developmental processes in *Arabidopsis*. For example, the FIS complex is involved in gametogenesis and endosperm development, the EMF complex is necessary for flower development and VRN-PRC2 complex mediates the vernalisation response (Hennig & Derkacheva, 2009).

As described earlier, VRN2, a homologue of Su(z)12, was identified in a vernalisation mutant screen (Chandler et al., 1996; Gendall et al., 2001), suggesting the involvement of a PRC2-like complex in the vernalisation-mediated *FLC* repression (Bastow et al., 2004; Gendall et al., 2001; Sung & Amasino, 2004b). Consistent with the role of PRC2 in vernalisation, the vernalisation-mediated enrichment of H3K9me2 and H3K27me3 at *FLC* chromatin is reduced in *vrn2* mutants compared to the wild type (Bastow et al., 2004; Sung & Amasino, 2004b). In *Arabidopsis*, the VRN–PRC2 complex is composed of the four core subunits, VRN2, CLF/SWN, FIE and MSI1 (Jiang, Wang, & He, 2008; Wood et al., 2006). CLF and SWN are homologues of E(z), histone methyltransferases, and SWN is partially redundant with CLF (Chanvivattana et al., 2004). The enrichment of CLF increases and mediates H3K27 trimethylation (H3K27me3) at *FLC* chromatin by vernalisation (Heo & Sung, 2011; Kim & Sung, 2013). Biochemical approaches revealed that the VRN–PRC2 complex includes VIN3, VIL1/VRN5 and VIL2/VEL1 during the cold exposure (De Lucia et al., 2008; Wood et al., 2006), suggesting that the VIN3 family of proteins functions together with VRN–PRC2 complex to trigger the chromatin changes at *FLC* by vernalisation.

In *Drosophila*, PRC2-mediated H3K27me3 is maintained by Polycomb Repressive Complex 1 (PRC1). PRC1 consists of four core protein subunits (POLYCOMB (PC), POSTERIOR SEX COMBS (PSC), POLYHOMEOTIC and RING) as well as associated proteins such as SEX COMBS ON MIDLEG, ZESTE, and TATA-BOX BINDING-ASSOCIATED FACTORS (Francis, Saurin, Shao, & Kingston, 2001). RING and PSC proteins have E3 ubiquitin ligase activity that mono-ubiquitinates histone H2A at lysine 119 (H2AK119ub1). PRC1 can recognise and bind to H3K27me3 through a chromodomain (CD) of the PC subunit. However, plant genomes do not contain clear PC homologues of PRC1 complex. In *Arabidopsis*, LIKE HETEROCHROMATIN PROTEIN 1 (LHP1) is the only homologue of the HETEROCHROMATIN PROTEIN 1 (HP1) and belongs to the large family of CD proteins, which includes HP1 and PC.

In animals and yeast, HP1 is predominantly localised in heterochromatin and plays a key role in the formation and maintenance of constitutive heterochromatin (Maison & Almouzni, 2004). HP1 binds to chromatin through the binding to methylated H3K9. Although LHP1 is a homologue of HP1, its function is similar to those of PC that binds to the PRC2-mediated H3K27me3 modification (Zhang et al., 2007). LHP1 preferentially binds to H3K27me3 through its CD (Turck et al., 2007; Zhang et al., 2007). By vernalisation, the enrichment of LHP1 at *FLC* chromatin increases (Figure 2.3 (D)) and the enrichment is maintained after cold (Sung, He, et al., 2006) (Figure 2.3 (E)). In *lhp1* mutants, the repression of *FLC* is still observed during the cold but *FLC* is de-repressed when plants are returned to warm temperature, suggesting that LHP1 functions to maintain the repressed state of *FLC*.

Taken together, these data indicate that an evolutionarily conserved gene repression system mediated by PRC2/PRC1 operates in the repression of *FLC*. Unlike other examples of PRC2/PRC1-mediated gene repressions, in which target genes are regulated in the context of developmental programming, vernalisation-mediated *FLC* repression is triggered by an external environmental stimulus, cold. The environmentally induced nature of the VRN–PRC2/LHP1 regulatory system provides a unique opportunity to study the mechanisms underlying PRC2/PRC1-mediated gene repressions.

2.6 NON-CODING RNAs: NEW PLAYERS IN VERNALISATION

As discussed earlier, the enrichment of VRN–PRC2 and LHP1 and corresponding histone modifications at *FLC* are required for the stable *FLC* repression by vernalisation (Bastow et al., 2004; De Lucia et al., 2008; Greb et al., 2007; Jiang et al., 2008; Sung & Amasino, 2004b; Sung, He, et al., 2006; Sung, Schmitz, et al., 2006). But how VRN–PRC2 is targeted to *FLC* in a cold-induced manner remained unclear. In *Drosophila*, *cis*-regulatory sequences, known as Polycomb Response Elements (PREs), mediate the recruitment of PcG proteins (Muller & Kassis, 2006; Simon, Chiang, Bender, Shimell, & O'Connor, 1993). PRE serves as a docking site for PcG proteins through DNA–protein and protein–protein interactions (Chan, Rastelli, & Pirrotta, 1994; Muller & Kassis, 2006). However, up to date no PRE has been discovered, nor direct DNA binding by Polycomb proteins has been reported in organisms other than *Drosophila* (Kohler & Villar, 2008). A PRE-like *cis*-element, named Vernalisation Response Element (VRE), was identified from the first intron of *FLC* (Sung, He, et al., 2006). In the absence of 238 bp element, the vernalisation-mediated *FLC* repression is not stable. Unstable *FLC* repression by vernalisation is also observed in PcG mutants (Bastow et al., 2004; De Lucia et al., 2008; Sung & Amasino, 2004b; Wood et al., 2006). However, how this VRE confers the PcG-mediated *FLC* repression was not clear.

In a human cell line, a long non-coding RNA (lncRNA) *HOTAIR* interacts with PRC2 and is required for the PRC2 enrichment and H3K27 methylation at target loci (Rinn et al., 2007). Following this finding, other groups reported similar interactions in different systems. For example, an ncRNA *Kcnq1ot1* interacts with the histone methyltransferase G9a and PRC2 in a lineage-specific manner during *Kcnq1* silencing (Pandey et al., 2008). Another ncRNA *RepA* recruits PRC2 in embryonic stem cells to initiate the X-chromosome inactivation (Zhao, Sun, Erwin, Song, & Lee, 2008). These findings offered a paradigm by which ncRNAs act as mediators for PRC2 targeting, suggesting that a similar mechanism may exist in the vernalisation-mediated *FLC* repression.

ncRNAs represent a set of RNAs that are transcribed from the genome, but not translated into proteins (Eddy, 2001). Based on the function, ncRNAs can be divided into two major groups: (1) classic ncRNAs, such as ribosomal RNAs, transfer RNAs, small nuclear RNAs, small nucleolar RNAs and telomere-associated RNAs; (2) regulatory ncRNAs, including microRNAs (miRNAs), small interfering RNAs (siRNAs), Piwi-interacting RNAs (piRNAs) and lncRNAs (Kung, Colognori, & Lee, 2013). Classic ncRNAs serve as structural molecules to maintain cell integrity and/or take part in fundamental cellular processes. Their genesis and function are relatively well understood. In contrast, the growing group of regulatory ncRNAs are engaged in various regulatory pathways that modulate gene expression and development. Identification and functional study of these regulatory ncRNAs have been fascinating topics in recent years.

Among regulatory ncRNAs, miRNAs, siRNAs and piRNAs are grouped as small ncRNAs because their lengths are relatively short, approximately 20–30 bases, whereas newly emerged lncRNAs range from approximately 200 to several thousand bases in length. Regulatory ncRNAs generally function by guiding different proteins to target loci to perform transcriptional or post-transcriptional regulations (Batista & Chang, 2013). Small RNAs are loaded onto a protein family called Argonautes (AGO). AGO-associated small ncRNAs guide AGO proteins to target regions through sequence complementarity, and subsequently other components are recruited through the interaction with AGOs (Czech & Hannon, 2011). Unlike small ncRNAs, lncRNAs do not need AGO proteins as mediators. Due to their longer length (>200 nt), lncRNAs tend to fold into thermodynamically stable secondary and higher order structures that serve as the interface for protein binding (Mercer & Mattick, 2013). In this way, lncRNAs are able to interact directly with various proteins and recruit them to specific loci through sequence complementarity.

A group of antisense transcripts, COOLAIR, was identified from the *FLC* region (Swiezewski, Liu, Magusin, & Dean, 2009). Transcription of COOLAIR initiates from the 3′ region of *FLC* and extends beyond the 5′ start site. These long antisense transcripts undergo alternative splicing and result in different isoforms. COOLAIR expression is induced by cold and peaks at 10–14 days of cold treatment when the level of *FLC* mRNA begins to decline (Figure 2.3 (B)). The cold-induced expression of COOLAIR suggests that it acts to repress *FLC* upon cold exposure. However, no direct interaction between COOLAIR and components of PRC2 has been observed. Instead, COOLAIR may repress *FLC* through its own transcription because the COOLAIR promoter alone is sufficient to induce cold-dependent silencing of sense genes. Therefore, the sequence of COOLAIR may not be relevant in the repression of *FLC* (Swiezewski et al., 2009).

Another lncRNA was identified from *FLC* named COLD Assisted Intronic non-coding RNA (COLDAIR) (Heo & Sung, 2011). COLDAIR is transcribed in a sense direction from the first intron of *FLC*. The promoter of COLDAIR and part of its transcripts overlap with VRE, a region required for the stable repression of *FLC* by vernalisation (Sung, He, et al., 2006). Its expression is also cold induced and peaks at around 20 days of cold treatment. The exciting feature of COLDAIR is its ability to bind CLF and SWN, the catalytic subunits of PRC2, and such interactions are enhanced during cold exposure (Figure 2.3 (C)). Consistent with its PRC2-binding property, COLDAIR knockdown using RNAi results in plants with reduced vernalisation responses and delayed flowering after cold exposure. This is due to a dramatic decrease in CLF enrichment and H3K27 trimethylation at *FLC*, consistent with the role of COLDAIR in recruiting PRC2 onto *FLC*. Taken together, the evolving model postulates that an lncRNA, COOLAIR, represses *FLC* expression at the early stage of vernalisation through its own transcription, in turn an lncRNA, COLDAIR, recruits PRC2 later to initiate the stable epigenetic changes at *FLC* chromatin (Figures 2.3 (B), (C) and (D)). As new players in vernalisation, lncRNAs provide insights on the mechanisms by which ncRNAs interface with chromatin modifications.

2.7 CHANGES AT *VIN3* CHROMATIN AND ITS TRANSCRIPTION BY VERNALISATION

It is possible that some components of the cold-sensing signal transduction pathway are shared between cold acclimation and vernalisation. For example, HOS1 is a RING-finger protein that acts as an E3 ligase required for ubiquitin-mediated protein degradation (Dong, Agarwal, Zhang, Xie, & Zhu, 2006). The mutation in *HOS1* leads to reduced levels of *CBF* expression and thus results in a reduced tolerance to freezing temperatures. Interestingly, *hos1* mutants exhibit reduced *FLC* expression and early flowering (Lee et al., 2001). On the basis of these data, it was proposed that *HOS1* functions in both cold acclimation and vernalisation. However, a recent study has shown that HOS1 is not likely to be involved in vernalisation because the mutation in *HOS1* does not affect the expression of *VIN3*, which is induced by vernalisation (Bond, Dennis, & Finnegan, 2011). So far, there is no known gene that is involved in both cold acclimation and vernalisation.

VIN3, a major player in the vernalisation-mediated *FLC* repression, undergoes a dramatic change in its expression level during the course of the vernalisation (Sung & Amasino, 2004b). Prior to cold and after cold, *VIN3* expression (both transcripts and protein) is barely detectable. When plants are exposed to the long period of cold (more than 2 weeks of cold at 4 °C), *VIN3* is induced at significantly higher levels. The induction kinetics of *VIN3* by vernalising cold is different from that of cold acclimation genes. For example, *CBF* genes are rapidly induced by cold (within several hours) (Gilmour et al., 1998; Medina et al., 1999; Zarka, Vogel, Cook, & Thomashow, 2003). Unlike its induction, the reduction of *VIN3* expression occurs rapidly when plants are returned to warm temperatures. Within a few days, *VIN3* expression completely diminishes (Kim & Sung, 2013; Kim, Zografos, & Sung, 2010; Sung & Amasino, 2004b). Therefore, the expression pattern of *VIN3* is a whole marker that reflects a long-term exposure to winter cold.

Both *VIN3* and *FLC* undergo changes in their chromatin landscape by vernalisation. Some chromatin remodelling complexes are involved in the regulation of both *VIN3* and *FLC* (Finnegan et al., 2011; Kim & Sung, 2013; Kim et al., 2010). EFS and ELF7 function to activate *VIN3* and *FLC*, whereas PRC2 and LHP1 function to repress *VIN3* and *FLC*. Given the opposite expression patterns of *VIN3* and *FLC* during the course of vernalisation, it is necessary for these chromatin remodelling complexes to coordinate their activities during the course of vernalisation. Before the cold exposure, LHP1 and CLF are enriched at *VIN3* chromatin (Figure 2.4 (A)).

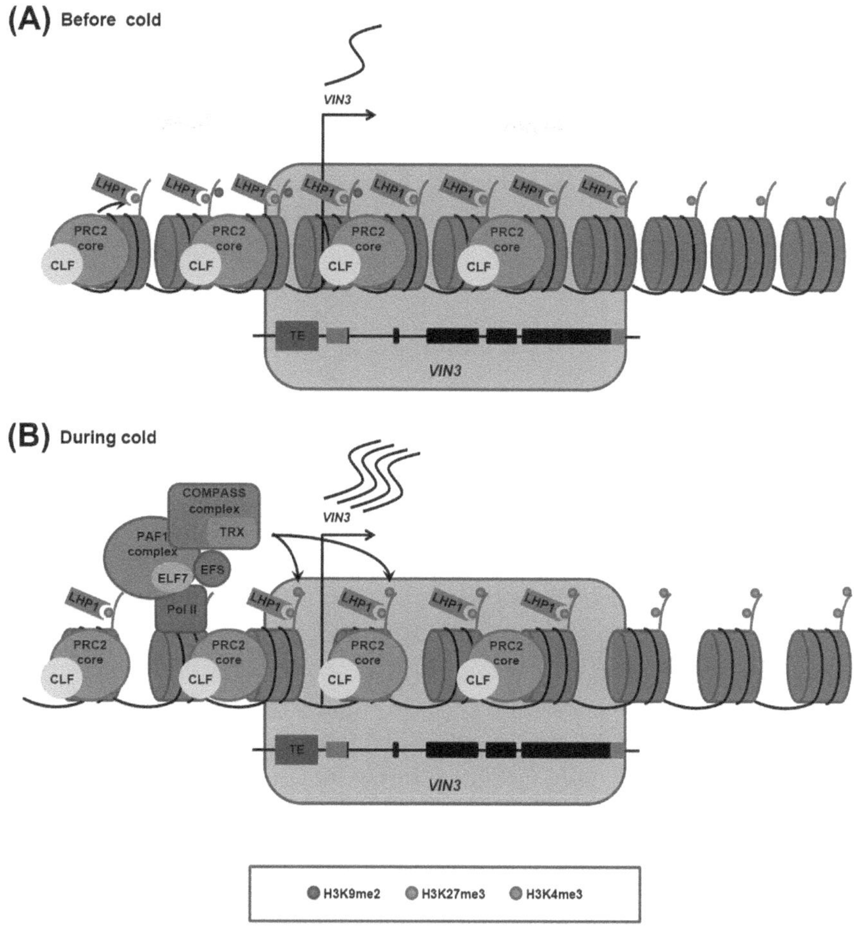

Figure 2.4 Changes at *VIN3* chromatin during the course of vernalisation. (A) Before cold, VIN3 is expressed at a very low level. Polycomb Repressive Complex 2 (PRC2) and LIKE HETEROCHROMATIN PROTEIN 1 (LHP1) are associated with VIN3 chromatin. VIN3 chromatin is enriched with repressive histone marks, H3 Lys9 dimethylation (H3K9me2) and H3 Lys 27 trimethylation (H3K27me3). In addition, a transposable element (TE)-derived sequence is present at the VIN3 promoter region and H3K9me2 is enriched around the TE region. (B) During cold, H3K9me2 decreases whereas an active histone mark, H3K4me3, increases at the transcription start site of VIN3. Polymerase II-associated factor 1 (PAF1) complex and *EARLY FLOWERING IN SHORT DAYS* (*EFS*) are necessary for the fullest extent of VIN3 induction. PRC2 and LHP1 are still associated with VIN3 chromatin when VIN3 is induced. After cold, H3K4me3 is decreased while H3K9me2 is increased again at VIN3 chromatin. (See the colour plate.)

lhp1 mutants show higher levels of *VIN3* expression compared to the wild type, suggesting that *VIN3* is de-repressed in *lhp1* mutants. Similarly, *clf;swn* double mutants show de-repression of *VIN3*. However, de-repressed levels of *VIN3* expression in *lhp1* and *clf;swn* mutants are significantly lower than the fully induced levels of *VIN3* expression by cold (Finnegan et al., 2011; Kim et al., 2010). This observation indicates that the removal of LHP1 and CLF/SWN from *VIN3* chromatin cannot mimic the induction of *VIN3* by vernalisation. Although LHP1 and CLF/SWN are necessary for the repressed state of *VIN3*, the enrichment of LHP1 and CLF at *VIN3* chromatin is maintained during the course of vernalisation (Figure 2.4 (B)) (Finnegan et al., 2011; Kim et al., 2010). Therefore, the induction of *VIN3* occurs independently of LHP1 and PRC2. This is consistent with the observation that the repressive mark, H3K27me3, is not changed at *VIN3* chromatin regardless of the cold exposure.

Transcriptional induction of *VIN3* by vernalisation accompanies changes in the chromatin landscape (Figure 2.4 (B)). An active histone mark H3K4me3 enriches at *VIN3* chromatin when *VIN3* is induced. Furthermore, ELF7 (a component of the PAF1 complex) and EFS (an H3K4/H3K36 methyltransferase) are necessary to achieve the fullest extent of *VIN3* induction by vernalisation (Kim et al., 2010). In *elf7* and *efs* mutants, the level of induced *VIN3* by vernalisation is lower than in the wild type, suggesting the role of these activating chromatin complexes is the induction of *VIN3*. Because *VIN3* is still induced by cold in *elf7* and *efs* mutants, it has been postulated that additional unknown components must operate to trigger the induction of *VIN3* independent of theses activating chromatin complexes.

2.8 CONCLUDING REMARKS

In recent years, significant progress has been made in understanding the molecular basis of vernalisation requirement and response by identifying several genes involved in the epigenetic regulation of *FLC*. Many studies have shown that the dynamic changes in histone modifications at *FLC* chromatin are a key aspect of vernalisation in *Arabidopsis*. PRC2 complex and lncRNAs play important roles in the epigenetic regulation of *FLC*. Although an lncRNA, COLDAIR, is required for the recruitment of PRC2 complex at *FLC* chromatin, the biochemical details explaining how COLDAIR operates to guide PRC2 to a specific locus, *FLC*, remain to be addressed. In addition, there are three known components in the vernalisation response that are known to be cold induced, *VIN3*, COOLAIR and COLDAIR. These three components exhibit different kinetics in their induction during cold, demanding an investigation to understand the mechanistic details of how cold coordinates the induction of their expressions. Despite recent progresses in understanding the molecular mechanisms underlying vernalisation, it is largely unknown how plants sense and measure the duration of cold. Efforts to identify upstream regulators for *VIN3*, COOLAIR and COLDAIR would be a solid ground to address the cold sensing mechanism in the vernalisation response.

ACKNOWLEDGEMENTS

Sung Lab is supported by grants from NIH (R01GM100108) and NSF (IOS0950785).

REFERENCES

Al-Shehbaz, I. A., & O'Kane, S. L., Jr. (2002). Taxonomy and phylogeny of Arabidopsis (brassicaceae). *Arabidopsis Book*, *1*, e0001. http://dx.doi.org/10.1199/tab.0001.

Alvarez-Venegas, R., Pien, S., Sadder, M., Witmer, X., Grossniklaus, U., & Avramova, Z. (2003). ATX-1, an Arabidopsis homolog of trithorax, activates flower homeotic genes. *Current Biology*, *13*, 627–637.

Amasino, R. (2004). Vernalization, competence, and the epigenetic memory of winter. *Plant Cell*, *16*, 2553–2559. http://dx.doi.org/10.1105/tpc.104.161070.

Andres, F., & Coupland, G. (2012). The genetic basis of flowering responses to seasonal cues. *Nature Reviews. Genetics*, *13*, 627–639. http://dx.doi.org/10.1038/nrg3291.

Bastow, R., Mylne, J. S., Lister, C., Lippman, Z., Martienssen, R. A., & Dean, C. (2004). Vernalization requires epigenetic silencing of FLC by histone methylation. *Nature*, *427*, 164–167. http://dx.doi.org/10.1038/nature02269.

Batista, P. J., & Chang, H. Y. (2013). Long noncoding RNAs: cellular address codes in development and disease. *Cell*, *152*, 1298–1307. http://dx.doi.org/10.1016/j.cell.2013.02.012.

Bienz, M. (2006). The PHD finger, a nuclear protein-interaction domain. *Trends in Biochemical Sciences*, *31*, 35–40. http://dx.doi.org/10.1016/j.tibs.2005.11.001.

Bond, D. M., Dennis, E. S., & Finnegan, E. J. (2011). The low temperature response pathways for cold acclimation and vernalization are independent. *Plant, Cell & Environment*, *34*, 1737–1748. http://dx.doi.org/10.1111/j.1365-3040.2011.02370.x.

Cao, Y., Dai, Y., Cui, S., & Ma, L. (2008). Histone H2B monoubiquitination in the chromatin of FLOWERING LOCUS C regulates flowering time in Arabidopsis. *Plant Cell*, *20*, 2586–2602. http://dx.doi.org/10.1105/tpc.108.062760.

Cao, R., Wang, L., Wang, H., Xia, L., Erdjument-Bromage, H., Tempst, P., et al. (2002). Role of histone H3 lysine 27 methylation in Polycomb-group silencing. *Science*, *298*, 1039–1043. http://dx.doi.org/10.1126/science.1076997.

Cao, R., & Zhang, Y. (2004). The functions of E(Z)/EZH2-mediated methylation of lysine 27 in histone H3. *Current Opinion in Genetics & Development*, *14*, 155–164. http://dx.doi.org/10.1016/j.gde.2004.02.001.

Chan, C. S., Rastelli, L., & Pirrotta, V. (1994). A Polycomb response element in the Ubx gene that determines an epigenetically inherited state of repression. *EMBO Journal*, *13*, 2553–2564.

Chandler, J., Wilson, A., & Dean, C. (1996). Arabidopsis mutants showing an altered response to vernalization. *Plant Journal*, *10*, 637–644.

Chanvivattana, Y., Bishopp, A., Schubert, D., Stock, C., Moon, Y. H., Sung, Z. R., et al. (2004). Interaction of Polycomb-group proteins controlling flowering in Arabidopsis. *Development, 131*, 5263–5276. http://dx.doi.org/10.1242/dev.01400.

Chinnusamy, V., Zhu, J., & Zhu, J. K. (2007). Cold stress regulation of gene expression in plants. *Trends in Plant Science, 12*, 444–451. http://dx.doi.org/10.1016/j.tplants.2007.07.002.

Choi, K., Kim, J., Hwang, H. J., Kim, S., Park, C., Kim, S. Y., et al. (2011). The FRIGIDA complex activates transcription of FLC, a strong flowering repressor in Arabidopsis, by recruiting chromatin modification factors. *Plant Cell, 23*, 289–303. http://dx.doi.org/10.1105/tpc.110.075911.

Choi, K., Kim, S., Kim, S. Y., Kim, M., Hyun, Y., Lee, H., et al. (2005). SUPPRESSOR OF FRIGIDA3 encodes a nuclear ACTIN-RELATED PROTEIN6 required for floral repression in Arabidopsis. *Plant Cell, 17*, 2647–2660. http://dx.doi.org/10.1105/tpc.105.035485.

Choi, K., Park, C., Lee, J., Oh, M., Noh, B., & Lee, I. (2007). Arabidopsis homologs of components of the SWR1 complex regulate flowering and plant development. *Development, 134*, 1931–1941. http://dx.doi.org/10.1242/dev.001891.

Chouard, P. (1960). Vernalization and its relations to dormancy. *Annual Review of Plant Physiology, 11*, 191–238. http://dx.doi.org/10.1146/annurev.pp.11.060160.001203.

Clarke, J. H., & Dean, C. (1994). Mapping FRI, a locus controlling flowering time and vernalization response in *Arabidopsis thaliana*. *Molecular & General Genetics, 242*, 81–89.

Corbesier, L., Vincent, C., Jang, S., Fornara, F., Fan, Q., Searle, I., et al. (2007). FT protein movement contributes to long-distance signaling in floral induction of Arabidopsis. *Science, 316*, 1030–1033. http://dx.doi.org/10.1126/science.1141752.

Czech, B., & Hannon, G. J. (2011). Small RNA sorting: matchmaking for Argonautes. *Nature Reviews. Genetics, 12*, 19–31. http://dx.doi.org/10.1038/nrg2916.

De Lucia, F., Crevillen, P., Jones, A. M., Greb, T., & Dean, C. (2008). A PHD-polycomb repressive complex 2 triggers the epigenetic silencing of FLC during vernalization. *Proceedings of the National Academy of Sciences of the United States of America, 105*, 16831–16836. http://dx.doi.org/10.1073/pnas.0808687105.

Deal, R. B., Kandasamy, M. K., McKinney, E. C., & Meagher, R. B. (2005). The nuclear actin-related protein ARP6 is a pleiotropic developmental regulator required for the maintenance of FLOWERING LOCUS C expression and repression of flowering in Arabidopsis. *Plant Cell, 17*, 2633–2646. http://dx.doi.org/10.1105/tpc.105.035196.

Deal, R. B., Topp, C. N., McKinney, E. C., & Meagher, R. B. (2007). Repression of flowering in Arabidopsis requires activation of FLOWERING LOCUS C expression by the histone variant H2A.Z. *Plant Cell, 19*, 74–83. http://dx.doi.org/10.1105/tpc.106.048447.

Deng, X., Gu, L., Liu, C., Lu, T., Lu, F., Lu, Z., et al. (2010). Arginine methylation mediated by the Arabidopsis homolog of PRMT5 is essential for proper pre-mRNA splicing. *Proceedings of the National Academy of Sciences of the United States of America, 107*, 19114–19119. http://dx.doi.org/10.1073/pnas.1009669107.

Ding, L., Kim, S.Y., & Michaels, S. D. (2013). FLOWERING LOCUS C EXPRESSOR family proteins regulate FLOWERING LOCUS C expression in both winter-annual and rapid-cycling Arabidopsis. *Plant Physiology, 163*, 243–252. http://dx.doi.org/10.1104/pp.113.223958.

Dong, C. H., Agarwal, M., Zhang, Y., Xie, Q., & Zhu, J. K. (2006). The negative regulator of plant cold responses, HOS1, is a RING E3 ligase that mediates the ubiquitination and degradation of ICE1. *Proceedings of the National Academy of Sciences of the United States of America, 103*, 8281–8286. http://dx.doi.org/10.1073/pnas.0602874103.

Eddy, S. R. (2001). Non-coding RNA genes and the modern RNA world. *Nature Reviews. Genetics, 2*, 919–929. http://dx.doi.org/10.1038/35103511.

Finnegan, E. J., Bond, D. M., Buzas, D. M., Goodrich, J., Helliwell, C. A., Tamada, Y., et al. (2011). Polycomb proteins regulate the quantitative induction of VERNALIZATION INSENSITIVE 3 in response to low temperatures. *Plant Journal, 65*, 382–391. http://dx.doi.org/10.1111/j.1365-313X.2010.04428.x.

Francis, N. J., Saurin, A. J., Shao, Z., & Kingston, R. E. (2001). Reconstitution of a functional core polycomb repressive complex. *Molecular Cell, 8*, 545–556.
Galvao, V. C., Horrer, D., Kuttner, F., & Schmid, M. (2012). Spatial control of flowering by DELLA proteins in *Arabidopsis thaliana*. *Development, 139*, 4072–4082. http://dx.doi.org/10.1242/dev.080879.
Gazzani, S., Gendall, A. R., Lister, C., & Dean, C. (2003). Analysis of the molecular basis of flowering time variation in Arabidopsis accessions. *Plant Physiol, 132*, 1107–1114. http://dx.doi.org/10.1104/pp.103.021212.
Gendall, A. R., Levy, Y. Y., Wilson, A., & Dean, C. (2001). The VERNALIZATION 2 gene mediates the epigenetic regulation of vernalization in Arabidopsis. *Cell, 107*, 525–535.
Gilmour, S. J., Zarka, D. G., Stockinger, E. J., Salazar, M. P., Houghton, J. M., & Thomashow, M. F. (1998). Low temperature regulation of the Arabidopsis CBF family of AP2 transcriptional activators as an early step in cold-induced COR gene expression. *Plant Journal, 16*, 433–442.
Greb, T., Mylne, J. S., Crevillen, P., Geraldo, N., An, H., Gendall, A. R., et al. (2007). The PHD finger protein VRN5 functions in the epigenetic silencing of Arabidopsis FLC. *Current Biology, 17*, 73–78. http://dx.doi.org/10.1016/j.cub.2006.11.052.
Gu, X., Jiang, D., Wang, Y., Bachmair, A., & He, Y. (2009). Repression of the floral transition via histone H2B monoubiquitination. *Plant Journal, 57*, 522–533. http://dx.doi.org/10.1111/j.1365-313X.2008.03709.x.
Gu, X., Le, C., Wang, Y., Li, Z., Jiang, D., & He, Y. (2013). Arabidopsis FLC clade members form flowering-repressor complexes coordinating responses to endogenous and environmental cues. *Nature Communications, 4*, 1947. http://dx.doi.org/10.1038/ncomms2947.
He, Y., Doyle, M. R., & Amasino, R. M. (2004). PAF1-complex-mediated histone methylation of FLOWERING LOCUS C chromatin is required for the vernalization-responsive, winter-annual habit in Arabidopsis. *Genes & Development, 18*, 2774–2784. http://dx.doi.org/10.1101/gad.1244504.
Hennig, L., & Derkacheva, M. (2009). Diversity of Polycomb group complexes in plants: same rules, different players? *Trends in Genetics, 25*, 414–423. http://dx.doi.org/10.1016/j.tig.2009.07.002.
Heo, J. B., & Sung, S. (2011). Vernalization-mediated epigenetic silencing by a long intronic noncoding RNA. *Science, 331*, 76–79. http://dx.doi.org/10.1126/science.1197349.
Hepworth, S. R., Valverde, F., Ravenscroft, D., Mouradov, A., & Coupland, G. (2002). Antagonistic regulation of flowering-time gene SOC1 by CONSTANS and FLC via separate promoter motifs. *EMBO Journal, 21*, 4327–4337.
Hong, S., Song, H. R., Lutz, K., Kerstetter, R. A., Michael, T. P., & McClung, C. R. (2010). Type II protein arginine methyltransferase 5 (PRMT5) is required for circadian period determination in Arabidopsis thaliana. *Proceedings of the National Academy of Sciences of the United States of America, 107*, 21211–21216. http://dx.doi.org/10.1073/pnas.1011987107.
Hsieh, T. F., Hakim, O., Ohad, N., & Fischer, R. L. (2003). From flour to flower: how Polycomb group proteins influence multiple aspects of plant development. *Trends in Plant Science, 8*, 439–445. http://dx.doi.org/10.1016/S1360-1385(03)00189-4.
Jiang, D., Gu, X., & He, Y. (2009). Establishment of the winter-annual growth habit via FRIGIDA-mediated histone methylation at FLOWERING LOCUS C in Arabidopsis. *Plant Cell, 21*, 1733–1746. http://dx.doi.org/10.1105/tpc.109.067967.
Jiang, D., Kong, N. C., Gu, X., Li, Z., & He, Y. (2011). Arabidopsis COMPASS-like complexes mediate histone H3 lysine-4 trimethylation to control floral transition and plant development. *PLoS Genetics, 7*, e1001330. http://dx.doi.org/10.1371/journal.pgen.1001330.
Jiang, D., Wang, Y., & He, Y. (2008). Repression of FLOWERING LOCUS C and FLOWERING LOCUS T by the Arabidopsis Polycomb repressive complex 2 components. *PLoS One, 3*, e3404. http://dx.doi.org/10.1371/journal.pone.0003404.

Johanson, U., West, J., Lister, C., Michaels, S., Amasino, R., & Dean, C. (2000). Molecular analysis of FRIGIDA, a major determinant of natural variation in Arabidopsis flowering time. *Science, 290*, 344–347.

Kamakaka, R. T., & Biggins, S. (2005). Histone variants: deviants? *Genes & Development, 19*, 295–310. http://dx.doi.org/10.1101/gad.1272805.

Kandasamy, M. K., Deal, R. B., McKinney, E. C., & Meagher, R. B. (2005). Silencing the nuclear actin-related protein AtARP4 in Arabidopsis has multiple effects on plant development, including early flowering and delayed floral senescence. *Plant Journal, 41*, 845–858. http://dx.doi.org/10.1111/j.1365-313X.2005.02345.x.

Kardailsky, I., Shukla, V. K., Ahn, J. H., Dagenais, N., Christensen, S. K., Nguyen, J. T., et al. (1999). Activation tagging of the floral inducer FT. *Science, 286*, 1962–1965.

Kim, D. H., Doyle, M. R., Sung, S., & Amasino, R. M. (2009). Vernalization: winter and the timing of flowering in plants. *Annual Review of Cell and Developmental Biology, 25*, 277–299. http://dx.doi.org/10.1146/annurev.cellbio.042308.113411.

Kim, D. H., & Sung, S. (2013). Coordination of the vernalization response through a VIN3 and FLC gene family regulatory network in Arabidopsis. *Plant Cell, 25*, 454–469. http://dx.doi.org/10.1105/tpc.112.104760.

Kim, D. H., Zografos, B. R., & Sung, S. (2010). Vernalization-mediated VIN3 Induction Overcomes the LIKE-HETEROCHROMATIN PROTEIN1/POLYCOMB REPRESSION COMPLEX2-mediated epigenetic repression. *Plant Physiolology, 154*, 949–957. http://dx.doi.org/10.1104/pp.110.161083.

Kim, S. Y., He, Y., Jacob, Y., Noh, Y. S., Michaels, S., & Amasino, R. (2005). Establishment of the vernalization-responsive, winter-annual habit in Arabidopsis requires a putative histone H3 methyl transferase. *Plant Cell, 17*, 3301–3310. http://dx.doi.org/10.1105/tpc.105.034645.

Ko, J. H., Mitina, I., Tamada, Y., Hyun, Y., Choi, Y., Amasino, R. M., et al. (2010). Growth habit determination by the balance of histone methylation activities in Arabidopsis. *EMBO Journal, 29*, 3208–3215. http://dx.doi.org/10.1038/emboj.2010.198.

Kobayashi, Y., Kaya, H., Goto, K., Iwabuchi, M., & Araki, T. (1999). A pair of related genes with antagonistic roles in mediating flowering signals. *Science, 286*, 1960–1962.

Kobor, M. S., Venkatasubrahmanyam, S., Meneghini, M. D., Gin, J. W., Jennings, J. L., Link, A. J., et al. (2004). A protein complex containing the conserved Swi2/Snf2-related ATPase Swr1p deposits histone variant H2A.Z into euchromatin. *PLoS Biology, 2*, E131. http://dx.doi.org/10.1371/journal.pbio.0020131.

Kohler, C., & Villar, C. B. (2008). Programming of gene expression by Polycomb group proteins. *Trends in Cell Biology, 18*, 236–243. http://dx.doi.org/10.1016/j.tcb.2008.02.005.

Krogan, N. J., Dover, J., Wood, A., Schneider, J., Heidt, J., Boateng, M. A., et al. (2003). The Paf1 complex is required for histone H3 methylation by COMPASS and Dot1p: linking transcriptional elongation to histone methylation. *Molecular Cell, 11*, 721–729.

Kung, J. T., Colognori, D., & Lee, J. T. (2013). Long noncoding RNAs: past, present, and future. *Genetics, 193*, 651–669. http://dx.doi.org/10.1534/genetics.112.146704.

Kuzmichev, A., Nishioka, K., Erdjument-Bromage, H., Tempst, P., & Reinberg, D. (2002). Histone methyltransferase activity associated with a human multiprotein complex containing the Enhancer of Zeste protein. *Genes & Development, 16*, 2893–2905. http://dx.doi.org/10.1101/gad.1035902.

Lachner, M., & Jenuwein, T. (2002). The many faces of histone lysine methylation. *Current Opinion in Cell Biology, 14*, 286–298.

Lang, A. (1965). Physiology of flower initiation. In A. Lang (Ed.), *Differenzierung und Entwicklung/Differentiation and Development* (Vol. 15, (pp. 1380–1536). Berlin Heidelberg: Springer.

Lazaro, A., Gomez-Zambrano, A., Lopez-Gonzalez, L., Pineiro, M., & Jarillo, J. A. (2008). Mutations in the Arabidopsis SWC6 gene, encoding a component of the SWR1 chromatin remodelling complex, accelerate flowering time and alter leaf and flower development. *Journal of Experimental Botany, 59*, 653–666. http://dx.doi.org/10.1093/jxb/erm332.

Lee, H., Xiong, L., Gong, Z., Ishitani, M., Stevenson, B., & Zhu, J. K. (2001). The Arabidopsis HOS1 gene negatively regulates cold signal transduction and encodes a RING finger protein that displays cold-regulated nucleo–cytoplasmic partitioning. *Genes & Development, 15*, 912–924. http://dx.doi.org/10.1101/gad.866801.

Lee, I., Bleecker, A., & Amasino, R. (1993). Analysis of naturally occurring late flowering in *Arabidopsis thaliana. Molecular & General Genetics, 237*, 171–176.

Levy, Y. Y., Mesnage, S., Mylne, J. S., Gendall, A. R., & Dean, C. (2002). Multiple roles of Arabidopsis VRN1 in vernalization and flowering time control. *Science, 297*, 243–246. http://dx.doi.org/10.1126/science.1072147.

Liu, J., Gilmour, S. J., Thomashow, M. F., & Van Nocker, S. (2002). Cold signalling associated with vernalization in *Arabidopsis thaliana* does not involve CBF1 or abscisic acid. *Physiologia Plantarum, 114*, 125–134.

Liu, Q., Kasuga, M., Sakuma, Y., Abe, H., Miura, S., Yamaguchi-Shinozaki, K., et al. (1998). Two transcription factors, DREB1 and DREB2, with an EREBP/AP2 DNA binding domain separate two cellular signal transduction pathways in drought- and low-temperature-responsive gene expression, respectively, in Arabidopsis. *Plant Cell, 10*, 1391–1406.

Maison, C., & Almouzni, G. (2004). HP1 and the dynamics of heterochromatin maintenance. *Nature Reviews. Molecular Cell Biology, 5*, 296–304. http://dx.doi.org/10.1038/nrm1355.

Martin-Trillo, M., Lazaro, A., Poethig, R. S., Gomez-Mena, C., Pineiro, M. A., Martinez-Zapater, J. M., et al. (2006). EARLY IN SHORT DAYS 1 (ESD1) encodes ACTIN-RELATED PROTEIN 6 (AtARP6), a putative component of chromatin remodelling complexes that positively regulates FLC accumulation in Arabidopsis. *Development, 133*, 1241–1252. http://dx.doi.org/10.1242/dev.02301.

Medina, J., Bargues, M., Terol, J., Perez-Alonso, M., & Salinas, J. (1999). The Arabidopsis CBF gene family is composed of three genes encoding AP2 domain-containing proteins whose expression is regulated by low temperature but not by abscisic acid or dehydration. *Plant Physiology, 119*, 463–470.

Mellor, J. (2006). It takes a PHD to read the histone code. *Cell, 126*, 22–24. http://dx.doi.org/10.1016/j.cell.2006.06.028.

Mercer, T. R., & Mattick, J. S. (2013). Structure and function of long noncoding RNAs in epigenetic regulation. *Nature Structural & Molecular Biology, 20*, 300–307. http://dx.doi.org/10.1038/nsmb.2480.

Metzger, J. D. (1988). Localization of the site of perception of thermoinductive temperatures in *Thlaspi arvense* L. *Plant Physiology, 88*, 424–428.

Michaels, S. D., & Amasino, R. M. (1999). FLOWERING LOCUS C encodes a novel MADS domain protein that acts as a repressor of flowering. *Plant Cell, 11*, 949–956.

Michaels, S. D., & Amasino, R. M. (2001). Loss of FLOWERING LOCUS C activity eliminates the late-flowering phenotype of FRIGIDA and autonomous pathway mutations but not responsiveness to vernalization. *Plant Cell, 13*, 935–941.

Michaels, S. D., Himelblau, E., Kim, S. Y., Schomburg, F. M., & Amasino, R. M. (2005). Integration of flowering signals in winter-annual Arabidopsis. *Plant Physiology, 137*, 149–156. http://dx.doi.org/10.1104/pp.104.052811.

Miura, K., & Furumoto, T. (2013). Cold signaling and cold response in plants. *International Journal of Molecular Sciences, 14*, 5312–5337. http://dx.doi.org/10.3390/ijms14035312.

Muller, J., Hart, C. M., Francis, N. J., Vargas, M. L., Sengupta, A., Wild, B., et al. (2002). Histone methyltransferase activity of a Drosophila Polycomb group repressor complex. *Cell, 111*, 197–208.
Muller, J., & Kassis, J. A. (2006). Polycomb response elements and targeting of Polycomb group proteins in Drosophila. *Current Opinion in Genetics & Development, 16*, 476–484. http://dx.doi.org/10.1016/j.gde.2006.08.005.
Napp-Zinn, K. (1987). Vernalization: environmental and genetic regulation. In J. G. Atherton (Ed.), *Manipulation of flowering* (pp. 123–132). London: Butterworths.
Noh, Y. S., & Amasino, R. M. (2003). PIE1, an ISWI family gene, is required for FLC activation and floral repression in Arabidopsis. *Plant Cell, 15*, 1671–1682.
Oh, S., Zhang, H., Ludwig, P., & van Nocker, S. (2004). A mechanism related to the yeast transcriptional regulator Paf1c is required for expression of the Arabidopsis FLC/MAF MADS box gene family. *Plant Cell, 16*, 2940–2953. http://dx.doi.org/10.1105/tpc.104.026062.
Pandey, R. R., Mondal, T., Mohammad, F., Enroth, S., Redrup, L., Komorowski, J., et al. (2008). Kcnq1ot1 antisense noncoding RNA mediates lineage-specific transcriptional silencing through chromatin-level regulation. *Molecular Cell, 32*, 232–246. http://dx.doi.org/10.1016/j.molcel.2008.08.022.
Pei, Y., Niu, L., Lu, F., Liu, C., Zhai, J., Kong, X., et al. (2007). Mutations in the Type II protein arginine methyltransferase AtPRMT5 result in pleiotropic developmental defects in Arabidopsis. *Plant Physiology, 144*, 1913–1923. http://dx.doi.org/10.1104/pp.107.099531.
Pien, S., Fleury, D., Mylne, J. S., Crevillen, P., Inze, D., Avramova, Z., et al. (2008). ARABIDOPSIS TRITHORAX1 dynamically regulates FLOWERING LOCUS C activation via histone 3 lysine 4 trimethylation. *Plant Cell, 20*, 580–588. http://dx.doi.org/10.1105/tpc.108.058172.
Porri, A., Torti, S., Romera-Branchat, M., & Coupland, G. (2012). Spatially distinct regulatory roles for gibberellins in the promotion of flowering of Arabidopsis under long photoperiods. *Development, 139*, 2198–2209. http://dx.doi.org/10.1242/dev.077164.
Ratcliffe, O. J., Kumimoto, R. W., Wong, B. J., & Riechmann, J. L. (2003). Analysis of the Arabidopsis MADS AFFECTING FLOWERING gene family: MAF2 prevents vernalization by short periods of cold. *Plant Cell, 15*, 1159–1169.
Ratcliffe, O. J., Nadzan, G. C., Reuber, T. L., & Riechmann, J. L. (2001). Regulation of flowering in Arabidopsis by an FLC homologue. *Plant Physiology, 126*, 122–132.
Rinn, J. L., Kertesz, M., Wang, J. K., Squazzo, S. L., Xu, X., Brugmann, S. A., et al. (2007). Functional demarcation of active and silent chromatin domains in human HOX loci by noncoding RNAs. *Cell, 129*, 1311–1323. http://dx.doi.org/10.1016/j.cell.2007.05.022.
Robzyk, K., Recht, J., & Osley, M. A. (2000). Rad6-dependent ubiquitination of histone H2B in yeast. *Science, 287*, 501–504.
Saksouk, N., Avvakumov, N., Champagne, K. S., Hung, T., Doyon, Y., Cayrou, C., et al. (2009). HBO1 HAT complexes target chromatin throughout gene coding regions via multiple PHD finger interactions with histone H3 tail. *Molecular Cell, 33*, 257–265. http://dx.doi.org/10.1016/j.molcel.2009.01.007.
Samach, A., Onouchi, H., Gold, S. E., Ditta, G. S., Schwarz-Sommer, Z., Yanofsky, M. F., et al. (2000). Distinct roles of CONSTANS target genes in reproductive development of Arabidopsis. *Science, 288*, 1613–1616.
Sanchez, S. E., Petrillo, E., Beckwith, E. J., Zhang, X., Rugnone, M. L., Hernando, C. E., et al. (2010). A methyl transferase links the circadian clock to the regulation of alternative splicing. *Nature, 468*, 112–116. http://dx.doi.org/10.1038/nature09470.
Schmitz, R. J., Sung, S., & Amasino, R. M. (2008). Histone arginine methylation is required for vernalization-induced epigenetic silencing of FLC in winter-annual *Arabidopsis thaliana. Proceedings of the National Academy of Sciences of the United States of America, 105*, 411–416. http://dx.doi.org/10.1073/pnas.0710423104.

Schmitz, R. J., Tamada, Y., Doyle, M. R., Zhang, X., & Amasino, R. M. (2009). Histone H2B deubiquitination is required for transcriptional activation of FLOWERING LOCUS C and for proper control of flowering in Arabidopsis. *Plant Physiology, 149*, 1196–1204. http://dx.doi.org/10.1104/pp.108.131508.

Schuettengruber, B., Chourrout, D., Vervoort, M., Leblanc, B., & Cavalli, G. (2007). Genome regulation by polycomb and trithorax proteins. *Cell, 128*, 735–745. http://dx.doi.org/10.1016/j.cell.2007.02.009.

Scortecci, K. C., Michaels, S. D., & Amasino, R. M. (2001). Identification of a MADS-box gene, FLOWERING LOCUS M, that represses flowering. *Plant Journal, 26*, 229–236.

Searle, I., He, Y., Turck, F., Vincent, C., Fornara, F., Krober, S., et al. (2006). The transcription factor FLC confers a flowering response to vernalization by repressing meristem competence and systemic signaling in Arabidopsis. *Genes & Development, 20*, 898–912. http://dx.doi.org/10.1101/gad.373506.

Seo, E., Lee, H., Jeon, J., Park, H., Kim, J., Noh, Y. S., et al. (2009). Crosstalk between cold response and flowering in Arabidopsis is mediated through the flowering-time gene SOC1 and its upstream negative regulator FLC. *Plant Cell, 21*, 3185–3197. http://dx.doi.org/10.1105/tpc.108.063883.

Sheldon, C. C., Conn, A. B., Dennis, E. S., & Peacock, W. J. (2002). Different regulatory regions are required for the vernalization-induced repression of FLOWERING LOCUS C and for the epigenetic maintenance of repression. *Plant Cell, 14*, 2527–2537.

Sheldon, C. C., Rouse, D. T., Finnegan, E. J., Peacock, W. J., & Dennis, E. S. (2000). The molecular basis of vernalization: the central role of FLOWERING LOCUS C (FLC). *Proceedings of the National Academy of Sciences of the United States of America, 97*, 3753–3758. http://dx.doi.org/10.1073/pnas.060023597.

Shilatifard, A. (2008). Molecular implementation and physiological roles for histone H3 lysine 4 (H3K4) methylation. *Current Opinion in Cell Biology, 20*, 341–348. http://dx.doi.org/10.1016/j.ceb.2008.03.019.

Simon, J., Chiang, A., Bender, W., Shimell, M. J., & O'Connor, M. (1993). Elements of the Drosophila bithorax complex that mediate repression by Polycomb group products. *Developmental Biology, 158*, 131–144. http://dx.doi.org/10.1006/dbio.1993.1174.

Southall, S. M., Wong, P. S., Odho, Z., Roe, S. M., & Wilson, J. R. (2009). Structural basis for the requirement of additional factors for MLL1 SET domain activity and recognition of epigenetic marks. *Molecular Cell, 33*, 181–191. http://dx.doi.org/10.1016/j.molcel.2008.12.029.

Srikanth, A., & Schmid, M. (2011). Regulation of flowering time: all roads lead to Rome. *Cellular and Molecular Life Sciences, 68*, 2013–2037. http://dx.doi.org/10.1007/s00018-011-0673-y.

Sung, S., & Amasino, R. M. (2004a). Vernalization and epigenetics: how plants remember winter. *Current Opinion in Plant Biology*, 7, 4–10.

Sung, S., & Amasino, R. M. (2004b). Vernalization in *Arabidopsis thaliana* is mediated by the PHD finger protein VIN3. *Nature, 427*, 159–164. http://dx.doi.org/10.1038/nature02195.

Sung, S., He, Y., Eshoo, T. W., Tamada, Y., Johnson, L., Nakahigashi, K., et al. (2006). Epigenetic maintenance of the vernalized state in *Arabidopsis thaliana* requires LIKE HETEROCHROMATIN PROTEIN 1. *Nature Genetics, 38*, 706–710. http://dx.doi.org/10.1038/ng1795.

Sung, S., Schmitz, R. J., & Amasino, R. M. (2006). A PHD finger protein involved in both the vernalization and photoperiod pathways in Arabidopsis. *Genes & Development, 20*, 3244–3248. http://dx.doi.org/10.1101/gad.1493306.

Swiezewski, S., Liu, F., Magusin, A., & Dean, C. (2009). Cold-induced silencing by long antisense transcripts of an Arabidopsis Polycomb target. *Nature, 462*, 799–802. http://dx.doi.org/10.1038/nature08618.

Takahashi, Y. H., Westfield, G. H., Oleskie, A. N., Trievel, R. C., Shilatifard, A., & Skiniotis, G. (2011). Structural analysis of the core COMPASS family of histone H3K4 methylases from yeast to human. *Proceedings of the National Academy of Sciences of the United States of America, 108*, 20526–20531. http://dx.doi.org/10.1073/pnas.1109360108.

Tamada, Y., Yun, J. Y., Woo, S. C., & Amasino, R. M. (2009). ARABIDOPSIS TRITHORAX-RELATED7 is required for methylation of lysine 4 of histone H3 and for transcriptional activation of FLOWERING LOCUS C. *Plant Cell, 21*, 3257–3269. http://dx.doi.org/10.1105/tpc.109.070060.

Thomashow, M. F. (1999). PLANT COLD ACCLIMATION: freezing tolerance genes and regulatory mechanisms. *Annual Review of Plant Physiology and Plant Molecular Biology, 50*, 571–599. http://dx.doi.org/10.1146/annurev.arplant.50.1.571.

Turck, F., Roudier, F., Farrona, S., Martin-Magniette, M. L., Guillaume, E., Buisine, N., et al. (2007). Arabidopsis TFL2/LHP1 specifically associates with genes marked by trimethylation of histone H3 lysine 27. *PLoS Genetics, 3*, e86. http://dx.doi.org/10.1371/journal.pgen.0030086.

Wang, X., Zhang, Y., Ma, Q., Zhang, Z., Xue, Y., Bao, S., et al. (2007). SKB1-mediated symmetric dimethylation of histone H4R3 controls flowering time in Arabidopsis. *EMBO Journal, 26*, 1934–1941. http://dx.doi.org/10.1038/sj.emboj.7601647.

Weigel, D. (2012). Natural variation in Arabidopsis: from molecular genetics to ecological genomics. *Plant Physiology, 158*, 2–22. http://dx.doi.org/10.1104/pp.111.189845.

Wellensiek, S. J. (1962). Dividing cells as the locus for vernalization. *Nature, 195*, 307–308.

Wood, A., Schneider, J., Dover, J., Johnston, M., & Shilatifard, A. (2003). The Paf1 complex is essential for histone monoubiquitination by the Rad6-Bre1 complex, which signals for histone methylation by COMPASS and Dot1p. *Journal of Biological Chemistry, 278*, 34739–34742. http://dx.doi.org/10.1074/jbc.C300269200.

Wood, C. C., Robertson, M., Tanner, G., Peacock, W. J., Dennis, E. S., & Helliwell, C. A. (2006). The *Arabidopsis thaliana* vernalization response requires a polycomb-like protein complex that also includes VERNALIZATION INSENSITIVE 3. *Proceedings of the National Academy of Sciences of the United States of America, 103*, 14631–14636. http://dx.doi.org/10.1073/pnas.0606385103.

Xu, L., Menard, R., Berr, A., Fuchs, J., Cognat, V., Meyer, D., et al. (2009). The E2 ubiquitin-conjugating enzymes, AtUBC1 and AtUBC2, play redundant roles and are involved in activation of FLC expression and repression of flowering in *Arabidopsis thaliana*. *Plant Journal, 57*, 279–288. http://dx.doi.org/10.1111/j.1365-313X.2008.03684.x.

Xu, L., Zhao, Z., Dong, A., Soubigou-Taconnat, L., Renou, J. P., Steinmetz, A., et al. (2008). Di- and tri- but not monomethylation on histone H3 lysine 36 marks active transcription of genes involved in flowering time regulation and other processes in *Arabidopsis thaliana*. *Molecular and Cell Biology, 28*, 1348–1360. http://dx.doi.org/10.1128/MCB.01607-07.

Yoo, S. K., Chung, K. S., Kim, J., Lee, J. H., Hong, S. M., Yoo, S. J., et al. (2005). CONSTANS activates SUPPRESSOR OF OVEREXPRESSION OF CONSTANS 1 through FLOWERING LOCUS T to promote flowering in Arabidopsis. *Plant Physiology, 139*, 770–778. http://dx.doi.org/10.1104/pp.105.066928.

Zarka, D. G., Vogel, J. T., Cook, D., & Thomashow, M. F. (2003). Cold induction of Arabidopsis CBF genes involves multiple ICE (inducer of CBF expression) promoter elements and a cold-regulatory circuit that is desensitized by low temperature. *Plant Physiology, 133*, 910–918. http://dx.doi.org/10.1104/pp.103.027169.

Zhang, H., Ransom, C., Ludwig, P., & van Nocker, S. (2003). Genetic analysis of early flowering mutants in Arabidopsis defines a class of pleiotropic developmental regulator required for expression of the flowering-time switch flowering locus C. *Genetics, 164*, 347–358.

Zhang, H., & van Nocker, S. (2002). The VERNALIZATION INDEPENDENCE 4 gene encodes a novel regulator of FLOWERING LOCUS C. *Plant Journal*, *31*, 663–673.

Zhang, X., Germann, S., Blus, B. J., Khorasanizadeh, S., Gaudin, V., & Jacobsen, S. E. (2007). The Arabidopsis LHP1 protein colocalizes with histone H3 Lys27 trimethylation. *Nature Structural & Molecular Biology*, *14*, 869–871. http://dx.doi.org/10.1038/nsmb1283.

Zhang, Z., Zhang, S., Zhang, Y., Wang, X., Li, D., Li, Q., et al. (2011). Arabidopsis floral initiator SKB1 confers high salt tolerance by regulating transcription and pre-mRNA splicing through altering histone H4R3 and small nuclear ribonucleoprotein LSM4 methylation. *Plant Cell*, *23*, 396–411. http://dx.doi.org/10.1105/tpc.110.081356.

Zhao, J., Sun, B. K., Erwin, J. A., Song, J. J., & Lee, J. T. (2008). Polycomb proteins targeted by a short repeat RNA to the mouse X chromosome. *Science*, *322*, 750–756. http://dx.doi.org/10.1126/science.1163045.

Zhao, Z., Yu, Y., Meyer, D., Wu, C., & Shen, W. H. (2005). Prevention of early flowering by expression of FLOWERING LOCUS C requires methylation of histone H3 K36. *Nature Cell Biology*, 7, 1256–1260. http://dx.doi.org/10.1038/ncb1329.

CHAPTER THREE

Regulation of Flowering by Endogenous Signals

Vinicius Costa Galvão* and Markus Schmid†,[1]
*Center for Integrative Genomics, Faculty of Biology and Medicine, University of Lausanne, Lausanne, Switzerland
†Max Planck Institute for Developmental Biology, Tuebingen, Germany
[1]Corresponding author: e-mail address: markus.schmid@tuebingen.mpg.de

Contents

Abstract

The transition from vegetative to reproductive development, or floral transition, is a crucial event in the life cycle of plants. Work carried out over the last decades has shown how environmental signals, such as seasonal changes in the day length and temperature, are perceived and accurately integrated into genetically defined pathways to properly time the induction of flowering. In addition to seasonal fluctuations, plants must cope with a vast array of often stressful conditions that greatly affect metabolism and physiology. In this context, plant hormones and sugars have emerged as important endogenous signalling molecules mediating the transition to

Advances in Botanical Research, Volume 72
ISSN 0065-2296
http://dx.doi.org/10.1016/B978-0-12-417162-6.00003-1

the reproductive phase. In this chapter we report the recent advances in understanding the molecular basis underlying the transition to flowering in response to these endogenous signals.

3.1 INTRODUCTION

The transition from vegetative growth to the reproductive development represents one of the most remarkable examples of how flowering plants are adapted to their environment. The correct timing of this important phase transition not only guarantees the appropriate development of the reproductive organs from the shoot apical meristem (SAM), but it also ensures that each individual will generate offspring under the most favourable conditions, thus maximising the reproductive success. Therefore it is not surprising that during their evolution, plants have developed sophisticated and highly sensitive mechanisms to cope with environmental cues such as temperature and photoperiod as detailed in chapters 1, 2 and 4.

Plants perceive day length in leaves and under inductive photoperiods produce a mobile long-distance signal, the so-called florigen, which induces flowering at the SAM. While its molecular nature remained elusive for many years, recent observations indicate that the small globular protein FLOWERING LOCUS T (FT) serves as a florigen in a variety of plant species. In *Arabidopsis thaliana*, expression of *FT* in the leaf vasculature is controlled by a complex regulatory network that involves the GIGANTEA (GI) and CONSTANS (CO) proteins (An, Roussot, Suarez-Lopez, Corbesier, Vincent, Pineiro, 2004; Imaizumi, Schultz, Harmon, Ho, & Kay, 2005; Samach et al., 2000; Sawa & Kay, 2011; Suarez-Lopez et al., 2001). At the SAM, FT activates flowering presumably through interaction with the b-Zip transcription factor FD and 14-3-3 proteins (Taoka et al., 2011; Wigge et al., 2005). Among the genes regulated by the FT–FD–14-3-3 activator complex, the MADS-box transcription factor *SUPPRESSOR OF OVEREXPRESSION OF CO 1* (*SOC1*) is an important gene integrating diverse environmental flowering signals. Genome-wide studies have shown that SOC1 binds to the regulatory region of several flowering-time genes to control their expression (Immink et al., 2012; Tao et al., 2012). Among them, *AGAMOUS-LIKE 24* (*AGL24*) seems to be particularly important for SOC1 activity during the flowering transition. Using expression and chromatin immunoprecipitation analyses, Liu and colleagues demonstrated that mutual transcriptional control between SOC1 and AGL24 creates a feed-forward regulatory loop, critical for the induction of flowering at the SAM (Liu et al., 2008).

Temperature has also been shown to affect flowering time in a variety of plant species. Vernalisation, the prolonged exposure to cold (overwintering), is required in many plant species, including certain natural accessions of *A. thaliana*, to ensure synchronous and rapid flowering in spring. In *A. thaliana*, vernalisation is mediated by the activity of the MADS-box transcription factor *FLOWERING LOCUS C* (*FLC*), which encodes a potent repressor of flowering, and *FRIGIDA* (*FRI*) (Michaels & Amasino, 1999a; Johanson et al., 2000). *FLC* is expressed at high levels in non-vernalised plants and epigenetically silenced upon exposure to cold (Bastow et al., 2004; Michaels & Amasino, 1999a) and Chapter 2. In contrast to vernalisation, which in *A. thaliana* is fairly well understood, the molecular mechanisms that regulate flowering in response to changes in ambient temperature are less clear. Recent analyses have reported that the temperature-dependent eviction of a special histone variant, H2A.Z, contributes to the regulation of the expression of flowering-time genes such as *FT* under non-inductive photoperiod (Kumar et al., 2012; Kumar & Wigge, 2010). In addition, accumulation of a repressor complex containing the MADS-box proteins FLC, SHORT VEGETATIVE PHASE (SVP) and FLOWERING LOCUS M seems to regulate temperature-dependent flowering in *A. thaliana* (Balasubramanian, Sureshkumar, Lempe, & Weigel, 2006; Lee et al., 2013; Lee et al., 2007; Li et al., 2008; Pose et al., 2013). For a more detailed analysis of the genetic basis of the flowering time regulation by environmental cues please see recent reviews (e.g. Andres & Coupland, 2012; Srikanth & Schmid, 2011) as well as chapters 1, 2 and 4.

In addition to the pathways that promote flowering in response to environmental signals, a recently discovered endogenous pathway, the so-called age pathway, ensures that flowering will eventually be initiated even under otherwise non-inductive conditions. This pathway relies on the activity of microRNA156 (miR156) and *SQUAMOSA PROMOTER BINDING PROTEIN LIKE* (*SPL*) transcription factors. *MIR156* genes are highly expressed during the early stages of vegetative development, and decrease gradually as plants age. Conversely, *SPL* transcript levels are at a minimum during early vegetative stages, and increase later in development (Wang & Weigel, 2009). Molecular analyses have shown that SPL proteins control flowering by directly binding to flowering-time flower patterning genes such as *LEAFY* (*LFY*), *APETALA1* (*AP1*), *SOC1*, *FRUITFUL* (*FUL*) and *MIR172* (Wang & Weigel, 2009; Wu et al., 2009; Yamaguchi et al., 2009).

While these studies provide us with a rather detailed understanding as to how flowering is regulated in response to environmental cues, how exactly endogenous signals, such as hormones and carbohydrates, contribute to the regulation of flowering is for the most part less well understood.

3.2 REGULATION OF FLOWERING BY PLANT HORMONES

Initial studies to investigate the effect of hormones on plant development in general and the regulation of flowering time in particular mostly depended on the exogenous application of these important growth regulators. Based on these experiments it became apparent that, rather than always acting as positive or negative regulator of flowering, the mode of action of a given hormone varied between plants, inducing flowering in one while having the opposite effect in another species. As a result, the mode of action by which plant hormones control the floral transition remained unresolved for many years. It was only after the identification and molecular characterisation of mutants defective in hormones biosynthesis and signalling that this somewhat unsatisfactory situation began to change. Here, we will review the recent advances that have been made in understanding the molecular bases of the regulation of flowering by plant hormones. Most progress in this field has been made in model plant species such as *A. thaliana*, but we will also refer to results from other species where appropriate.

3.2.1 Gibberellic Acid

Among the five classical plant hormones abscisic acid (ABA), auxin, cytokinin, ethylene and gibberellic acid (GA), the latter has attracted the greatest attention among scientists investigating the mechanisms of reproductive transitions because of its absolute requirement for flowering under non-inductive short days (SD) in *A. thaliana* (Wilson, Heckman, & Sommerville, 1992). Therefore, it is not surprising that the molecular mechanisms by which GA regulates the transition to flowering have been studied extensively and is now relatively well understood. One outcome of these analyses was that GA signalling, rather than acting independently and endogenously, is linked to genetic pathways that regulate flowering in response to environmental stimuli, such as photoperiod and temperature. In addition, other hormonal signalling pathways are tightly interconnected with GA signalling as well. Therefore, it is possible that other hormones control the flowering transition indirectly by modulating GA biosynthesis and signalling. We will present a general overview on the regulation of flowering by GA and briefly introduce the most important components of the GA signalling pathway before we continue with a discussion of how these proteins function to regulate flowering.

3.2.1.1 Exogenous Hormone Application and Mutant Analysis Reveal a Role for GA in Flowering

First identified in rice infected with the fungus *Gibberella fujikuroi* (Kurosawa, 1926), gibberellins encompass a class of chemically related compounds, of which only a minor proportion is biologically active (Olszewski, Sun, & Gubler, 2002). After the isolation and discovery of its diterpenoid nature (Curtis & Cross, 1954; Takahashi et al., 1955), several reports demonstrated that GA plays a critical role in several developmental processes such as seed germination, internode elongation, fruit and flower development and control of flowering time (reviewed in Davies, 2004).

Experiments performed by Anton Lang in *Hyoscyamus niger* and several other species for the first time demonstrated the inductive effect of GA on flowering (Lang, 1956a, 1956b, 1957). Later it was shown that GA inhibited flowering in citrus and strawberry (Guardiola, Monerri, & Agusti, 1982), while other species, such as soybean grown under non-inductive long day (LD) photoperiods, did not respond to GA application or showed a mild effect, as in the case of *Daucus carota* (Lang, 1957). These results indicated that GA affects flowering in a species-dependent manner (reviewed in Pharis & King, 1985; Zeevaart, 1976).

Despite the initial excitement caused by these findings, it took several decades before the basic molecular mechanisms underlying the regulation of flowering by GA became clear. Most of the genetic and molecular data accumulated so far came after the identification and characterisation of mutants defective in GA biosynthesis and signalling in pioneering studies performed by Maarten Koornneef in the model plant *A. thaliana* (Koornneef et al., 1985; Koornneef & van der Veen, 1980). Among the mutants identified in these genetic screens was a recessive loss-of-function mutation in the *GA1* gene (*ga1-3*), which encodes an ent-copalyl diphosphate synthase that catalyses the first rate-limiting step of GA biosynthesis (Sun & Kamiya, 1994), resulting in a reduction in the levels of bioactive GA (Silverstone et al., 2001). *ga1-3* mutant plants display a pleiotropic phenotype, including dark green colour, dwarfism and compromised flower development (Figure 3.1). Strikingly, *ga1-3* mutant completely failed to flower under SD conditions, but flowered at approximately the same time as wild type plants under inductive LD, indicating that GA plays a critical role controlling the induction of flowering in *A. thaliana* only under non-inductive

Figure 3.1 Effect of mutations in gibberellic acid (GA) biosynthesis and signalling genes on flowering in *Arabidopsis thaliana.* Wild type *Arabidopsis thaliana* (L*er*-1), GA signalling mutant lacking four DELLA proteins (*ga1-3 gai-t6 rga-t2 rgl1-1 rgl2-1*) and the GA biosynthesis mutant *ga1-3* grown under long day photoperiod at 23 °C. The *ga1-3* mutant flowers slightly late compared to L*er*-1 plants, while *ga1-3 gai-t6 rga-t2 rgl1-1 rgl2-1* flowers early. (See the colour plate.)

photoperiods (Wilson, Heckman, & Sommerville, Wilson, 1992). More recently, however, a *ga1* allele that displayed severe late flowering in LD has been described in the Col-0 accession (Richter et al., 2013a), suggesting that GA signalling might contribute to the control of flowering time in *A. thaliana*, irrespective of day length. The reason for the distinct flowering behaviour observed in *ga1-3* (L*er*-1) relative to *ga1* (Col-0) is still unclear (Richter et al., 2013a; Wilson et al., 1992). However, it has been shown that GA biosynthesis itself is regulated by photoperiod in several species, which could indicate a role for GA in the regulation of flowering in response to photoperiod (Garcia-Martinez & Gil, 2001; Kamiya & Garcia-Martinez, 1999; Weller, Hecht, Vander Schoor, Davidson, & Ross, 2009).

3.2.1.2 GA Perception and Signalling

GA is perceived by the GIBBERELLIN INSENSITIVE DWARF1 (GID1) receptor, which was originally identified in rice (Ueguchi-Tanaka et al., 2005). *A. thaliana* contains three highly redundant GA receptor genes,

GID1a-c (Griffiths et al., 2006; Nakajima et al., 2006; Ueguchi-Tanaka et al., 2005; Willige et al., 2007). Loss of individual *GID1* genes has only a minor or no effect at all on growth and development, whereas the triple *gid1a-c* mutant displays all phenotypes commonly observed in GA biosynthesis mutants such as *ga1-3*, including dwarfism, dark green colour, impaired seed germination and severely delayed flowering, irrespective of the photoperiod (Griffiths et al., 2006; Iuchi et al., 2007; Willige et al., 2007). In contrast to mutants defective in GA biosynthesis, *gid1a-c* plants are completely insensitive to treatments with exogenous bioactive GA, confirming the importance of GID1 in GA perception (Griffiths et al., 2006; Ueguchi-Tanaka et al., 2005; Willige et al., 2007). Structural analyses have highlighted the importance of conformational changes of GID1 after binding to bioactive GA as a key event in GA signalling (Murase, Hirano, Sun, & Hakoshima, 2008; Shimada et al., 2008). After binding, bioactive GA is locked in a pocket of GID1 through its N-terminal region (Murase et al., 2008), which creates a hydrophobic surface on the GID1–GA complex that facilitates the interaction with a class of growth repressors called DELLA proteins (Griffiths et al., 2006; Nakajima et al., 2006; Willige et al., 2007).

DELLA proteins are important negative regulators of GA signalling, which were first identified in *A. thaliana* in a genetic screen for GA insensitive mutants (Koornneef et al., 1985; Silverstone, Mak, Martinez, & Sun, 1997). This screen recovered a dominant mutation in the *GIBBERELLIC ACID INSENSITIVE* (*GAI*) gene, *gai-1*, which resembled GA biosynthesis mutants but could not be rescued by exogenous GA (Koornneef et al., 1985). In contrast, loss-of-function DELLA mutants display a constitutive GA response and strongly suppress the GA-deficient phenotype of *ga1-3* (Figure 3.1) (Cheng et al., 2004; Dill & Sun, 2001). Cloning of *GAI* and analysis of the *gai-1* mutant allele revealed an in-frame 17 amino acids deletion that removes a conserved five amino acids motif (DELLA) in the N-terminal region of GAI (Peng et al., 1997). Further analyses demonstrated that the DELLA motif is required for interaction between GAI and the GID1 receptor in the presence of GA. This interaction further stabilises the GID1–GA–DELLA complex, and promotes its interaction with the F-box protein SLEEPY1, which is part of the SKP1-CUL1-F-BOX PROTEIN (SCF) E3 ubiquitin ligase complex (Dill, Jung, & Sun, 2001; Dill & Sun, 2001; Dill, Thomas, Hu, Steber, & Sun, 2004; Griffiths et al., 2006; Murase et al., 2008; Peng et al., 1997; Peng & Harberd, 1993; Willige et al., 2007; McGinnis et al., 2003). Ultimately, GA binding to GID1 results in the ubiquitination and degradation of DELLA proteins via the proteasome.

DELLA proteins have a conserved role in GA signalling in several species, including the *DWARF-8* gene in maize and *REDUCED HEIGHT 1* (*rht-B1*) and *rht-D1* genes in wheat, the latter of which were instrumental in the green revolution (Peng et al., 1999). Transcriptome analyses in *A. thaliana* revealed that GA regulates the expression of a large number of genes involved in a wide range of biological processes in a DELLA-dependent manner (Willige et al., 2007; Zentella et al., 2007). However, DELLA proteins do not contain any canonical DNA binding domain, suggesting that they function as co-factors rather than direct transcriptional regulators (Bolle, 2004). A ground-breaking observation on DELLA function came from the study of PHYTOCHROME INTERACTING FACTOR (PIF) proteins during photomorphogenesis. Two independent studies described the direct interaction between DELLA proteins and PIF transcription factors (de Lucas et al., 2008; Feng et al., 2008). Interestingly, EMSA assay demonstrated that binding to DELLA proteins impaired the capacity of PIF4 to bind to DNA, therefore blocking its activity (de Lucas et al., 2008). More recent studies have shown that in addition to preventing transcription factors from binding to DNA, DELLA proteins can bind to DNA when associated with BOTRYTIS SUSCEPTIBLE 1 INTERACTOR (BOI) and BOI-RELATED GENE 1 (BRG1), BRG2 and BRG3 (collectively referred to as BOI transcription factors) (Park, Nguyen, Park, Jeon, & Choi, 2013). In addition, DELLA proteins have been shown to control the sub-cellular localisation of prefoldin to regulate cortical microtubules organisation (Locascio, Blazquez, & Alabadi, 2013), and directly regulate the chromatin remodelling protein SWITCH/SUCROSE NONFERMENTING 3C (Sarnowska et al., 2013). Based on these findings it is now believed that DELLA proteins activity is dependent on direct protein–protein interactions.

Another important player in the GA signalling pathway is *SPINDLY* (*SPY*), which encodes an O-linked N-acetylglucosamine transferase that regulates DELLA function through post-translational modifications. *spy* mutants flower early under LD conditions (Jacobsen & Olszewski, 1993), and are partially epistatic to strong GA biosynthesis mutants and dominant DELLA mutants, despite the high levels of DELLA proteins found in the latter (Jacobsen & Olszewski, 1993; Silverstone et al., 2007).

3.2.1.3 Integration of GA Signalling at the SAM

As mentioned before, the transition to flowering is mainly controlled in two separate tissues, the leaf phloem companion cells, in which photoperiod is

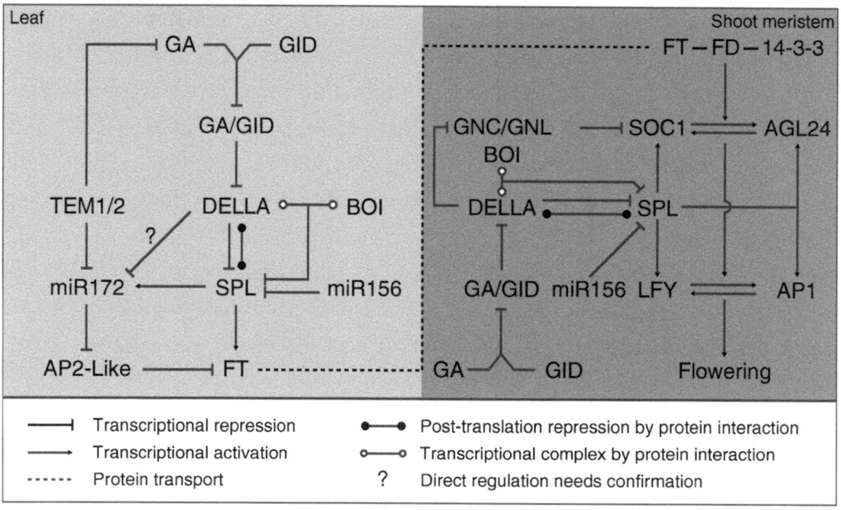

Figure 3.2 Spatial separation of gibberellic acid (GA) signalling events in leaves and at the shoot apical meristem (SAM). Depicted is the integration of GA signalling into the photoperiod pathway in leaves (light grey) and at the SAM (dark grey).

perceived, and the shoot meristem. While the findings discussed above provide a framework for the general mechanism of GA perception and signalling, they do not explain how exactly GA contributes to the temporal and spatial control of flowering (Figure 3.2).

This question has recently been addressed by tissue-specific misexpression of dominant (degradation insensitive) versions of DELLA proteins (*dellaΔ17*) and GA catabolic enzymes, which reduce the availability of bioactive GA (Galvão, Horrer, Kuttner, & Schmid, 2012; Porri, Torti, Romera-Branchat, & Coupland, 2012; Yu et al., 2012). Expression of these genes at the SAM consistently delayed flowering under both SD and LD (Galvão et al., 2012; Porri et al., 2012). In contrast, expression in the leaf vasculature from the phloem companion cell specific *SUCROSE-PROTON SYMPORTER 2* (*SUC2*) promoter had hardly any effect on flowering in SD, suggesting that GA regulates flowering under SD predominantly at the SAM (Galvão et al., 2012; Porri et al., 2012). This is in agreement with other studies that reported an increase in GA levels at the SAM both prior to flowering under SD and shortly after exposure to inductive LD in both *A. thaliana* and *Lolium* (Eriksson, Bohlenius, Moritz, & Nilsson, 2006; King, Moritz, Evans, Junttila, & Herlt, 2001; MacMillan, Blundell, & King, 2005).

At the SAM, GA seems to activate flowering at least in part through the miR156-targeted *SPL* transcription factors (Figure 3.2) (Yu et al., 2012).

Several reports demonstrated that accumulation of DELLA proteins at the SAM impairs the gradual increase of *SPL* transcripts normally associated with the age-dependent decline in miR156, thus repressing flowering (Galvão et al., 2012; Jung et al., 2012; Porri et al., 2012; Yu et al., 2012). How exactly GA regulates expression of *SPL* genes has not been determined, but apparently depends on miR156-independent direct interaction between DELLA proteins and BOI transcription factors (Galvão et al., 2012; Jung et al., 2012; Park et al., 2013). BOI proteins have been shown to affect several typical GA-responses, such as germination, juvenile-to-adult transition and flowering. BOI proteins can interact with DELLA proteins to form a complex and directly repress GA-responsive genes (Park et al., 2013). The quadruple *boi* mutant (*boi-Q*) flowers early in both SD and LD and shows increased *SPL3*, *SPL4* and *SPL5* expression (Park et al., 2013). Conversely, BOI overexpressing lines flower significantly late in SD and LD. Supporting a role of BOI proteins in the regulation of flowering, *boi-Q* greatly suppresses the late flowering of *gai-1* mutant in LD (Park et al., 2013). However, at the current stage it is not entirely clear whether the BOI–DELLA complex regulates *SPL* expression through direct binding to *SPL* promoters or whether this regulation occurs indirectly. In addition, DELLA proteins have been shown to directly interact with certain SPL proteins, adding SPL proteins to the list of post-transcriptional DELLA targets (Yu et al., 2012).

Another important integrator of diverse flowering-time signals, including GA, is the MADS-box transcription factor *SOC1*, whose expression is induced by GA, and accelerates flowering in SD (Moon et al., 2003). In addition, loss of *SOC1* attenuates the early flowering normally observed in plants treated with exogenous GA, whereas the gain-of-function *soc1-101D* line was found to be largely insensitive to the GA biosynthesis inhibitor paclobutrazol (Moon et al., 2003). It has recently been proposed that *SOC1* and the MADS-box transcription factor *FUL* regulate flowering in SD in response to GA downstream of the age pathway (Figure 3.2) (Yu et al., 2012). According to this model, the reduction of GA levels and consequently higher DELLA accumulation result in the transcriptional repression of *SPL3, SPL4* and *SPL5*, and post-transcriptional repression of SPL9 through direct interaction with DELLA proteins. In turn, the reduced SPL activity causes a reduction in *SOC1* and *FUL* expression, delaying flowering (Yu et al., 2012). Moreover, it has been shown that the inductive effect of GA on *AGL24* and *SOC1* expression was nearly abolished in *soc1* or *agl24* single mutants, respectively, indicating that the interaction between these two genes is important for the amplification of the GA signal (Liu et al., 2008). Nevertheless, treatment of the non-flowering *soc1 agl24* double mutant with GA is sufficient to induce flowering in SD, indicating that other genes contribute to the induction of flowering in response to GA at the SAM.

In contrast to its well-established role in SD, whether *SOC1* participates in mediating GA-dependent flowering under LD is still under debate. On the one hand, analysis of dissected meristems and *in situ* hybridisation of *pKNAT1::GA2ox7* and *pFD::dellaΔ17* lines indicated that GA has only a very mild or no effect on *SOC1* expression at the SAM under inductive LD. Instead, the late flowering observed in these lines was attributed to the reduced expression of *SPL* genes downstream of *SOC1* (Galvão et al., 2012; Porri et al., 2012). On the other hand, a recent report attributes the severe late flowering phenotype of the Col-0 *ga1* allele under LD to reduced *SOC1* and *FT* expression (Richter et al., 2013a). The authors demonstrated that the reduction of *SOC1* expression in the *ga1* background was due to the activity of the GATA transcription factors *GATA/NITRATE-INDUCIBLE/CARBON-METABOLISM INVOLVED* (*GNC*) and *GNC-LIKE/CYTOKININ-RESPONSIVE GATA FACTOR1* (*GNL/CGA1*) (Figure 3.2) (Richter et al., 2013a; Richter, Behringer, Muller, & Schwechheimer, 2010). GNC/GNL have been shown to act as negative regulators of GA signalling and are repressed by GA in a DELLA-dependent manner (Richter, Behringer, Muller, & Schwechheimer, 2010). Mutations in *GNC/GNL* have a very mild flowering phenotype under LD, while their overexpression results in late flowering and reduced *SOC1* expression independently of FT (Richter et al., 2013a). In addition, GNC/GNL directly bind to the *SOC1* promoter to regulate its expression, presumably accounting for the observed changes in flowering time (Richter et al., 2013a). It should be noted, however, that the *gnc gnl* double mutant slightly suppresses the extreme late flowering phenotype of the *ga1* mutant under LD. Taken together, these results strongly indicate that the regulation of *SOC1* by GA has a minor effect on flowering regulation under LD and suggest that a *GNC/GNL*-independent pathway contributes to the regulation of flowering transition at the SAM. Additional experiments addressing the genetic interaction between *GNC/GNL* and SPL proteins are clearly required to solve this question.

The floral meristem identity gene *LFY* constitutes another important hub of GA signal integration (Blazquez, Soowal, Lee, & Weigel, 1997; Weigel, Alvarez, Smyth, Yanofsky, & Meyerowitz, 1992). Application of GA has been shown to enhance *LFY* expression and the activity of a *pLFY::GUS* reporter was reduced in the non-flowering *ga1-3* background

(Blazquez, Green, Nilsson, Sussman, & Weigel, 1998; Blazquez et al., 1997). Analysis of the *LFY* promoter has identified a small GA-responsive *cis* element, which contained a potential MYB (myeloblastosis) transcription factor binding site (Blazquez & Weigel, 2000; Gocal et al., 2001). GA has been shown to induce the expression of so-called *GAMYB* genes in *Hordeum vulgare* and *Lolium temulentum* (Gocal et al., 1999). Also in *A. thaliana*, expression of *MYB33*, which is closely related to *HvGAMYB*, was strongly promoted at the SAM in response to GA application (Gocal et al., 1999; Gocal et al., 2001), suggesting that this class of transcription factors might fulfil an evolutionary conserved role in GA signalling. Interestingly, *GAMYB* genes are direct targets of miR159 (Rhoades et al., 2002). Achard and colleagues have shown that GA regulates miR159 levels in *A. thaliana*, suggesting a possible role in regulating *MYB33* and *LFY* expression in response to GA (Achard, Herr, Baulcombe, & Harberd, 2004). In agreement with LFY functioning downstream of GA, constitutive *LFY* expression was found sufficient to restore flowering in *ga1-3* in SD (Blazquez et al., 1998). More recently, ChIP-seq experiments demonstrated that LFY binds to the regulatory elements of several GA biosynthesis and signalling genes, which might contribute to the increase in GA levels previously observed to occur at the SAM at the time of floral transition (Eriksson et al., 2006; Moyroud et al., 2011). Taken together these studies provide compelling evidence that GA signalling contributes to the regulation of key flowering time and floral meristem identity genes at the SAM and constitutes an important regulatory node in the control of flowering.

3.2.1.4 GA Signalling Modulates Flowering under Inductive Photoperiod in Leaves

Besides the SAM, leaves have been shown to play an important role in the regulation of flowering in many plant species. Inductive photoperiods are perceived in leaves and ultimately result in the production of the FT protein, which has been shown to have florigenic function in a variety of plants (Corbesier et al., 2007; Jaeger & Wigge, 2007; Kojima et al., 2002; Mathieu, Warthmann, Kuttner, & Schmid, 2007; Srikanth & Schmid, 2011).

As mentioned above, the *A. thaliana ga1-3* mutant (L*er*-1 background) flowered only slightly later than control plants under LD, but failed to flower under SD. This was initially considered as evidence that GA did not contribute to the regulation of flowering under inductive photoperiods (Wilson et al., 1992). However, several lines of evidence suggest that GA contributes substantially to the regulation of flowering in *A. thaliana* under

LD. In particular, the late or non-flowering phenotype of the triple *gid1a-c* mutant clearly indicated a role for GA signalling in the control of flowering time under LD conditions (Griffiths et al., 2006; Willige et al., 2007). Similarly, depletion of bioactive GA by overexpression of the GA catabolic enzymes *GIBBERELLIN-2-OXIDASE 7* (*GA2OX7*) and *GA2OX8* has been shown to delay flowering under LD (Galvão et al., 2012; Porri et al., 2012; Schomburg, Bizzell, Lee, Zeevaart, & Amasino, 2003), while the overexpression of the GA metabolic enzyme *GIBBERELLIN-20-OXIDASE1* (*GA20OX1/GA5*) gene induced flowering (Domagalska, Sarnowska, Nagy, & Davis, 2010). Therefore, it is now clear that GA contributes to the control of flowering both under inductive and non-inductive photoperiods.

More detailed analyses revealed that, in addition to its effects at the SAM, GA promoted flowering through transcriptional activation of *FT* in leaves (Figure 3.2) (Galvão et al., 2012; Hisamatsu & King, 2008; Porri et al., 2012; Richter et al., 2013a; Yu et al., 2012). *FT* expression was found to be significantly reduced in late flowering *pSUC2::dellaΔ17* and *pSUC2::GA2ox7* lines, whereas exogenous GA_3 treatment strongly induced *FT* in the vasculature (Galvão et al., 2012; Porri et al., 2012; Yu et al., 2012). Similarly, increased DELLA accumulation in the triple *gid1a-c* and *ga1* mutant resulted in a strong reduction of *FT* expression (Galvão et al., 2012; Richter et al., 2013a), whereas, *FT* expression was increased in the early flowering *ga1-3 gai-t6 rga-t2 rgl1-1 rgl2-1* quintuple mutant (Galvão et al., 2012). Interestingly, the regulation of *FT* by GA appears to be independent of CO and GI, which act upstream of *FT* in the photoperiod pathway, since transcription of these two genes was not affected in either the triple *gid1a-c* mutant or the transgenic *pSUC2::dellaΔ17* and *pSUC2::GA2ox7* lines (Galvão et al., 2012; Porri et al., 2012; Yu et al., 2012).

Regulation of *FT* by GA seems also to occur largely independently of the potent floral repressor *FLC*, because increased DELLA accumulation, induced upon paclobutrazol application, resulted in a marked repression of *FT* in the *flc* mutant (Porri et al., 2012). In addition, expression of the floral repressor *SVP* remained nearly unchanged in LD (Porri et al., 2012; Yu et al., 2012). Whether the floral repressors *TEMPRANILLO1* (*TEM1*) and TEM2 (Castillejo & Pelaz, 2008) participate in the regulation of flowering in response to GA is still under debate. It has been reported that expression of these genes was not significantly changed in *pSUC2::dellaΔ17* and *pSUC2:GA2ox7* (Porri et al., 2012; Yu et al., 2012). However, a recent report demonstrated that TEM1/2 regulates GA levels by directly binding to the promoters of the GA metabolic genes *GIBBERELLIN-3-OXIDASE1*

(*GA3OX1/GA4*) and GIBBERELLIN-3-OXIDASE2 (*GA3OX2*) (Osnato, Castillejo, Matias-Hernandez, & Pelaz, 2012). Based on these results the authors postulated that *TEM1/2* acts upstream of the GA pathway by regulating its biosynthesis in response to photoperiod (Castillejo & Pelaz, 2008; Osnato et al., 2012). Therefore, instead of acting as transcriptional regulators downstream of GA signalling, TEM1/2 seems to regulate flowering responses by controlling GA biosynthesis.

In contrast, miR172 and its targets, six AP2-like transcription factors that have been shown to regulate phase transitions in *A. thaliana* (reviewed in Huijser & Schmid, 2011), likely contribute to the regulation of *FT* by GA. Induction of flowering in response to GA application was attenuated in transgenic lines with reduced miR172 (Franco-Zorrilla et al., 2007; Galvão et al., 2012; Todesco, Rubio-Somoza, Paz-Ares, & Weigel, 2010). In addition, accumulation of DELLA proteins was shown to result in reduced mature miR172 levels in the vasculature of plants grown under LD (Galvão et al., 2012; Yu et al., 2012). Taken together, these results suggest that modulation of miR172, and consequently AP2-like transcripts, may contribute to the GA-dependent regulation of flowering under LD.

3.2.1.5 The Role of GA in Regulating Flowering in Response to Temperature

Flowering time is greatly affected by temperature. In many species, flowering is accelerated after a long exposure to winter cold in a process called vernalisation. GA has long been suspected to be linked to vernalisation since some species that need to experience prolonged cold prior to flowering display features of GA-deficient mutants, with dark green leaves and reduced internode length (Rood, Pearce, Williams, & Pharis, 1989; Zanewich, Rood, & Williams, 1990). Furthermore, the observation that vernalisation is strictly necessary for some plant species to bolt and flower, events also controlled by GA, leads to the hypothesis that prolonged cold exposure might induce GA biosynthesis (Lang, 1957). Indeed, several studies indicate that GA biosynthesis is induced in response to vernalisation in certain species. For example, it has been shown that GA levels are higher in vernalised compared to non-vernalised radish plants (Suge, 1970). Interestingly, GA biosynthesis seems to be controlled by vernalisation in a tissue-specific manner. In *Thlaspi arvense* kaurenoic acid metabolism is dramatically increased specifically at the SAM after vernalisation, while it remains unaltered in leaves (Hazebroek, Metzger, & Mansager, 1993). Similarly, GA metabolism was found to be increased at the shoot tip after vernalisation treatments in *Brassica napus* (Zanewich &

Rood, 1995). In addition, application of bioactive GA has been shown to be sufficient to bypass the vernalisation requirement in several biennial plants (Lang, 1957).

Genome-wide ChIP experiments showed that FLC, which mediates the vernalisation response in *A. thaliana*, directly binds to the GA biosynthesis gene *ENT-KAURENE OXIDASE 1* (GA3) and the GA receptor gene *GID1-C* (Deng et al., 2011), suggesting that FLC might contribute to the control of GA biosynthesis and signalling in *A. thaliana*. However, GA apparently does not affect *FLC* expression in *A. thaliana*, since application of exogenous GA had no effect on *FLC* levels (Moon et al., 2003). Furthermore, despite the fact that *FLC* is repressed in *ga1-3 FRI FLC* in response to vernalisation (Moon et al., 2003), prolonged exposure to cold of *ga1-3* and *ga1-3 FRI FLC* does not result in flowering under SD condition (Chandler, Martinez-Zapater, & Dean, 2000; Michaels & Amasino, 1999b; Moon et al., 2003; Wilson et al., 1992). These results indicate that GA may be connected to the vernalisation pathway downstream of *FLC*, possibly by regulating common targets. This idea is supported by the finding that *SOC1*, which is regulated by both FLC and the GA pathway, only responds to prolonged cold exposure in the presence of GA (Moon et al., 2003). Other genes targeted by both the vernalisation and the GA signalling pathways include *FT* and *SPL3* (Deng et al., 2011; Galvão et al., 2012; Porri et al., 2012; Yu et al., 2012). In addition, treatments with exogenous GA have been shown to bypass the requirement for cold to induce flowering in autonomous pathway mutants, which display increased *FLC* expression (Chandler & Dean, 1994; Sheldon et al., 1999).

In contrast to the vernalisation insensitive phenotype observed in SD, *ga1-3* plants promptly responded to vernalisation when grown under LD (Michaels & Amasino, 1999b; Moon et al., 2003). Nevertheless, it has been shown that *ga1-3* only moderately affect *FT* expression and flowering time under inductive condition (Galvão et al., 2012; Wilson et al., 1992). Therefore, additional experiments using stronger *ga1* or *gid1a-c* alleles are necessary to confirm this result. Interestingly, Porri and colleagues have shown that *FLC* expression is moderately increased in *pSUC2::GA2ox7* under LD, which could explain the reduced *FT* expression (Porri et al., 2012). However, FLC is apparently not repressing the GA-mediated induction of *FT* since treatments with the GA biosynthesis inhibitor paclobutrazol were found to repress *FT* expression similarly in wild type and *flc* mutant plants (Porri et al., 2012).

In addition to vernalisation, ambient temperature has been shown to have a strong effect on the control of flowering time in many plant species. Exposure to high ambient temperature (27 °C) drastically accelerates flowering in *A. thaliana* under SD (Balasubramanian et al., 2006). GA seems to play a critical role in flowering regulation in response to temperature. The *ga1-3* mutant completely fails to flower under SD at 27 °C, while the quintuple DELLA mutant flowers significantly earlier under SD at 12 °C (Balasubramanian et al., 2006; Kumar et al., 2012). At the molecular level, induction of flowering in response to higher temperatures has been proposed to be mediated through temperature-dependent deposition/eviction of a histone variant, H2A.Z, at the promoters of temperature regulated genes (Kumar & Wigge, 2010). One potential target of H2A.Z-mediated transcriptional control is *FT*, based on the observation that the *ft-10* mutant showed a reduced response to thermal induction (Balasubramanian et al., 2006). Indeed, it has recently been shown that PIF4 binds to regulatory elements in the *FT* promoter after eviction of H2A.Z in response to elevated temperatures (Kumar et al., 2012). Interestingly, DELLA proteins directly interact with and impair the ability of PIF4 to bind to DNA (de Lucas et al., 2008). Therefore, it seems likely that increased DELLA accumulation in *ga1-3* would impair the capacity of PIF4 to bind to *FT* in response to high temperatures (Kumar et al., 2012). Another possible mechanism recently proposed to regulate *FT* in response to ambient temperature involves *SPL3*, whose expression is regulated by GA (Galvão et al., 2012; Kim et al., 2012; Porri et al., 2012; Yu et al., 2012). However, several aspects of the link between ambient temperature and endogenous GA levels in the control flowering are still not fully understood. Most notably, it is currently unclear whether GA-dependent regulation of *SPL3* affects *FT* expression.

3.2.2 Brassinosteroids

Brassinosteroids (BR) are a class of plant hormones derived from sterols that regulate diverse developmental processes in plants (Clouse, 2011). Signal transduction involves the perception of BR by the BRASSINOSTEROID INSENSITIVE 1 (BRI1) – BRI1-ASSOCIATED KINASE co-receptor complex at the plasma membrane and a subsequent (auto-) phosphorylation cascade that ultimately leads to the activation of the BRASSINAZOLE-RESISTANT 1 (BZR1) and BRI1-EMS-SUPPRESSOR 1 (BES1) transcription factors. In the presence of BR, BZR1 and BES1 proteins are dephosphorylated and enter into the nucleus to regulate the expression of BR-responsive genes either directly, or indirectly via interaction with other transcription factors. At low BR levels, BZR1 and BES1 are phosphorylated and retained in the cytoplasm (reviewed in Kim & Wang, 2010).

Analyses of mutants defective in *BRI1* and the BR deficient mutants *constitutive photomorphogenic and dwarf* (*cpd*), *de-etiolated 2* (*det2*) and *dwarf4* (*dwf4*) indicate that BR contribute to the regulation of flowering in *A. thaliana* (Azpiroz, Wu, Locascio, & Feldmann, 1998; Domagalska et al., 2010; Li & Chory, 1997). These mutants present severe dwarfism and dark green leaves, flower slightly late under LD but completely fail to bolt under SD (Azpiroz et al., 1998; Chory, Nagpal, & Peto, 1991; Domagalska et al., 2007; Li & Chory, 1997). In addition, the finding that the BR receptor *bri1* and the BR biosynthesis mutants *cpd* and *det2* flower moderately late under LD, provides additional evidence for the importance of BR signalling in the control of flowering (Chory et al., 1991; Domagalska et al., 2010; Domagalska et al., 2007). In contrast, increased BR levels have only a weak or no effect on flowering, depending on the photoperiod. Transgenic plants overexpressing the BR biosynthesis enzyme *DWF4* are mildly early flowering under LD but showed no significant effect under SD (Domagalska et al., 2010). Similarly, single *PHYB activation tagged suppressor 1* and *suppressor of phyB-4* 7 mutants, which are impaired in BR catabolism, flowered at the same time as Col-0 plants under LD, whereas the double mutant flowered only slightly earlier (Turk et al., 2005).

Interestingly, the positive effect of higher BR level on flowering is greatly enhanced by GA. Constitutive co-expression of the BR biosynthesis enzyme *DWF4* and *GA20OX1* (*GIBBERELLIN 20-OXIDASE 1/GA5*) significantly accelerated flowering under SD when compared to lines overexpressing *GA5*. This result indicates that GA is necessary for the effective induction of flowering by BR (Domagalska et al., 2010), and can be explained by the recent finding that DELLA proteins directly interact with and impair the DNA binding capacity of BZR1 (Bai et al., 2012; Oh, Zhu, & Wang, 2012). In this scenario, the activity of BZR1 depends both on its dephosphorylation and activation in response to BR as well as its release from DELLA-mediated repression in response to GA (Bai et al., 2012; Oh et al., 2012). In addition, the GA-dependent and additive effects of BR indicate that despite their direct interaction, GA and BR signalling pathways may at least in part target different downstream genes to control flowering. Indeed, recent findings indicate that the negative effect of BR on flowering results from the upregulation of the flowering repressor protein FLC (Domagalska et al., 2007), which does not play a major role in the GA-mediated regulation of flowering.

3.2.3 Auxin

Auxin plays an important role in basically all aspects of plant development. Initial evidences that auxin participates in the regulation of flowering came from exogenous hormone application. These experiments demonstrated that auxin acts as a repressor of flowering in the SD plant *Pharbitis nil* (Wijayanti, Fujioka, Kobayashi, & Sakurai, 1997). Further analyses revealed that auxin levels were strongly reduced at the SAM of SD-grown *P. nil* when compared to plants grown under continuous light (Wijayanti et al., 1997), suggesting that light-mediated regulation of auxin synthesis at the SAM may contribute to the regulation of flowering.

Auxin signalling begins with the perception of the hormone by the F-box TRANSPORT INHIBITOR RESPONSE 1 (TIR1) receptor. The binding of auxin induces conformational changes in the TIR1/SCF complex that favour its interaction with members of the AUXIN-REPONSIVE PROTEIN/INDOLEACETIC ACID INDUCED PROTEIN (Aux/IAA) protein family, which ultimately results in their ubiquitination and degradation via the 26S proteasome (Dharmasiri, Dharmasiri, & Estelle, 2005; Kepinski & Leyser, 2005). At low auxin concentrations transcriptional repressors of the Aux/IAA family accumulate and directly interact with a class of transcriptional regulators, the AUXIN RESPONSE FACTORS (ARF), repressing their activity. Increased auxin levels induce TIR1/SCF-dependent Aux/IAA degradation, releasing ARF proteins to directly bind to auxin response elements in the promoters of their target genes, and act as either transcriptional activator or repressors (reviewed in Mockaitis & Estelle, 2008).

In *A. thaliana*, the role of auxin in flowering-time control has been demonstrated through the characterisation of auxin signalling and transport mutants. For instance, loss-of-function mutations in the *TIR3/BIG* gene, which participates in polar auxin transport, slightly delayed flowering under both SD and LD conditions (Kanyuka et al., 2003). Interestingly, polar auxin transport seems to be important for the induction of flowering in shade conditions because early flowering of *phyA phyB* double mutants and of plants exposed to a low red:far-red ratio was consistently suppressed in *tir3/big* mutants (Kanyuka et al., 2003). In contrast, the auxin biosynthesis mutant *shade avoidance 2* suppresses classical shade-related phenotypes (i.e. leaf shape and hypocotyl growth), but has no effect on flowering time (Tao et al., 2008). Which genes are regulated by TIR3/BIG in response to low red:far-red ratios remains to be determined.

More recently, the molecular mechanisms involved in the auxin-mediated regulation of flowering have been investigated in closer detail. Mai and colleagues reported that dominant mutations in *INDOLE-3-ACETIC ACID7/ AUXIN RESISTANT 2* (*IAA7/AXR2*) delayed flowering under SD in a GA-dependent manner, presumably due to reduced *SOC1* and *LFY* expression. Consistent with this idea, the artificial reduction of free auxin by overexpressing the enzyme indoleacetic acid lysine synthetase resulted in a late flowering phenotype in *A. thaliana* under SD (Mai, Wang, & Yang, 2011). The authors propose that the delay in flowering observed in *iaa7/axr2* can in part be explained by reduced expression of the GA biosynthesis genes *GA20OX1 and GA20OX2*, which would result in a reduction of bioactive GA and increased DELLA accumulation. Genetic analyses of several single and higher order mutants suggest that *ARF7* and *ARF19* also contribute to this regulatory circuit (Mai et al., 2011). However, more detailed genetic analyses are required to place these genes genetically downstream of *IAA7/AXR2* and to test whether they are involved in the regulation of GA biosynthesis.

Besides ARF7 and ARF19, ARF1 and ARF2 have also been demonstrated to participate in the regulation of flowering (Ellis et al., 2005; Okushima, Mitina, Quach, & Theologis, 2005). While *arf1* flowered at about the same time as wild type plants, the *arf2* single mutant flowered late under inductive LD. This phenotype was further enhanced in the double *arf1 arf2*, indicating functional redundancy between the two genes (Ellis et al., 2005). In contrast to *iaa7/axr2*, which apparently modulate GA biosynthesis (as described above), recent data indicate that ARF2, along with ARF7 and ARF19, directly regulates the GNC/GNL transcription factors to control flowering (Richter et al., 2013b). Therefore, two conflicting models have been proposed, one of which suggests that *iaa7/axr2* regulates flowering in a GA-dependent manner, whereas the second suggests that GA and auxin regulate GNC and GNL independently of each other (Mai et al., 2011; Richter et al., 2013b). Additionally, it has been reported that ARF2 can be regulated independently of the auxin-Aux/IAA pathway (Ulmasov, Hagen, & Guilfoyle, 1999; Vernoux et al., 2011). Clearly, more work is required to dissect the role of individual auxin signalling components and their interaction with the GA pathway in the regulation of flowering.

3.2.4 Cytokinins

Cytokinins comprise a class of plant hormones derived from adenine that were first identified as factors controlling cell proliferation. In addition to the control of cell division, cytokinins are involved in other events during the plant life cycle, such as photomorphogenesis and regulation of organ growth (Kieber, 2002). A phosphorylation cascade similar to the bacterial two-component system lies at the heart of cytokinin signalling. Binding of cytokinin to the histidine kinase receptors ARABIDOPSIS HISTIDINE KINASE-2 (AHK2), AHK3 and AHK4 induces auto-phosphorylation of their kinase domain. Subsequently, the phosphate group is transferred to the histidine phosphotransfer proteins, which move to the nucleus to phosphorylate response regulators (ARR). Type-B ARR transcription factors in turn regulate the expression of cytokinin responsive genes (reviewed in Argueso, Raines, & Kieber 2010).

Several lines of evidence indicate that cytokinin has an inductive role during the flowering transition in *A. thaliana*. Increased levels of cytokinin in response to either exogenous application of the hormone or a mutation in *ALTERED MERISTEM PROGRAM 1*, which encodes a glutamate carboxypeptidase, significantly induced flowering (Chaudhurz, Letham, Craig, & Dennis, 1993; He & Loh, 2002). Conversely, reduced hormone levels in transgenic plants overexpressing the cytokinin catabolic enzyme *CYTOKININ OXIDASE 1* (*CKX1*) or *CKX3* flowered late and even died before flowering (Werner et al., 2003). Likewise, *ahk2 ahk3 ahk4* triple mutants flowered late or completely failed to flower (Riefler, Novak, Strnad, & Schmulling, 2006). In addition, cytokinin has been shown to promote flowering in the LD plant *Sinapis alba* (Bernier, Havelange, Houssa, Petitjean, & Lejeune, 1993; Bonhomme, Kurz, Melzer, Bernier, & Jacqmard, 2000).

Similar to GA, cytokinin signalling seems to regulate flowering both in leaves and at the SAM. In leaves, cytokinin is thought to regulate flowering mostly through the *FT* homologue *TWIN SISTER OF FT* (*TSF*). This conclusion is based on the finding that in hydroponically grown *A. thaliana* seedlings, cytokinin promoted flowering in wild type and *ft-10* mutant plants, while it failed to induce flowering in *tsf-1* (D'Aloia et al., 2011). In addition, *TSF* expression was found to respond rapidly to application of cytokinin, whereas *FT* expression remained unchanged. Interestingly, as for *TSF*, expression of the bZIP transcription factor *FD* increased at the SAM shortly after application of cytokinin and an *fd* loss-of-function mutation partially suppressed the induction of flowering by cytokinin (D'Aloia et al., 2011). Taken together, these results suggest a model in which *TSF* is induced in leaves and, presumably after movement to the SAM, interacts at the shoot meristem with FD to induce flowering in response to cytokinin. Cytokinin treatments could also induce *SOC1* expression at the SAM in

A. thaliana and the *SOC1* homologue in *S. alba*, *SaMADS A* (Bonhomme et al., 2000). *SOC1* likely mediates between cytokinins and flowering as loss-of-function *soc1* mutants failed to flower upon hormone application (D'Aloia et al., 2011). However, it is not yet clear if *SOC1* is a direct target of cytokinin signalling at the SAM or if the induction of *SOC1* is mediated through a TSF–FD–(14-3-3) complex (D'Aloia et al., 2011).

Exposure to a single inductive LD increased the level of cytokinin in the leaf exudate and subsequently at the SAM of *A. thaliana* (Corbesier et al., 2003), indicating that cytokinin levels are controlled by photoperiod. Nevertheless, additional experiments are required to unravel the molecular details of the spatial separation of cytokinin effects on flowering in leaves and at the SAM, and to identify the underlying genes.

3.2.5 Ethylene

Despite its simple chemical structure, ethylene plays an important role throughout the entire plant life cycle, from germination and photomorphogenesis to senescence and fruits ripening. At the molecular level ethylene is perceived by the receptors ETHYLENE RESPONSE 1 (ETR1), ETR2, ETHYLENE RESPINSE SENSOR 1 (ERS1), ERS2 and ETHYLENE INSENSITIVE 4 (EIN4), which negatively regulate ethylene signalling by activating the Ser/Thr kinase CONSTITUTIVE TRIPLE RESPONSE (CTR1) (Merchante, Alonso, & Stepanova, 2013). Active CTR1 suppresses the response to ethylene by inactivating the protein EIN2 through direct phosphorylation at the C-terminal position. By contrast, binding of ethylene inactivates the receptors and CTR1, preventing the phosphorylation of EIN2, which in turn moves to the nucleus and stabilises the ETHYLENE INSENSITIVE3/EIN3-LIKE (EIN3/EIL1) transcription factors. Therefore, the transcription factors EIN3/EIL1 directly regulate the expression of ethylene response genes (Merchante et al., 2013).

Similar to what has been observed for other plant hormones, ethylene has either a positive or negative effect on flowering, depending on the species analysed. For instance, in *Xanthium pensylvanicum, P. nil,* rice, Lotus and *A. thaliana* ethylene has been shown to repress flowering (Abeles, 1967; Chan, Biswas, & Gresshoff, 2013; Suge, 1972; Wang, Zhang, Yin, & Wen, 2013; Wuriyanghan et al., 2009). In contrast, ethylene has been used to induce flowering to synchronise fruit production in pineapple cultures (Trusov & Botella, 2006; Wang, Hsu, Bartholomew, Maruthasalam, & Lin, 2007). In *A. thaliana*, the repressive role of ethylene on flowering has been demonstrated by exogenous application of the hormone as well as by mutants affected in ethylene biosynthesis and signalling. For example, mutations in the ethylene receptor, which results in constitutive ethylene responses, caused extreme late flowering under inductive LD photoperiod (Hall & Bleecker, 2003; Ogawara, Higashi, Kamada, & Ezura, 2003). Corroborating this result, fumigation with ethylene or treatment with ethylene precursors delayed flowering in *A. thaliana* both under inductive and non-inductive photoperiods (Achard et al., 2007).

While the mechanism underlying the regulation of flowering in LD remains elusive, under SD ethylene seems to control flowering at least in part through regulation of GA metabolism (Achard et al., 2007). Late flowering in *ctr1* under SD has been attributed to reduced expression of the flowering-time genes *LFY* and *SOC1*, which are also targets of GA signalling at the shoot meristem. Expression of these genes and the late flowering of *ctr1* could be rescued by GA treatment or loss-of-function mutations in DELLA genes, suggesting that ethylene might act upstream of GA biosynthesis. In agreement with this hypothesis, the levels of bioactive GA_1 and GA_4 were found to be strongly decreased in *ctr1*, while GA precursors were increased. Based on these findings it has been proposed that ethylene might regulate DELLA proteins accumulation, which in turn regulate flowering time through *SOC1* and *LFY* (Achard et al., 2007).

Considering that its synthesis is highly modulated by diverse environmental cues such as temperature and salt stress, it seems likely that ethylene also functions in mediating flowering in response to environmental fluctuations (Wang, Li, & Ecker, 2002).

3.2.6 Salicylic Acid

Salicylic acid (SA) is best known for its prominent role in stress and pathogen defence responses. Besides this, SA has also been shown to positively regulate flowering. This was first reported in *Lemna gibba* and *Lemna paucicostata*, in which SA application induced flowering specifically under inductive photoperiods (Cleland & Tanaka, 1979).

In *A. thaliana*, however, SA seems to have only a minor or no effect on plants growing under inductive LD conditions since mutants with reduced SA levels flower at the same time as wild type (Li et al., 2012; Martinez, Pons, Prats, & Leon, 2004). Nevertheless, SA is apparently modulating the transition to flowering of plants growing under stressful conditions. For instance, high UV-C irradiation leads to higher SA accumulation and early flowering. Interestingly, this phenotype is largely suppressed in mutants with reduced SA, such as *nahG*, which encodes salicylate hydroxylase, an enzyme that converts SA to catechol, indicating that SA is necessary for flowering induction in response to stress (Martinez et al., 2004). Corroborating this idea, it has been shown that SA is necessary for flowering in *P. nil* growing under mineral deprivation (Wada, Yamada, Shiraya, & Takeno, 2010), and exogenous application of the hormone accelerates flowering in wild type and SA-deficient *A. thaliana* mutants (Martinez et al., 2004).

The molecular mechanisms underlying the SA-mediated control of flowering under LD have not been completely elucidated but apparently involve transcriptional regulation of *FT*. UV-C treated *A. thaliana* showed increased *FT* expression, which was completely suppressed in the SA-deficient *nahG* mutant (Martinez et al., 2004). In addition, SA application induced *FT* expression in *A. thaliana*, sunflower, and in *P. nil* grown under mineral deprivation (Dezar et al., 2011; Martinez et al., 2004; Yamada & Takeno, 2014). Interestingly, as observed for the ABA-mediated regulation of flowering under drought stress (see below), SA is apparently regulating flowering through a GI-dependent mechanism because SA completely failed to induce flowering in a loss-of-function *gi* mutant, and double *gi nahG* mutants flowered at the same time as *gi* single mutant (Martinez et al., 2004; Riboni, Galbiati, Tonelli, & Conti, 2013). However, weather mutations in *gi* suppress SA-mediated *FT* induction still needs to be tested.

The integration of SA signalling and regulation of *FT* expression appears to be in part mediated by *CO*. *CO* expression is weakly induced upon UV-C irradiation and slightly reduced in the SA-deficient *nahG* mutant, and strongly reduced at dawn in *enhanced disease susceptibility 5* and *salicylic acid induction deficient 2* (Martinez et al., 2004; Segarra, Mir, Martinez, & Leon, 2010). Despite this, *co* mutants were still responsive to SA treatments, indicating that the regulation of *FT* expression in response to SA was not mediated exclusively by CO (Martinez et al., 2004). More recently, it has been suggested that a protein called PATHOGEN AND CIRCADIAN CONTROLLED 1 (PCC1), which is strongly regulated in response to UV-C irradiation, presumably due to higher SA levels, contributes to the regulation of *FT* in LD (Segarra et al., 2010). In agreement with this, knocking down *PCC1* by RNAi impaired the UV-C mediated flowering induction, possibly because of reduced *FT* expression (Segarra et al., 2010). However, the molecular mechanism by which PCC1 controls *FT* expression is largely unclear and still needs to be investigated further.

SA has also been shown to regulate expression of the potent floral repressor *FLC* both in SD and LD. SA treatments resulted in reduced *FLC* transcription, while its expression was increased in SA-deficient mutants (Martinez et al., 2004). The effect of SA seemed to be largely independent of the autonomous pathway since SA treatment of *fve* and *fca* loss-of-function mutants still repressed *FLC*. Interestingly, a more detailed genetic analysis indicated that SA-mediated flowering did not require FLC in LD (Martinez et al., 2004). *nahG flc* double mutant plants were indistinguishable from the *nahG* single mutant, indicating that FLC was not responsible for the late flowering phenotype of the latter. In addition, the *nahG* mutant was found to be fully responsive to vernalisation, indicating that SA was not necessary for seasonal cold-induced flowering (Martinez et al., 2004).

In contrast to the mild phenotype observed in LD-grown plants, SA-deficient mutants flowered consistently late under SD. The data accumulated so far indicate that *FLC* is at least partially mediating this response because late flowering of *nahG* mutants was considerably suppressed by mutations in *FLC*. However, known FLC target genes, such as *FT* and *SOC1* were apparently not responsible for this phenotype, since *SOC1* transcript levels remained unchanged in *nahG* and expression of FT was only moderately reduced (Martinez et al., 2004).

3.2.7 Abscisic Acid

Another plant hormone known for its prominent role in adaptation to environmental stress is ABA, which is synthesised from carotenoids and is perceived by the soluble PYRABACTIN RESISTANT 1 (PYR1) and PYR-LIKE (PYLs) receptors. Binding of ABA leads to the inhibition of negative regulators of the group A PP2C Ser/Thr phosphatases, such as ABA-INSENSITIVE 1 (ABI1) and ABI2. Active PP2C phosphatases are believed to inactivate positive downstream regulators, such as the *Snf1-related protein kinase 2*. In contrast, in the absence of ABA, PYR/PYL receptors cannot bind to PP2Cs, preventing SNF1-related protein kinases2 (SnRK2) activation (reviewed in Cutler, Rodriguez, Finkelstein, & Abrams, 2010).

Based on the early flowering phenotype of ABA-deficient mutants it has been proposed that ABA acts as a repressor of flowering in *A. thaliana* (Domagalska et al., 2010; Martinez-Zapater, Coupland, Dean, & Koornneef, 1994). Moreover, the *abi3* mutant flowers significantly earlier under both SD and LD, whereas the ABA hypersensitive mutant *hyponastic leaves 1* delays flowering in SD (Lu & Fedoroff, 2000), further supporting the role of ABA as a repressor of flowering. ABA has been proposed to repress flowering under unfavourable conditions by modulating DELLA signalling (Achard et al., 2006). *A. thaliana* plants treated with high salt concentrations were found to flower late compared to control plants and presented increased, ABA-dependent DELLA accumulation (Achard et al., 2006). This accumulation of DELLA proteins in response to ABA is consistent with the epistatic effect of ABA suppressing *LFY* expression after GA treatment at the SAM in *A. thaliana* (Blazquez et al., 1998).

A recent report by Riboni and colleagues proposed a novel mechanism mediating the ABA-dependent induction of flowering in response to severe drought (Riboni et al., 2013). The drought escape response is believed to accelerate flowering to ensure that plants will complete their life cycle when exposed to severe stress conditions (Verslues & Juenger, 2011). Using several flowering-time mutants, the authors demonstrated that *FT* and *TSF* are necessary for the induction of flowering through a *GI*-dependent mechanism, but independently of *CO*, specifically under inductive LD (Riboni et al., 2013). However, and in contrast to the previously reported early flowering phenotype of ABA-deficient mutants, which had been interpreted as ABA functioning to repress flowering, these authors observed that ABA had a positive effect on flowering in LD. In particular, *ABA-deficient 1 (aba1)* and *aba2* mutants were found to flower later than control plants specifically under LD condition, while no differences in flowering time could be observed under SD. In addition, the signalling mutants *hypersensitive to aba 1 (hab1)*, *abi2* and the *hab1 abi1 pp2ca* triple mutant flowered significantly earlier under LD and later under SD. To explain these somewhat controversial results, the authors proposed that ABA acts as a positive regulator of flowering through the regulation of *GI* and *FT/TSF* under inductive conditions. In addition, induction of flowering by ABA was suggested to occur independently of GA since *ga1-3* plants still flowered earlier under drought stress (Riboni et al., 2013). However, it would be interesting to test this response using a strong *ga1* allele since it has been shown that *ga1-3* has only a mild flowering phenotype in LD (Galvão et al., 2012; Richter et al., 2013a; Wilson et al., 1992).

3.3 REGULATION OF FLOWERING BY SUGARS

One of the hallmarks of plants is their photoautotroph life style, which enables them to fix carbon dioxide and synthesise, as a first product, simple sugars. These monosaccharides are subsequently converted into a wide range of complex oligo- and polysaccharides that function, for example, as structural components in the cell wall and as short- and long-term energy sources. In addition, sugars have been shown to function as signalling molecules, an area of research that has lately attracted great attention. For example, it has been shown that sugars participate in the entrainment of circadian rhythms in *A. thaliana* (Haydon, Mielczarek, Robertson, Hubbard, & Webb, 2013) and control root meristem activity through the regulation of the TARGET-OF-RAPAMYCIN signalling pathway (Xiong et al., 2013). In addition, fluctuation in glucose concentration is perceived by the bifunctional enzyme HEXOKINASE, which both catalyses the first step of glycolysis, regulating the conversion of glucose into glucose-6-phosphate, and functions as a transcriptional regulator in the nucleus (Cho, Yoo, & Sheen, 2006; Moore et al., 2003). In the context of this review, it is important to note that sugar signalling has also been shown to contribute to the regulation of diverse developmental processes, including flowering (reviewed in Paul, Primavesi, Jhurreea, & Zhang, 2008; Ponnu, Wahl, & Schmid, 2011).

The analysis of several mutants impaired in starch metabolism strongly suggested that carbohydrates play a prominent role in regulating the transition to flowering. Starch is the main energy storage form in plants, fuelling metabolism and growth when plants are unable to synthesise sugar through photosynthesis (Streb & Zeeman, 2012). Starch deficient mutants, such as *phosphoglucomutase1* (*pgm1*), *ADP glucose phosphorylase 1* and *phosphoglucose isomerase*, and mutants deficient in the mobilisation of starch during the night, such as *starch in excess 1* (*sex1*), *sex4* and *like sex four 1*, have been shown to generally develop more slowly and to flower later than wild type plants (Caspar, Huber, & Somerville, 1985; Caspar et al., 1991; Corbesier, Lejeune, & Bernier, 1998; Eimert, Wang, Lue, & Chen, 1995; Lin, Caspar, Somerville, & Preiss, 1988; Paparelli et al., 2013; Yu, Lue, Wang, & Chen, 2000). Interestingly, while these phenotypes are mostly observed in shorter photoperiod, the growth rate and flowering time of these mutants are restored when grown under constant light (Corbesier et al., 1998; Eimert et al., 1995). The most plausible conclusion for this observation is that prolonged/constant exposure to light apparently results in sufficient sugars to be produced through photosynthesis to compensate for the limited or no sugars available from starch mobilisation.

Interestingly, exogenous application of sucrose reverted the late flowering phenotype of the starch-deficient *pgm* mutant, indicating that sucrose acts as positive regulator of flowering in *A. thaliana* (Yu et al., 2000). In addition, it has been shown that photoperiodic induction by exposure to either a single LD or displaced SD rapidly and transiently increases sucrose

Figure 3.3 *Arabidopsis thaliana* plants impaired in trehalose-6-phosphate synthesis are late flowering. The *tps1* mutant carrying the chemically inducible rescue construct *PGVG::TPS1* flowers extremely late under inductive photoperiodic conditions compared to control plants. Depicted are 20 (Col-0) and 50 (*tps1 PGVG::TPS1*)-day-old plants grown under long day at 23 °C. (See the colour plate.) *Picture credit: Jathish Ponnu.*

production and export from leaves to the shoot apex of *S. alba* and *A. thaliana* (Bernier et al., 1993; Corbesier et al., 1998; Eriksson et al., 2006). However, it should be noted that sucrose and glucose were found to repress flowering when applied exogenously at very high concentrations (Ohto et al., 2001; Zhou, Jang, Jones, & Sheen, 1998). *A. thaliana* plants grown under constant light on MS medium supplemented with 6% glucose flowered 16 days later than plants grown on 2% glucose medium (Zhou et al., 1998). Similarly, *A. thaliana* plants flowered considerably later when grown on a growth medium supplemented with 5% sucrose than on the same medium supplemented with 2% sucrose (Ohto et al., 2001).

Another striking example demonstrating the role of sugars on flowering comes from the analysis of loss-of-function alleles of TREHALOSE-6-PHOSPHATE SYNTHASE 1 (TPS1), which catalyses the conversion of glucose-6-phosphate and uridine-diphosphate (UDP) glucose into trehalose-6-phosphate (T6P). T6P is a disaccharide found at very low concentration in most plants and is believed to function as a signalling molecule, conveying the information on carbohydrate availability to other signalling pathways (Lunn et al., 2006). Loss of *TPS1* results in impaired embryo development and late embryo lethality (Eastmond et al., 2002), which can be overcome by expression of *TPS1* from either the seed-specific *ABI3* promoter or the dexamethasone-inducible GAL4-VP16-GR (GVG) construct (Gomez, Gilday, Feil, Lunn, & Graham, 2010; van Dijken, Schluepmann, & Smeekens, 2004). Strikingly, rescued transgenic *tps1 GVG::TPS1* plants display pleiotropic developmental defects, including growth arrest, and late or non-flowering phenotype (Figure 3.3) (van Dijken et al., 2004; Wahl et al., 2013).

Several lines of evidence indicate that the extreme late flowering phenotype of *tps1 GVG::TPS1* under inductive photoperiods is at least in part due to reduced expression of *FT* in leaf phloem companion cells (Wahl et al., 2013). Diurnal expression analyses showed a dramatic reduction of *FT* and its close homologue *TSF* in *tps1 GVG::TPS1* compared to control plants in LD. Furthermore, treatment of *tps1 GVG::TPS1* with dexamethasone significantly induced *FT* expression and accelerated flowering (Wahl et al., 2013). The finding that expression of *FT* from a constitutive or phloem companion cell specific promoter almost completely suppressed the late flowering of an artificial microRNA *TPS1* (*p35S::amiR-TPS1*) line provides additional evidence that *FT* constitutes an important node of T6P signal integration. In contrast to the reduced *FT* expression in *tps1 GVG::TPS1, CO* and *GI* transcripts remained unchanged, suggesting that T6P regulates *FT* independently of these genes (Wahl et al., 2013). This is in agreement with the observation that flowering can be induced in dark-grown *co* and *gi* but not in *ft* mutants by exogenous sucrose (Roldan, Gomez-Mena, Ruiz-Garcia, Salinas, & Martinez-Zapater, 1999).

In addition to the control of *FT* in leaves, T6P seems to also regulate flowering by an *FT*-independent mechanism, as evidenced by the fact that *tps1 GVG::TPS1* flowers much later than *ft* mutants. Several lines of evidence suggest that, similar to GA and cytokinin, the SAM plays an important role in regulating flowering in response to T6P signalling. For example, shifting wild type plants from non-inductive SD to inductive LD photoperiods boosts T6P at the SAM (Wahl et al., 2013). Interestingly, while sucrose is synthesised in source leaves, T6P seems to also be produced at the SAM. RNA in situ hybridisation detected strong *TPS1* expression at the SAM in *A. thaliana* plants growing under LD very early after germination (Wahl et al., 2013). Functional data supporting the role of T6P at the SAM came from the expression of *TPS1* and the *E. coli* trehalose-6-phosphate phosphatase *otsB*, which catalyses the conversion of T6P to trehalose, from the meristem-specific *CLAVATA3* (*CLV3*) promoter. Increased *TPS1* expression at the SAM (*pCLV3::TPS1*) resulted in early flowering both under SD and LD, while *pCLV3::otsB* had the opposite phenotype (Wahl et al., 2013). Since T6P has been suggested to serve as a proxy for carbohydrate availability (Lunn et al., 2006), it is tempting to speculate that the observed increase in T6P reflects an increase in sucrose availability at the SAM at the time of floral transition.

At the molecular level, T6P signalling at the SAM seems to connect to the canonical flowering-time pathways at the level of the miR156/SPL module, which has been shown to play a central role in the regulation of flowering depending on plant age (Wang & Weigel, 2009; Yamaguchi et al., 2009). Expression analyses performed in dissected apices of *tps1 GVG::TPS1* revealed that *SPL3, SPL4* and *SPL5* genes were significantly reduced in the mutant compared to control plants, whereas levels of mature miR156 were initially increased (Wahl et al., 2013). Interestingly, regulation of the miR156/SPL module by carbohydrates has also been implicated in the regulation of the juvenile-to-adult vegetative phase transition (Yang, Xu, Koo, He, & Poethig, 2013; Yu et al., 2013).

3.4 CONCLUSIONS

Despite the great advances that have been made over the last decade in understanding the mechanisms that regulate flowering in response to endogenous signals, many questions still remain. Even in *A. thaliana*, for which by far the most data have been accumulated over the years, in many instances it is still not possible to precisely pinpoint the site at which endogenous signals are integrated into the genetic pathways that regulate flowering in response to environmental cues such as light and temperature. The issue is further complicated by the fact that the different regulatory pathways are not strictly separated but regulate each other in an elaborate cross-regulatory network, which makes extremely challenging to dissect the role of individual signals (Figure 3.4). Moreover, the molecular circuits that regulate flowering in response to endogenous factors seem to vary widely

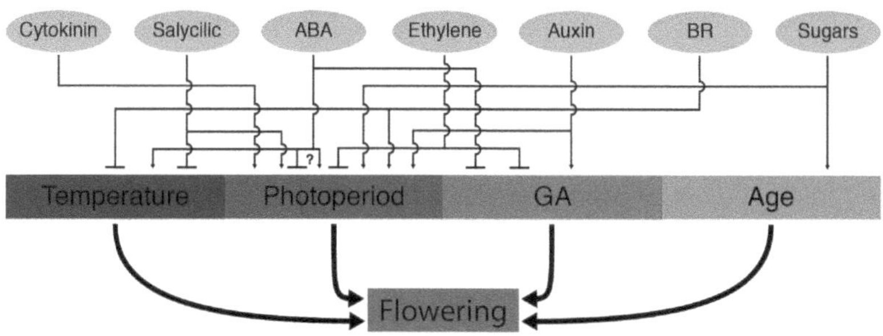

Figure 3.4 Regulatory interactions between flowering time signalling pathways. Endogenous hormone and sugar signalling pathways are intimately linked to each other and modulate flowering time in response to environmental stimuli.

between species, and it is not uncommon that a given growth regulator has opposite effects in different species. This makes the transfer of knowledge from model species, such as *A. thaliana*, to species with agronomical importance often difficult, if not outright impossible. Thus, while results from model species can serve as guidelines, clearly additional efforts are needed to unravel the gene regulatory networks underlying the regulation of flowering in a wider variety of species.

ACKNOWLEDGEMENTS

V.C.G. is supported by a long-term fellowship from European Molecular Biology Organisation (EMBO). Work on flowering time regulation in the Schmid lab is supported by the Deutsche Forschungsgemeinschaft (DFG) and the Max Planck Society.

REFERENCES

Abeles, F. B. (1967). Inhibition of flowering in Xanthium pensylvanicum Walln. by ethylene. *Plant Physiology*, *42*(4), 608–609.

Achard, P., Baghour, M., Chapple, A., Hedden, P., Van Der Straeten, D., et al. (2007). The plant stress hormone ethylene controls floral transition via DELLA-dependent regulation of floral meristem-identity genes. *Proceeding of the National Academy of Sciences of the United States of America*, *104*(15), 6484–6489.

Achard, P., Cheng, H., De Grauwe, L., Decat, J., Schoutteten, H., Moritz, T., et al. (2006). Integration of plant responses to environmentally activated phytohormonal signals. *Science*, *311*(5757), 91–94.

Achard, P., Herr, A., Baulcombe, D. C., & Harberd, N. P. (2004). Modulation of floral development by a gibberellin-regulated microRNA. *Development*, *131*(14), 3357–3365.

An, H., Roussot, C., Suarez-Lopez, P., Corbesier, L., Vincent, C., Pineiro, M., et al. (2004). CONSTANS acts in the phloem to regulate a systemic signal that induces photoperiodic flowering of *Arabidopsis*. *Development*, *131*(15), 3615–3626.

Andres, F., & Coupland, G. (2012). The genetic basis of flowering responses to seasonal cues. *Nature Reviews Genetics*, *13*(9), 627–639.

Argueso, C. T., Raines, T., & Kieber, J. J. (2010). Cytokinin signaling and transcriptional networks. *Current Opinion in Plant Biology*, *13*(5), 533–539.

Azpiroz, R., Wu, Y. W., Locascio, J. C., & Feldmann, K. A. (1998). An *Arabidopsis* brassinosteroid-dependent mutant is blocked in cell elongation. *Plant Cell*, *10*(2), 219–230.

Bai, M.Y., Shang, J. X., Oh, E., Fan, M., Bai, Y., Zentella, R., et al. (2012). Brassinosteroid, gibberellin and phytochrome impinge on a common transcription module in *Arabidopsis*. *Nature Cell Biology*, *14*(8), 810–817.

Balasubramanian, S., Sureshkumar, S., Lempe, J., & Weigel, D. (2006). Potent induction of *Arabidopsis thaliana* flowering by elevated growth temperature. *PLoS Genetics*, *2*(7), e106.

Bastow, R., Mylne, J. S., Lister, C., Lippman, Z., Martienssen, R. A., & Dean, C. (2004). Vernalization requires epigenetic silencing of FLC by histone methylation. *Nature*, *427*(6970), 164–167.

Bernier, G., Havelange, A., Houssa, C., Petitjean, A., & Lejeune, P. (1993). Physiological signals that induce flowering. *Plant Cell*, *5*(10), 1147–1155.

Blazquez, M. A., Green, R., Nilsson, O., Sussman, M. R., & Weigel, D. (1998). Gibberellins promote flowering of *Arabidopsis* by activating LEAFY promoter. *Plant Cell*, *10*(5), 791–800.

Blazquez, M. A., Soowal, L. N., Lee, I., & Weigel, D. (1997). LEAFY expression and flower initiation in *Arabidopsis*. *Development*, *124*(19), 3835–3844.
Blazquez, M., & Weigel, D. (2000). Integration of floral inductive signals in *Arabidopsis*. *Nature*, *404*, 889–892.
Bolle, C. (2004). The role of GRAS proteins in plant signal transduction and development. *Planta*, *218*(5), 683–692.
Bonhomme, F., Kurz, B., Melzer, S., Bernier, G., & Jacqmard, A. (2000). Cytokinin and gibberellin activate SaMADS A, a gene apparently involved in regulation of the floral transition in *Sinapis alba*. *Plant Journal*, *24*(1), 103–111.
Caspar, T., Huber, S. C., & Somerville, C. (1985). Alterations in growth, photosynthesis, and respiration in a starchless mutant of Arabidopsis-thaliana (L) deficient in chloroplast phosphoglucomutase activity. *Plant Physiology*, *79*(1), 11–17.
Caspar, T., Lin, T. P., Kakefuda, G., Benbow, L., Preiss, J., & Somerville, C. (1991). Mutants of *Arabidopsis* with altered regulation of starch degradation. *Plant Physiology*, *95*(4), 1181–1188.
Castillejo, C., & Pelaz, S. (2008). The balance between CONSTANS and TEMPRANILLO activities determines FT expression to trigger flowering. *Current Biology*, *18*(17), 1338–1343.
Chan, P. K., Biswas, B., & Gresshoff, P. M. (2013). Classical ethylene insensitive mutants of the *Arabidopsis* EIN2 orthologue lack the expected 'hypernodulation' response in Lotus japonicus. *Journal of Integrative Plant Biology*, *55*(4), 395–408.
Chandler, J., & Dean, C. (1994). Factors influencing the vernalization response and flowering time of late flowering mutants of *Arabidopsis thaliana* (L.) Heynh. *Journal of Experimental Botany*, *45*(278), 1279–1288.
Chandler, J., Martinez-Zapater, J. M., & Dean, C. (2000). Mutations causing defects in the biosynthesis and response to gibberellins, abscisic acid and phytochrome B do not inhibit vernalization in Arabidopsis fca-1. *Planta*, *210*(4), 677–682.
Chaudhurz, A. M., Letham, S., Craig, S., & Dennis, E. S. (1993). amp1-a mutant with high cytokinin levels and altered embryonic pattern faster vegetative growth constitutive photomorphogenesis and precocious flowering. *Plant Journal*, *4*(6), 907–916.
Cheng, H., Qin, L., Lee, S., Fu, X., Richards, D. E., Cao, D., et al. (2004). Gibberellin regulates *Arabidopsis* floral development via suppression of DELLA protein function. *Development*, *131*(5), 1055–1064.
Cho, Y. H., Yoo, S. D., & Sheen, J. (2006). Regulatory functions of nuclear hexokinase1 complex in glucose signaling. *Cell*, *127*(3), 579–589.
Chory, J., Nagpal, P., & Peto, C. A. (1991). Phenotypic and genetic-analysis of Det2, a new mutant that affects light-regulated seedling development in *Arabidopsis*. *Plant Cell*, *3*(5), 445–459.
Cleland, C. F., & Tanaka, O. (1979). Effect of daylength on the ability of salicylic acid to induce flowering in the long-day plant Lemna gibba G3 and the short-day plant Lemna paucicostata 6746. *Plant Physiology*, *64*(3), 421–424.
Clouse, S. D. (2011). Brassinosteroids. *The Arabidopsis Book*, *9*, e0151.
Corbesier, L., Lejeune, P., & Bernier, G. (1998). The role of carbohydrates in the induction of flowering in *Arabidopsis thaliana*: comparison between the wild type and a starchless mutant. *Planta*, *206*(1), 131–137.
Corbesier, L., Prinsen, E., Jacqmard, A., Lejeune, P., Onckelen, H. V., Perilleux, C., et al. (2003). Cytokinin levels in leaves leaf exudate and shoot apical meristem of *Arabidopsis thaliana* during floral transition. *Journal of Experimental Botany*, *54*(392), 2511–2517.
Corbesier, L., Vincent, C., Jang, S. H., Fornara, F., Fan, Q. Z., Searle, I., et al. (2007). FT protein movement contributes to long-distance signaling in floral induction of *Arabidopsis*. *Science*, *316*(5827), 1030–1033.
Curtis, P. J., & Cross, B. E. (1954). Gibberellic acid. A new metabolite from the culturefiltrates of Gibberella fujikuroi. *Chemical & Industry*, 1066.

Cutler, S. R., Rodriguez, P. L., Finkelstein, R. R., & Abrams, S. R. (2010). Abscisic acid: emergence of a core signaling network. *Annual Review in Plant Biology, 61*, 651–679.

D'Aloia, M., Bonhomme, D., Bouche, F., Tamseddak, K., Ormenese, S., Torti, S., et al. (2011). Cytokinin promotes flowering of *Arabidopsis* via transcriptional activation of the FT paralogue TSF. *Plant Journal, 65*(6), 972–979.

Davies, P. J. (2004). *Plant hormones: biosynthesis, signal transduction, action!* (3rd ed). Dordrecht, the Netherlands: Springer.

de Lucas, M., Daviere, J. M., Rodriguez-Falcon, M., Pontin, M., Iglesias-Pedraz, J. M., Lorrain, S., et al. (2008). A molecular framework for light and gibberellin control of cell elongation. *Nature, 451*(7177), 480–484.

Deng, W., Ying, H., Helliwell, C. A., Taylor, J. M., Peacock, W. J., & Dennis, E. S. (2011). FLOWERING LOCUS C (FLC) regulates development pathways throughout the life cycle of *Arabidopsis*. *Proceeding of the National Academy of Sciences of the United States of America, 108*(16), 6680–6685.

Dezar, C. A., Giacomelli, J. I., Manavella, P. A., Re, D. A., Alves-Ferreira, M., Baldwin, I. T., et al. (2011). HAHB10, a sunflower HD-Zip II transcription factor, participates in the induction of flowering and in the control of phytohormone-mediated responses to biotic stress. *Journal of Experimental Botany, 62*(3), 1061–1076.

Dharmasiri, N., Dharmasiri, S., & Estelle, M. (2005). The F-box protein TIR1 is an auxin receptor. *Nature, 435*(7041), 441–445.

Dill, A., Jung, H. S., & Sun, T. P. (2001). The DELLA motif is essential for gibberellin-induced degradation of RGA. *Proceeding of the National Academy of Sciences of the United States of America, 98*(24), 14162–14167.

Dill, A., & Sun, T.-P. (2001). Synergistic derepression of gibberellin signaling by removing RGA and GAI function in *Arabidopsis thaliana*. *Genetics, 159*, 777–785.

Dill, A., Thomas, S. G., Hu, J., Steber, C. M., & Sun, T. P. (2004). The *Arabidopsis* F-box protein SLEEPY1 targets gibberellin signaling repressors for gibberellin-induced degradation. *Plant Cell, 16*(6), 1392–1405.

Domagalska, M. A., Sarnowska, E., Nagy, F., & Davis, S. J. (2010). Genetic analyses of interactions among gibberellin, abscisic acid, and brassinosteroids in the control of flowering time in *Arabidopsis thaliana*. *PLoS ONE, 5*(11).

Domagalska, M. A., Schomburg, F. M., Amasino, R. M., Vierstra, R. D., Nagy, F., & Davis, S. J. (2007). Attenuation of brassinosteroid signaling enhances FLC expression and delays flowering. *Development, 134*(15), 2841–2850.

Eastmond, P. J., van Dijken, A. J., Spielman, M., Kerr, A., Tissier, A. F., Dickinson, H. G., et al. (2002). Trehalose-6-phosphate synthase 1, which catalyses the first step in trehalose synthesis, is essential for *Arabidopsis* embryo maturation. *Plant Journal, 29*(2), 225–235.

Eimert, K., Wang, S. M., Lue, W. L., & Chen, J. C. (1995). Monogenic recessive mutations causing both late floral initiation and excess starch accumulation in *Arabidopsis*. *Plant Cell*, 7(10), 1703–1712.

Ellis, C. M., Nagpal, P., Young, J. C., Hagen, G., Guilfoyle, T. J., & Reed, J. W. (2005). AUXIN RESPONSE FACTOR1 and AUXIN RESPONSE FACTOR2 regulate senescence and floral organ abscission in *Arabidopsis thaliana*. *Development, 132*(20), 4563–4574.

Eriksson, S., Bohlenius, H., Moritz, T., & Nilsson, O. (2006). GA4 is the active gibberellin in the regulation of LEAFY transcription and *Arabidopsis* floral initiation. *Plant Cell, 18*(9), 2172–2181.

Feng, S., Martinez, C., Gusmaroli, G., Wang, Y., Zhou, J., Wang, F., et al. (2008). Coordinated regulation of *Arabidopsis thaliana* development by light and gibberellins. *Nature, 451*(7177), 475–479.

Franco-Zorrilla, J. M., Valli, A., Todesco, M., Mateos, I., Puga, M. I., Rubio-Somoza, I., et al. (2007). Target mimicry provides a new mechanism for regulation of microRNA activity. *Nature Genetics, 39*(8), 1033–1037.

Galvão, V. C., Horrer, D., Kuttner, F., & Schmid, M. (2012). Spatial control of flowering by DELLA proteins in *Arabidopsis thaliana*. *Development*, *139*(21), 4072–4082.
Garcia-Martinez, J. L., & Gil, J. (2001). Light regulation of gibberellin biosynthesis and mode of action. *Journal of Plant Growth Regulation*, *20*(4), 354–368.
Gocal, G. F., Poole, A. T., Gubler, F., Watts, R. J., Blundell, C., & King, R. W. (1999). Long-day up-regulation of a GAMYB gene during Lolium temulentum inflorescence formation. *Plant Physiology*, *119*(4), 1271–1278.
Gocal, G. F. W., Sheldon, C. C., Gubler, F., Moritz, T., Bagnall, D. J., MacMillan, C. P., et al. (2001). GAMYB-like genes, flowering, and gibberellin signaling in *Arabidopsis*. *Plant Physiology*, *127*(4), 1682–1693.
Gomez, L. D., Gilday, A., Feil, R., Lunn, J. E., & Graham, I. A. (2010). AtTPS1-mediated trehalose 6-phosphate synthesis is essential for embryogenic and vegetative growth and responsiveness to ABA in germinating seeds and stomatal guard cells. *Plant Journal*, *64*(1), 1–13.
Griffiths, J., Murase, K., Rieu, I., Zentella, R., Zhang, Z. L., Powers, S. J., et al. (2006). Genetic characterization and functional analysis of the GID1 gibberellin receptors in *Arabidopsis*. *Plant Cell*, *18*(12), 3399–3414.
Guardiola, J. L., Monerri, C., & Agusti, M. (1982). The inhibitory effect of GA on flowering in citrus. *Physiologia Plantarum*, *55*, 136–142.
Hall, A. E., & Bleecker, A. B. (2003). Analysis of combinatorial loss-of-function mutants in the *Arabidopsis* ethylene receptors reveals that the ers1 etr1 double mutant has severe developmental defects that are EIN2 dependent. *Plant Cell*, *15*(9), 2032–2041.
Haydon, M. J., Mielczarek, O., Robertson, F. C., Hubbard, K. E., & Webb, A. A. (2013). Photosynthetic entrainment of the *Arabidopsis thaliana* circadian clock. *Nature*, *502*, 689–692.
Hazebroek, J. P., Metzger, J. D., & Mansager, E. R. (1993). Thermoinductive regulation of gibberellin metabolism in *Thlaspi arvense* L. (II. cold induction of enzymes in gibberellin biosynthesis). *Plant Physiology*, *102*(2), 547–552.
He, Y. W., & Loh, C. S. (2002). Induction of early bolting in *Arabidopsis thaliana* by triacontanol, cerium and lanthanum is correlated with increased endogenous concentration of isopentenyl adenosine. *Journal of Experimental Botany*, *53*(368), 505–512.
Hisamatsu, T., & King, R. W. (2008). The nature of floral signals in *Arabidopsis*. II. Roles for FLOWERING LOCUS T (FT) and gibberellin. *Journal of Experimental Botany*, *59*(14), 3821–38219.
Huijser, P., & Schmid, M. (2011). The control of developmental phase transitions in plants. *Development*, *138*(19), 4117–4129.
Imaizumi, T., Schultz, T. F., Harmon, F. G., Ho, L. A., & Kay, S. A. (2005). FKF1 F-box protein mediates cyclic degradation of a repressor of CONSTANS in *Arabidopsis*. *Science*, *309*(5732), 293–297.
Immink, R. G., Pose, D., Ferrario, S., Ott, F., Kaufmann, K., Valentim, F. L., et al. (2012). Characterization of SOC1's central role in flowering by the identification of its upstream and downstream regulators. *Plant Physiology*, *160*(1), 433–449.
Iuchi, S., Suzuki, H., Kim, Y. C., Iuchi, A., Kuromori, T., Ueguchi-Tanaka, M., et al. (2007). Multiple loss-of-function of *Arabidopsis* gibberellin receptor AtGID1s completely shuts down a gibberellin signal. *Plant Journal*, *50*(6), 958–966.
Jacobsen, S. E., & Olszewski, N. E. (1993). Mutations at the SPINDLY locus of *Arabidopsis* alter gibberellin signal transduction. *Plant Cell*, *5*(8), 887–896.
Jaeger, K. E., & Wigge, P. A. (2007). FT protein acts as a long-range signal in *Arabidopsis*. *Current Biology*, *17*(12), 1050–1054.
Johanson, U., West, J., Lister, C., Michaels, S., Amasino, R., & Dean, C. (2000). Molecular analysis of FRIGIDA, a major determinant of natural variation in *Arabidopsis* flowering time. *Science*, *290*(5490), 344–347.

Jung, J. H., Ju, Y., Seo, P. J., Lee, J. H., & Park, C. M. (2012). The SOC1-SPL module integrates photoperiod and gibberellic acid signals to control flowering time in *Arabidopsis*. *Plant Journal, 69*(4), 577–588.

Kamiya, Y., & Garcia-Martinez, J. L. (1999). Regulation of gibberellin biosynthesis by light. *Current Opinion in Plant Biology, 2*(5), 398–403.

Kanyuka, K., Praekelt, U., Franklin, K. A., Billingham, O. E., Hooley, R., Whitelam, G. C., et al. (2003). Mutations in the huge *Arabidopsis* gene BIG affect a range of hormone and light responses. *Plant Journal, 35*(1), 57–70.

Kepinski, S., & Leyser, O. (2005). The *Arabidopsis* F-box protein TIR1 is an auxin receptor. *Nature, 435*(7041), 446–451.

Kieber, J. J. (2002). Cytokinins. *The Arabidopsis Book, 1*, e0063.

Kim, J. J., Lee, J. H., Kim, W., Jung, H. S., Huijser, P., & Ahn, J. H. (2012). The microRNA156-SQUAMOSA PROMOTER BINDING PROTEIN-LIKE3 module regulates ambient temperature-responsive flowering via FLOWERING LOCUS T in *Arabidopsis*. *Plant Physiology, 159*(1), 461–478.

Kim, T. W., & Wang, Z. Y. (2010). Brassinosteroid signal transduction from receptor kinases to transcription factors. *Annual Review of Plant Biology, 61*, 681–704.

King, R. W., Moritz, T., Evans, L. T., Junttila, O., & Herlt, A. J. (2001). Long-day induction of flowering in *Lolium temulentum* involves sequential increases in specific gibberellins at the shoot apex. *Plant Physiology, 127*(2), 624–632.

Kojima, S., Takahashi, Y., Kobayashi, Y., Monna, L., Sasaki, T., Araki, T., et al. (2002). Hd3a, a rice ortholog of the Arabidopsis FT gene, promotes transition to flowering downstream of Hd1 under short-day conditions. *Plant & Cell Physiology, 43*(10), 1096–1105.

Koornneef, M., Elgersma, A., Hanhart, C. J., Vanloenenmartinet, E. P., Vanrijn, L., & Zeevaart, J. A. D. (1985). A gibberellin insensitive mutant of *Arabidopsis thaliana*. *Physiologia Plantarum, 65*(1), 33–39.

Koornneef, M., & van der Veen, J. (1980). Induction and analysis of gibberellin sensitive mutants in *Arabidopsis thaliana* (L.) Heynh. *Theoretical and Applied Genetics, 58*(6), 257–263.

Kumar, S. V., Lucyshyn, D., Jaeger, K. E., Alos, E., Alvey, E., Harberd, N. P., et al. (2012). Transcription factor PIF4 controls the thermosensory activation of flowering. *Nature, 484*(7393). 242–U127.

Kumar, S. V., & Wigge, P. A. (2010). H2A.Z-containing nucleosomes mediate the thermosensory response in *Arabidopsis*. *Cell, 140*(1), 136–147.

Kurosawa, E. (1926). Experimental studies on the nature of the substance secreted by the "bakanae" fungus. *Natural History Society of Formosa, 16*, 213–227.

Lang, A. (1956a). Gibberellin and flower formation. *Naturwissenschaften, 43*(23). 544–544.

Lang, A. (1956b). Induction of flower formation in biennial Hyoscyamus by treatment with gibberellin. *Naturwissenschaften, 43*(12), 284–285.

Lang, A. (1957). The effect of gibberellin upon flower formation. *Proceeding of the National Academy of Sciences of the United States of America, 43*, 709–717.

Lee, J. H., Ryu, H. S., Chung, K. S., Pose, D., Kim, S., Schmid, M., et al. (2013). Regulation of temperature-responsive flowering by MADS-box transcription factor repressors. *Science, 342*(6158), 628–632.

Lee, J. H., Yoo, S. J., Park, S. H., Hwang, I., Lee, J. S., & Ahn, J. H. (2007). Role of SVP in the control of flowering time by ambient temperature in *Arabidopsis*. *Genes & Development, 21*(4), 397–402.

Li, D., Liu, C., Shen, L., Wu, Y., Chen, H., Robertson, M., et al. (2008). A repressor complex governs the integration of flowering signals in *Arabidopsis*. *Developmental Cell, 15*(1), 110–120.

Li, J. M., & Chory, J. (1997). A putative leucine-rich repeat receptor kinase involved in brassinosteroid signal transduction. *Cell, 90*(5), 929–938.

Li, W., Ahn, I. P., Ning, Y. S., Park, C. H., Zeng, L. R., Whitehill, J. G. A., et al. (2012). The U-Box/ARM E3 ligase PUB13 regulates cell death, defense, and flowering time in *Arabidopsis*. *Plant Physiology*, *159*(1), 239–250.

Lin, T. P., Caspar, T., Somerville, C., & Preiss, J. (1988). Isolation and characterization of a starchless mutant of *Arabidopsis-thaliana* (L.) Heynh lacking adpglucose pyrophosphorylase Activity. *Plant Physiology*, *86*(4), 1131–1135.

Liu, C., Chen, H., Er, H. L., Soo, H. M., Kumar, P. P., Han, J. H., et al. (2008). Direct interaction of AGL24 and SOC1 integrates flowering signals in *Arabidopsis*. *Development*, *135*(8), 1481–1491.

Locascio, A., Blazquez, M. A., & Alabadi, D. (2013). Dynamic regulation of cortical microtubule organization through prefoldin-DELLA interaction. *Current Biology*, *23*(9), 804–809.

Lu, C., & Fedoroff, N. (2000). A mutation in the arabidopsis HYL1 gene encoding a dsRNA binding protein affects responses to abscisic acid, auxin, and cytokinin. *Plant Cell*, *12*(12), 2351–2365.

Lunn, J. E., Feil, R., Hendriks, J. H. M., Gibon, Y., Morcuende, R., Osuna, D., et al. (2006). Sugar-induced increases in trehalose 6-phosphate are correlated with redox activation of ADPglucose pyrophosphorylase and higher rates of starch synthesis in *Arabidopsis thaliana*. *Biochemical Journal*, *397*, 139–148.

MacMillan, C. P., Blundell, C. A., & King, R. W. (2005). Flowering of the grass *Lolium perenne*. Effects of vernalization and long days on gibberellin biosynthesis and signaling. *Plant Physiology*, *138*(3), 1794–1806.

Mai, Y. X., Wang, L., & Yang, H. Q. (2011). A gain-of-function mutation in IAA7/AXR2 confers late flowering under short-day light in *Arabidopsis*. *Journal of Integrative Plant Biology*, *53*(6), 480–492.

Martinez, C., Pons, E., Prats, G., & Leon, J. (2004). Salicylic acid regulates flowering time and links defence responses and reproductive development. *Plant Journal*, *37*(2), 209–217.

Martinez-Zapater, J. M., Coupland, G., Dean, C., & Koornneef, M. (1994). *The transition to flowering in Arabidopsis thaliana*. Cold Spring Harbor, NY: Cold Spring Harbor Laboratory Press.

Mathieu, J., Warthmann, N., Kuttner, F., & Schmid, M. (2007). Export of FT protein from phloem companion cells is sufficient for floral induction in *Arabidopsis*. *Current Biology*, *17*(12), 1055–1060.

McGinnis, K. M., Thomas, S. G., Soule, J. D., Strader, L. C., Zale, J. M., Sun, T. P., et al. (2003). The *Arabidopsis* SLEEPY1 gene encodes a putative F-box subunit of an SCF E3 ubiquitin ligase. *Plant Cell*, *15*(5), 1120–1130.

Merchante, C., Alonso, J. M., & Stepanova, A. N. (2013). Ethylene signaling: simple ligand, complex regulation. *Current Opinion in Plant Biology*, *16*(5), 554–560.

Michaels, S. D., & Amasino, R. M. (1999a). FLOWERING LOCUS C encodes a novel MADS domain protein that acts as a repressor of flowering. *Plant Cell*, *11*(5), 949–956.

Michaels, S. D., & Amasino, R. M. (1999b). The gibberellic acid biosynthesis mutant ga1-3 of *Arabidopsis thaliana* is responsive to vernalization. *Developmental Genetics*, *25*(3), 194–198.

Mockaitis, K., & Estelle, M. (2008). Auxin receptors and plant development: a new signaling paradigm. *Annual Review of Cell and Developmental Biology*, *24*, 55–80.

Moon, J., Suh, S. S., Lee, H., Choi, K. R., Hong, C. B., Paek, N. C., et al. (2003). The SOC1 MADS-box gene integrates vernalization and gibberellin signals for flowering in *Arabidopsis*. *Plant Journal*, *35*, 613–623.

Moore, B., Zhou, L., Rolland, F., Hall, Q., Cheng, W. H., Liu, Y. X., et al. (2003). Role of the *Arabidopsis* glucose sensor HXK1 in nutrient, light, and hormonal signaling. *Science*, *300*(5617), 332–336.

Moyroud, E., Minguet, E. G., Ott, F., Yant, L., Pose, D., Monniaux, M., et al. (2011). Prediction of regulatory interactions from genome sequences using a biophysical model for the *Arabidopsis* LEAFY transcription factor. *Plant Cell, 23*(4), 1293–1306.
Murase, K., Hirano, Y., Sun, T. P., & Hakoshima, T. (2008). Gibberellin-induced DELLA recognition by the gibberellin receptor GID1. *Nature, 456*(7221), 459–463.
Nakajima, M., Shimada, A., Takashi, Y., Kim, Y. C., Park, S. H., Ueguchi-Tanaka, M., et al. (2006). Identification and characterization of *Arabidopsis* gibberellin receptors. *Plant Journal, 46*(5), 880–889.
Ogawara, T., Higashi, K., Kamada, H., & Ezura, H. (2003). Ethylene advances the transition from vegetative growth to flowering in *Arabidopsis thaliana. Journal of Plant Physiology, 160*(11), 1335–1340.
Oh, E., Zhu, J. Y., & Wang, Z. Y. (2012). Interaction between BZR1 and PIF4 integrates brassinosteroid and environmental responses. *Nature Cell Biology, 14*(8), 802–U864.
Ohto, M., Onai, K., Furukawa, Y., Aoki, E., Araki, T., & Nakamura, K. (2001). Effects of sugar on vegetative development and floral transition in *Arabidopsis. Plant Physiology, 127*(1), 252–261.
Okushima, Y., Mitina, I., Quach, H. L., & Theologis, A. (2005). Auxin response factor 2 (Arf2): A pleiotropic developmental regulator. *Plant Journal, 43*(1), 29–46.
Olszewski, N., Sun, T. P., & Gubler, F. (2002). Gibberellin signaling: biosynthesis, catabolism, and response pathways. *Plant Cell, 14*(Suppl), S61–S80.
Osnato, M., Castillejo, C., Matias-Hernandez, L., & Pelaz, S. (2012). TEMPRANILLO genes link photoperiod and gibberellin pathways to control flowering in *Arabidopsis. Nature Communications, 3*(808).
Paparelli, E., Parlanti, S., Gonzali, S., Novi, G., Mariotti, L., Ceccarelli, N., et al. (2013). Night-time sugar starvation orchestrates gibberellin biosynthesis and plant growth in *Arabidopsis. Plant Cell, 25*(10), 3760–3769.
Park, J., Nguyen, K. T., Park, E., Jeon, J. S., & Choi, G. (2013). DELLA proteins and their interacting RING finger proteins repress gibberellin responses by binding to the promoters of a subset of gibberellin-responsive genes in *Arabidopsis. Plant Cell, 25*(3), 927–943.
Paul, M. J., Primavesi, L. F., Jhurreea, D., & Zhang, Y. (2008). Trehalose metabolism and signaling. *Annual Review of Plant Biology, 59*, 417–441.
Peng, J. R., Carol, P., Richards, D. E., King, K. E., Cowling, R. J., Murphy, G. P., et al. (1997). The *Arabidopsis* GAI gene defines a signaling pathway that negatively regulates gibberellin responses. *Genes & Development, 11*(23), 3194–3205.
Peng, J. R., & Harberd, N. P. (1993). Derivative alleles of the *Arabidopsis* gibberellin-insensitive (gai) mutation confer a wild-type phenotype. *Plant Cell, 5*(3), 351–360.
Peng, J. R., Richards, D. E., Hartley, N. M., Murphy, G. P., Devos, K. M., Flintham, J. E., et al. (1999). 'Green revolution' genes encode mutant gibberellin response modulators. *Nature, 400*(6741), 256–261.
Pharis, R. P., & King, R. W. (1985). Gibberellins and reproductive development in seed plants. *Annual Review of Plant Physiology and Plant Molecular Biology, 36*, 517–568.
Ponnu, J., Wahl, V., & Schmid, M. (2011). Trehalose-6-phosphate: connecting plant metabolism and development. *Frontiers in Plant Science, 2*, 1–6.
Porri, A., Torti, S., Romera-Branchat, M., & Coupland, G. (2012). Spatially distinct regulatory roles for gibberellins in the promotion of flowering of *Arabidopsis* under long photoperiods. *Development, 139*(12), 2198–2209.
Pose, D., Verhage, L., Ott, F., Yant, L., Mathieu, J., Angenent, G. C., et al. (2013). Temperature-dependent regulation of flowering by antagonistic FLM variants. *Nature, 503*(7476), 414–417.
Rhoades, M. W., Reinhart, B. J., Lim, L. P., Burge, C. B., Bartel, B., & Bartel, D. P. (2002). Prediction of plant microRNA targets. *Cell, 110*(4), 513–520.

Riboni, M., Galbiati, M., Tonelli, C., & Conti, L. (2013). GIGANTEA enables drought escape response via abscisic acid-dependent activation of the florigens and SUPPRESSOR OF OVEREXPRESSION OF CONSTANS. *Plant Physiology*, *162*(3), 1706–1719.

Richter, R., Bastakis, E., & Schwechheimer, C. (2013a). Cross-repressive interactions between SOC1 and the GATAs GNC and GNL/CGA1 in the control of greening, cold tolerance, and flowering time in *Arabidopsis*. *Plant Physiology*, *162*(4), 1992–2004.

Richter, R., Behringer, C., Muller, I. K., & Schwechheimer, C. (2010). The GATA-type transcription factors GNC and GNL/CGA1 repress gibberellin signaling downstream from DELLA proteins and PHYTOCHROME-INTERACTING FACTORS. *Genes & Development*, *24*(18), 2093–2104.

Richter, R., Behringer, C., Zourelidou, M., & Schwechheimer, C. (2013b). Convergence of auxin and gibberellin signaling on the regulation of the GATA transcription factors GNC and GNL in *Arabidopsis thaliana*. *Proceeding of the National Academy of Sciences of the United States of America*, *110*(32), 13192–13197.

Riefler, M., Novak, O., Strnad, M., & Schmulling, T. (2006). *Arabidopsis* cytokinin receptor mutants reveal functions in shoot growth, leaf senescence, seed size, germination, root development, and cytokinin metabolism. *Plant Cell*, *18*(1), 40–54.

Roldan, M., Gomez-Mena, C., Ruiz-Garcia, L., Salinas, J., & Martinez-Zapater, J. M. (1999). Sucrose availability on the aerial part of the plant promotes morphogenesis and flowering of *Arabidopsis* in the dark. *Plant Journal*, *20*(5), 581–590.

Rood, S. B., Pearce, D., Williams, P. H., & Pharis, R. P. (1989). A gibberellin-deficient Brassica mutant-rosette. *Plant Physiology*, *89*(2), 482–487.

Samach, A., Onouchi, H., Gold, S. E., Ditta, G. S., Schwarz-Sommer, Z., Yanofsky, M. F., et al. (2000). Distinct roles of CONSTANS target genes in reproductive development of *Arabidopsis*. *Science*, *288*(5471), 1613–1616.

Sarnowska, E. A., Rolicka, A. T., Bucior, E., Cwiek, P., Tohge, T., Fernie, A. R., et al. (2013). DELLA-interacting SWI3C core subunit of switch/sucrose nonfermenting chromatin remodeling complex modulates gibberellin responses and hormonal cross talk in *Arabidopsis*. *Plant Physiology*, *163*(1), 305–317.

Sawa, M., & Kay, S. A. (2011). GIGANTEA directly activates Flowering Locus T in *Arabidopsis thaliana*. *Proceeding of the National Academy of Sciences of the United States of America*, *108*(28), 11698–11703.

Schomburg, F. M., Bizzell, C. M., Lee, D. J., Zeevaart, J. A., & Amasino, R. M. (2003). Overexpression of a novel class of gibberellin 2-oxidases decreases gibberellin levels and creates dwarf plants. *Plant Cell*, *15*(1), 151–163.

Segarra, S., Mir, R., Martinez, C., & Leon, J. (2010). Genome-wide analyses of the transcriptomes of salicylic acid-deficient versus wild-type plants uncover Pathogen and Circadian Controlled 1 (PCC1) as a regulator of flowering time in *Arabidopsis*. *Plant Cell and Environment*, *33*(1), 11–22.

Sheldon, C. C., Burn, J. E., Perez, P. P., Metzger, J., Edwards, J. A., Peacock, W. J., et al. (1999). The FLF MADS box gene: a repressor of flowering in *Arabidopsis* regulated by vernalization and methylation. *Plant Cell*, *11*(3), 445–458.

Shimada, A., Ueguchi-Tanaka, M., Nakatsu, T., Nakajima, M., Naoe, Y., Ohmiya, H., et al. (2008). Structural basis for gibberellin recognition by its receptor GID1. *Nature*, *456*(7221), 520–523.

Silverstone, A. L., Jung, H.-S., Dill, A., Kawaide, H., Kamiya, Y., & Sun, T.-P. (2001). Repressing a repressor - Gibberellin induced rapid reduction of the RGA protein in *Arabidopsis*. *Plant Cell*, *13*, 1555–1565.

Silverstone, A. L., Mak, P. Y., Martinez, E. C., & Sun, T. P. (1997). The new RGA locus encodes a negative regulator of gibberellin response in *Arabidopsis thaliana*. *Genetics*, *146*(3), 1087–1099.

Silverstone, A. L., Tseng, T. S., Swain, S. M., Dill, A., Jeong, S.Y., Olszewski, N. E., et al. (2007). Functional analysis of SPINDLY in gibberellin signaling in *Arabidopsis*. *Plant Physiology*, *143*(2), 987–1000.

Srikanth, A., & Schmid, M. (2011). Regulation of flowering time: all roads lead to Rome. *Cellular and Molecular Life Sciences*, *68*(12), 2013–2037.

Streb, S., & Zeeman, S. C. (2012). Starch metabolism in *Arabidopsis*. *The Arabidopsis Book*, *10*, e0160.

Suarez-Lopez, P., Wheatley, K., Robson, F., Onouchi, H., Valverde, F., & Coupland, G. (2001). CONSTANS mediates between the circadian clock and the control of flowering in *Arabidopsis*. *Nature*, *410*(6832), 1116–1120.

Suge, H. (1970). Changes of endogenous gibberellins in vernalized radish plants. *Plant and Cell Physiology*, *11*(5), 729–735.

Suge, H. (1972). Inhibition of photoperiodic floral induction in *Pharbitis nil* by ethylene. *Plant and Cell Physiology*, *13*(6), 1031–1038.

Sun, T. P., & Kamiya, Y. (1994). The *Arabidopsis* GA1 locus encodes the cyclase ent-kaurene synthetase A of gibberellin biosynthesis. *Plant Cell*, *6*(10), 1509–1518.

Takahashi, N., Kitamura, H., Kawarada, A., Seta, Y., Takai, M., Tamura, S., et al. (1955). Isolation of gibberellins and their properties. *Bulletin of Agricultural Chemical Society, Japan*, *19*, 267–277.

Tao, Y., Ferrer, J. L., Ljung, K., Pojer, F., Hong, F. X., Long, J. A., et al. (2008). Rapid synthesis of auxin via a new tryptophan-dependent pathway is required for shade avoidance in plants. *Cell*, *133*(1), 164–176.

Tao, Z., Shen, L., Liu, C., Liu, L., Yan, Y., & Yu, H. (2012). Genome-wide identification of SOC1 and SVP targets during the floral transition in *Arabidopsis*. *Plant Journal*, *70*(4), 549–561.

Taoka, K., Ohki, I., Tsuji, H., Furuita, K., Hayashi, K., Yanase, T., et al. (2011). 14-3-3 proteins act as intracellular receptors for rice Hd3a florigen. *Nature*, *476*(7360), 332–335.

Todesco, M., Rubio-Somoza, I., Paz-Ares, J., & Weigel, D. (2010). A collection of target mimics for comprehensive analysis of microRNA function in *Arabidopsis thaliana*. *PLoS Genetics*, *6*(7), e1001031.

Trusov, Y., & Botella, J. R. (2006). Silencing of the ACC synthase gene ACACS2 causes delayed flowering in pineapple [*Ananas comosus* (L.) Merr.]. *Journal of Experimental Botany*, *57*(14), 3953–3960.

Turk, E. M., Fujioka, S., Seto, H., Shimada, Y., Takatsuto, S., Yoshida, S., et al. (2005). BAS1 and SOB7 act redundantly to modulate Arabidopsis photomorphogenesis via unique brassinosteroid inactivation mechanisms. *Plant Journal*, *42*(1), 23–34.

Ueguchi-Tanaka, M., Ashikari, M., Nakajima, M., Itoh, H., Katoh, E., Kobayashi, M., et al. (2005). GIBBERELLIN INSENSITIVE DWARF1 encodes a soluble receptor for gibberellin. *Nature*, *437*(7059), 693–698.

Ulmasov, T., Hagen, G., & Guilfoyle, T. J. (1999). Activation and repression of transcription by auxin-response factors. *Proceeding of the National Academy of Sciences of the United States of America*, *96*(10), 5844–5849.

van Dijken, A. J. H., Schluepmann, H., & Smeekens, S. C. M. (2004). Arabidopsis trehalose-6-phosphate synthase 1 is essential for normal vegetative growth and transition to flowering. *Plant Physiology*, *135*(2), 969–977.

Vernoux, T., Brunoud, G., Farcot, E., Morin, V., Van den Daele, H., Legrand, J., et al. (2011). The auxin signalling network translates dynamic input into robust patterning at the shoot apex. *Molecular Systems Biology*, 7, 508.

Verslues, P. E., & Juenger, T. E. (2011). Drought, metabolites, and *Arabidopsis* natural variation: a promising combination for understanding adaptation to water-limited environments. *Current Opinion in Plant Biology*, *14*(3), 240–245.

Wada, K. C., Yamada, M., Shiraya, T., & Takeno, K. (2010). Salicylic acid and the flowering gene FLOWERING LOCUS T homolog are involved in poor-nutrition stress-induced flowering of *Pharbitis nil*. *Journal of Plant Physiology*, *167*(6), 447–452.

Wahl, V., Ponnu, J., Schlereth, A., Arrivault, S., Langenecker, T., Franke, A., et al. (2013). Regulation of flowering by trehalose-6-phosphate signaling in *Arabidopsis thaliana*. *Science*, *339*(6120), 704–707.

Wang Czech, B., & Weigel, D. (2009). miR156-regulated SPL transcription factors define an endogenous flowering pathway in *Arabidopsis thaliana*. *Cell*, *138*(4), 738–749.

Wang, K. L., Li, H., & Ecker, J. R. (2002). Ethylene biosynthesis and signaling networks. *Plant Cell*, *14*(Suppl), S131–S151.

Wang, Q., Zhang, W., Yin, Z., & Wen, C. K. (2013). Rice CONSTITUTIVE TRIPLE-RESPONSE2 is involved in the ethylene-receptor signalling and regulation of various aspects of rice growth and development. *Journal of Experimental Botany*, *64*(16), 4863–4875.

Wang, R. H., Hsu, Y. M., Bartholomew, D. P., Maruthasalam, S., & Lin, C. H. (2007). Delaying natural flowering in pineapple through foliar application of aviglycine, an inhibitor of ethylene biosynthesis. *Hortscience*, *42*(5), 1188–1191.

Weigel, D., Alvarez, J., Smyth, D. R., Yanofsky, M. F., & Meyerowitz, E. M. (1992). LEAFY controls floral meristem identity in *Arabidopsis*. *Cell*, *69*(5), 843–859.

Weller, J. L., Hecht, V., Vander Schoor, J. K., Davidson, S. E., & Ross, J. J. (2009). Light regulation of gibberellin biosynthesis in pea is mediated through the COP1/HY5 pathway. *Plant Cell*, *21*(3), 800–813.

Werner, T., Motyka, V., Laucou, V., Smets, R., Van Onckelen, H., & Schmulling, T. (2003). Cytokinin-deficient transgenic *Arabidopsis* plants show multiple developmental alterations indicating opposite functions of cytokinins in the regulation of shoot and root meristem activity. *Plant Cell*, *15*(11), 2532–2550.

Wigge, P. A., Kim, M. C., Jaeger, K. E., Busch, W., Schmid, M., Lohmann, J. U., et al. (2005). Integration of spatial and temporal information during floral induction in *Arabidopsis*. *Science*, *309*(5737), 1056–1059.

Wijayanti, L., Fujioka, S., Kobayashi, M., & Sakurai, A. (1997). Involvement of abscisic acid and indole-3-acetic acid in the flowering of *Pharbitis nil*. *Journal of Plant Growth Regulation*, *16*(2), 115–119.

Willige, B. C., Ghosh, S., Nill, C., Zourelidou, M., Dohmann, E. M., Maier, A., et al. (2007). The DELLA domain of GA INSENSITIVE mediates the interaction with the GA INSENSITIVE DWARF1A gibberellin receptor of *Arabidopsis*. *Plant Cell*, *19*(4), 1209–1220.

Wilson, R. N., Heckman, J. W., & Sommerville, C. R. (1992). Gibberellin is required for flowering in *Arabidopsis thaliana* under short days. *Plant Physiology*, *100*, 403–408.

Wu, G., Park, M. Y., Conway, S. R., Wang, J. W., Weigel, D., & Poethig, R. S. (2009). The sequential action of miR156 and miR172 regulates developmental timing in *Arabidopsis*. *Cell*, *138*(4), 750–759.

Wuriyanghan, H., Zhang, B., Cao, W. H., Ma, B., Lei, G., Liu, Y. F., et al. (2009). The ethylene receptor ETR2 delays floral transition and affects starch accumulation in rice. *Plant Cell*, *21*(5), 1473–1494.

Xiong, Y., McCormack, M., Li, L., Hall, Q., Xiang, C., & Sheen, J. (2013). Glucose-TOR signalling reprograms the transcriptome and activates meristems. *Nature*, *496*(7444), 181–186.

Yamada, M., & Takeno, K. (2014). Stress and salicylic acid induce the expression of PnFT2 in the regulation of the stress-induced flowering of Pharbitis nil. *Journal of Plant Physiology*, *171*(3–4), 205–212.

Yamaguchi, A., Wu, M. F., Yang, L., Wu, G., Poethig, R. S., & Wagner, D. (2009). The MicroRNA-regulated SBP-Box transcription factor SPL3 is a direct upstream activator of LEAFY, FRUITFULL, and APETALA1. *Developmental Cell*, *17*(2), 268–278.

Yang, L., Xu, M., Koo, Y., He, J., & Poethig, R. S. (2013). Sugar promotes vegetative phase change in *Arabidopsis thaliana* by repressing the expression of MIR156A and MIR156C. *eLife*, *2*, e00260.

Yu, S., Cao, L., Zhou, C.-M., Zhang, T.-Q., Lian, H., Sun, Y., et al. (2013). Sugar is an endogenous cue for juvenile to adult phase transition in plants. *eLife*, *2*, e00269.

Yu, S., Galvão, V. C., Zhang, Y. C., Horrer, D., Zhang, T. Q., Hao, Y. H., et al. (2012). Gibberellin regulates the *Arabidopsis* floral transition through miR156-targeted SQUAMOSA promoter binding-like transcription factors. *Plant Cell*, *24*(8), 3320–3332.

Yu, T. S., Lue, W. L., Wang, S. M., & Chen, J. C. (2000). Mutation of Arabidopsis plastid phosphoglucose isomerase affects leaf starch synthesis and floral initiation. *Plant Physiology*, *123*(1), 319–325.

Zanewich, K. P., & Rood, S. B. (1995). Vernalization and gibberellin physiology of winter canola (endogenous gibberellin (GA)) content and metabolism of [3H]GA1 and [3H] GA20. *Plant Physiology*, *108*(2), 615–621.

Zanewich, K. P., Rood, S. B., & Williams, P. H. (1990). Growth and development of Brassica genotypes differing in endogenous gibberellin content. I. Leaf and reproductive development. *Physiologia Plantarum*, *79*(4), 673–678.

Zeevaart, J. A. D. (1976). Physiology of flower formation. *Annual Review of Plant Physiology and Plant Molecular Biology*, *27*, 321–348.

Zentella, R., Zhang, Z. L., Park, M., Thomas, S. G., Endo, A., Murase, K., et al. (2007). Global analysis of DELLA direct targets in early gibberellin signaling in *Arabidopsis*. *Plant Cell*, *19*(10), 3037–3057.

Zhou, L., Jang, J. C., Jones, T. L., & Sheen, J. (1998). Glucose and ethylene signal transduction crosstalk revealed by an Arabidopsis glucose-insensitive mutant. *Proceedings of the National Academy of Sciences of United States of America*, *95*(17), 10294–10299.

CHAPTER FOUR

Critical Gates in Day-Length Recognition to Control the Photoperiodic Flowering

Asami Osugi and Takeshi Izawa[1]
Functional Plant Research Unit, National Institute of Agrobiological Sciences, Tsukuba, Ibaraki, Japan
[1]Corresponding author: e-mail address: tizawa@nias.affrc.go.jp

Contents

Advances in Botanical Research, Volume 72
ISSN 0065-2296
http://dx.doi.org/10.1016/B978-0-12-417162-6.00004-3

Abstract

Many plants flower in response to specific day lengths, which is crucial for successful reproduction, indicating that plants have developed sophisticated molecular mechanisms for day-length recognition. The external coincidence model proposes a crosstalk between the internal circadian clock and external light stimuli. This review focuses on recent progress in understanding the molecular mechanisms underlying the 'gating effect', by which any biological responses to environmental stimuli are restricted to certain phases of the circadian clock. For instance, in rice, two distinct predominant gating mechanisms recognise the critical day length. The gating mechanism for red light induces morning expression of the floral repressor *Grain number, plant height and heading date7*, but only under long-day conditions. The gating mechanism for blue light induces morning expression of the floral promoter *Early heading date 1*. Here, we summarise various gating mechanisms that recognise day length in several plant species, and also discuss the existence of nongating mechanisms involved in the photoperiodic control of flowering.

4.1 INTRODUCTION

4.1.1 Florigen as an Indicator of Day-Length Recognition

Plants sense environmental conditions, which enables them to flower during the appropriate season and reproduce successfully. Among various environmental cues, day length is suitable to anticipate seasonal changes because of its predictable changes during the year. About a century ago, Garner and Allard first described that plants can flower in response to day length (Garner & Allard, 1920); this phenomenon is now termed photoperiodism. These authors classified many plants as long-day plants, i.e. which flower when day length is increasing, and short-day plants, which flower when day length is decreasing. Short-day plants often have a clear day-length threshold, called critical day length (critical photoperiod length), after which flower formation is induced. In *Arabidopsis thaliana*, a long-day plant, the molecular mechanisms of photoperiodic flowering are well characterised. Among short-day plants, such molecular mechanisms have been well studied in rice, especially those that confer critical day-length recognition.

Chailakhyan (1936) proposed that a compound termed 'florigen' is produced in leaves and can induce floral organ differentiation at the shoot apex, and thus acts similarly to plant hormones. Several recent reports suggested that the *FLOWERING LOCUS T* (*FT*) gene product is a candidate for florigen in *Arabidopsis*, since *FT* is expressed in the leaf and promotes flowering at the shoot apex (Abe et al., 2005; Kardailsky et al., 1999; Kobayashi, Kaya, Goto, Iwabuchi, & Araki, 1999). Subsequently, it was reported that *FT* and its rice orthologue Heading date 3a (*Hd3a*) can move from leaves to the shoot apex (Corbesier et al., 2007; Kojima et al., 2002; Tamaki, Matsuo, Wong, Yokoi, & Shimamoto, 2007), indicating that the *FT* family genes encode the florigens. A florigen gene has the following characteristics: (1) its mRNA is induced in the leaf under inductive day-length conditions; (2) the gene product moves from the leaf to the shoot apex; (3) the gene product can induce floral organ formation at the shoot apex (4) and florigens from different species are evolutionarily and functionally conserved. Long-day and short-day plants produce florigen under inductive long- and short-day conditions, respectively. Thus, the levels of florigen mRNA in leaves provide critical information about the status of floral transitions.

4.1.2 Models for Day-Length Recognition Based on Physiological Evidences

Today, it is widely recognised that both light signal transduction cascades and the circadian clock are involved in the molecular recognition of day length. This idea was first suggested by Dr. Erwin Bünning (Figure 4.1(A);

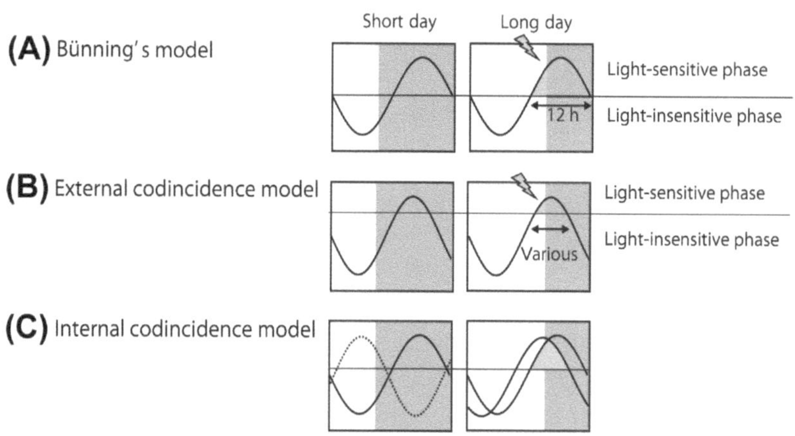

Figure 4.1 The models for day-length recognition. (A) Bünning's model. The light-sensitive phase is set by the circadian clock during the second half of the day (evening and night). When the light signal is present during the light-sensitive phase, plants perceive it as long day. (B) The external coincidence model. As in Bünning's model, the light-sensitive phase is set by the circadian clock, and the light signal during the light-sensitive phase is perceived as long day, but the light-sensitive phase is considerably shorter than in Bünning's model. (C) The internal coincidence model. Two distinct diurnal rhythms are formed by the circadian clock, or circadian clock and other diurnal rhythms coexist. When the two rhythms overlap, the day length is recognised. (See the colour plate.)

Bünning, 1936, 1950). According to Bünning's model, often called 'Bünning's hypothesis', plants recognise day length by perceiving (or not perceiving) the light during the light-sensitive phase of circadian clocks. The light-sensitive phase occurs in the evening and during the night and corresponds to the second half of the day and is reset at dawn by the circadian clock. However, Bünning's historical pioneer work was unable to explain the variability in the critical day length among plant species. Thirty years later, Dr. Colin Pittendrigh proposed the 'external coincidence model', which was likely inspired by Bünning's model (Figure 4.1(B); Pittendrigh & Minis, 1964). Similarly to Bünning's model, the new model proposed that light perceived during the light-sensitive phase can provide information about day length. However, in this model the light-sensitive phase is substantially shorter than 12 h, and is entrained by the light stimulus. As a result, the light-sensitive phase can shift in response to the changing day length. Later, Pittendrigh suggested the 'internal coincidence model' (Figure 4.1(C); Pittendrigh, 1966), according to which an interaction between two differentially entrained circadian (or diurnal) rhythms contributes to day-length recognition. Recent studies have discovered the molecular mechanisms that support the external coincidence model, although some other mechanisms support the internal coincidence model.

4.1.3 Gating Mechanisms in Day-Length Recognition

Many biological processes are diurnally regulated, either directly by the circadian clock or by the coincidence of environmental stimuli with a specific time of the day. The 'gating effect' concept has been proposed to explain the latter case (Pittendrigh, 1966). According to this concept, an external stimulus can be perceived only when the gate is 'open'. The gate has a diurnal rhythm, regulated mainly by the circadian clock, and biological processes can be induced by external stimuli only at a specific time of the day. The timing of such gate opening would be altered by light signals (either light-on or light-off signals) in response to a given day length, and the presence of light during the opening time of the gate depends on day length. When taking into account the expression dynamics of many genes involved in day-length recognition and responsive to the fluctuations of the gate, the opening time is often set around dawn or dusk, when light conditions change greatly within a few hours (Nagano et al., 2012). The molecular nature of such gates may include transcription, protein abundance, or enzymatic activity at a specific time of the day.

4.1.4 Photoreceptors Involved in Day-Length Recognition

Various light wavelengths are detected by several distinct plant photoreceptors, and plants use this information for day-length recognition. In rice, blue light accelerates flowering and red light delays it (Hirose, Shinomura, Tanabata, Shimada, & Takano, 2006; Itoh, Nonoue, Yano, & Izawa, 2010; Izawa, Oikawa, Tokutomi, Okuno, & Shimamoto, 2000; Takano et al., 2005). Far-red light at the 'end of the day' accelerates flowering, although these effects are poorly characterised (Takano et al., 2005). Photoreceptors in higher plants include phytochromes, cryptochromes, phototropins and some special F-box proteins (reviewed in Kami, Lorrain, Hornitschek, & Fankhauser (2010), Wang (2005)). Phytochromes are able to absorb a broad wavelength range, but are most sensitive to red and far-red light. Red light transforms the inactive Pr form into the active Pfr form, which activates a specific signalling cascade to induce downstream gene expression (reviewed in Quail (2002), Nagatani (2010), Chen and Chory (2011)). Rice has three phytochrome genes, *PHYA*, *PHYB* and *PHYC* (Takano et al., 2005). *Photoperiod sensitivity 5* (*se5*), a phytochrome chromophore-deficient mutant, has almost no functional phytochrome molecules (Izawa et al., 2000). This mutant is almost insensitive to day-length changes, indicating that phytochromes are critically required for day-length recognition in rice (Izawa et al., 2000). Distinct flowering time phenotypes of phytochrome single and double mutants indicate that all three rice phytochromes play different roles in the flowering time control (Takano et al., 2005). Cryptochromes, phototropins and those F-box proteins (the latter include *ZEITLUPE, FLAVIN-BINDING, KELCH REPEAT, F-BOX 1* (*FKF1*) and *LOV KELCH PROTEIN 2* in *Arabidopsis thaliana*), are blue light receptors. Less is known about the role of blue light as compared to red light in the regulation of rice flowering time. *Arabidopsis* cryptochrome 2 significantly promotes flowering under long-day conditions (Liu et al., 2008; Zuo, Liu, Liu, Liu, & Lin, 2011), whereas rice cryptochromes accelerate flowering time independently of day length (Hirose et al., 2006). Importantly, the expression of *Early heading date 1* (*Ehd1*), a floral promoter in rice, is blue light inducible, although its induction by blue light is also regulated by the circadian clock, which requires *OsGigantea (OsGI)*, a circadian clock core gene in rice (Doi et al., 2004; Itoh et al., 2010; Izawa et al., 2011). The F-box protein FKF1 plays a major role in *Arabidopsis* flowering time regulation, although such role of the *FKF1*-type blue light photoreceptors in rice has not yet been reported (Imaizumi, Schultz, Harmon, Ho, & Kay, 2005; Imaizumi, Tran, Swartz, Briggs, & Kay, 2003; Sawa, Nusinow, Kay, & Imaizumi, 2007; Song, Smith, To, Millar, & Imaizumi, 2012).

4.2 GATE-REGULATED EXPRESSION OF FLORIGEN GENES IN RESPONSE TO DAY LENGTH

4.2.1 Two Regulators of Florigen Genes in Rice

Rice has several florigen homologue genes (Izawa et al., 2002). Among them, *Hd3a* induces flowering under inductive short-day conditions, and *RICE FLOWERING LOCUS T1* (*RFT1*) induces flowering under noninductive long-day conditions (Kojima et al., 2002; Komiya, Ikegami, Tamaki, Yokoi, & Shimamoto, 2008). A number of genes encoding *Hd3a* and *RFT1* regulators have been identified. Almost all reports suggest that those regulators control *Hd3a* and *RFT1* via *Heading date 1* (*Hd1*) or *Ehd1* (or both) (Doi et al., 2004; Itoh & Izawa, 2013; Izawa, 2007; Yano et al., 2000). *Hd1* is a rice homologue of the *Arabidopsis CONSTANS* (*CO*) gene, which encodes a well-conserved plant transcription factor (Putterill, Robson, Lee, Simon, & Coupland, 1995; Griffiths, Dunford, Coupland, & Laurie, 2003; Nemoto, Kisaka, Fuse, Yano, & Ogihara, 2003; Yano et al., 2000). *Hd1* promotes the expression of *Hd3a* and *RFT1* under short-day conditions, but represses their expression under long-day conditions, whereas *CO* induces the expression of the *FT* gene (a florigen gene in *Arabidopsis*) only under long-day conditions (Izawa et al., 2002; Kojima et al., 2002; Samach et al., 2000). *Ehd1* encodes a putative B-type response regulator (Doi et al., 2004) that may include his-phosphorylation as a response regulator and specific DNA-binding ability as a transcription factor. However, any biochemical analysis of *Ehd1* activity is still to be reported. Although the *Arabidopsis* genome encodes several B-type response regulators, it has no apparent *Ehd1* orthologue. In young rice seedlings, *Ehd1* induces *Hd3a* and *RFT1* only under short-day conditions (Doi et al., 2004; Itoh et al., 2010). This day-length-specific function correlates well with the *Ehd1* mRNA level (Doi et al., 2004), indicating a critical role of *Ehd1* transcriptional regulation for its correct function. However, ectopic *Ehd1* expression still promotes floral transition only under short-day conditions (Osugi, Itoh, Ikeda-Kawakatsu, Takano, & Izawa, 2011), indicating that *Ehd1* activity can also respond to day length posttranscriptionally. At the mature stage, *Ehd1* induces *RFT1* even under long-day conditions (Komiya, Yokoi, & Shimamoto, 2009). This is consistent with the reports indicating that rice plants bearing nonfunctional *ehd1* alleles flowered much later than those with functional *Ehd1* alleles under long-day conditions, in genetic backgrounds where functional *Hd1* alleles are present (Doi et al., 2004; Izawa, 2007).

Several genes are involved in *Ehd1* upregulation at the mature stage (Gao et al., 2013; Komiya et al., 2009; Matsubara et al., 2011). For example, *Early Heading date 2 (Ehd2)/OsINDETERMINATE 1/RICE INDETERMINATE 1*, encoding a possible transcription factor, upregulates *Ehd1* regardless of day length (Matsubara et al., 2008; Park et al., 2008; Wu et al., 2008). Although severe mutations in *Ehd2* result in never-flowering phenotypes under long-day conditions but not under short-day conditions, the function of *Ehd2* may be day-length independent, and other photoperiod-responsive genes might cause flowering under short-day conditions in *ehd2*. The function or expression of *Ehd2* may depend on plant age but not on day length. Thus, *Ehd1* is believed to promote floral induction at the mature stage in response to both age and day length, whereas even at the young stage plants very clearly recognise day length regardless of the age background effect.

In following sections, we will focus on the regulation of *Hd3a* in young rice seedlings in terms of photoperiodic day-length recognition. We have demonstrated that the expression level of *Hd3a* is precisely regulated by two gating mechanisms of *Grain number, plant height and heading date7* (*Ghd7*) and *Ehd1* responding to at least two distinct light signalling pathways (Itoh et al., 2010). These mechanisms set up the critical day-length threshold to switch on *Hd3a* expression. This molecular switch can recognise subtle differences in day length, and only a 30 min difference is sufficient to toggle the switch of florigen gene expression on and off.

4.2.2 Circadian Gating Mechanisms with Photoperiodic Gate Shifts in Rice

Ghd7 plays a central role in day-length recognition (Itoh et al., 2010). Although Ghd7 is a nuclear protein with a constans, constans-like, toc1 (CCT) domain, its function as a transcription factor remains to be established (Xue et al., 2008). Under long-day conditions, *Ghd7* expression is induced, leading to a delay in flowering because of a strong repression of *Ehd1* and subsequently *Hd3a* (Xue et al., 2008; Itoh et al., 2010). In the *ghd7* mutant, flowering occurs very early independently of the day length (Xue et al., 2008).

The day-length response of *Ghd7* mRNA is achieved by a complex gating mechanism (Figure 4.2(A); Itoh et al., 2010). Phytochrome-mediated red light signalling induces *Ghd7* only at a specific time of the day when 'the *Ghd7* gate is open' (Itoh et al., 2010). Under long-day conditions, the *Ghd7* gate is set to open around dawn, whereas under short-day conditions its opening is shifted to midnight, when it is usually dark (Itoh et al., 2010). Under

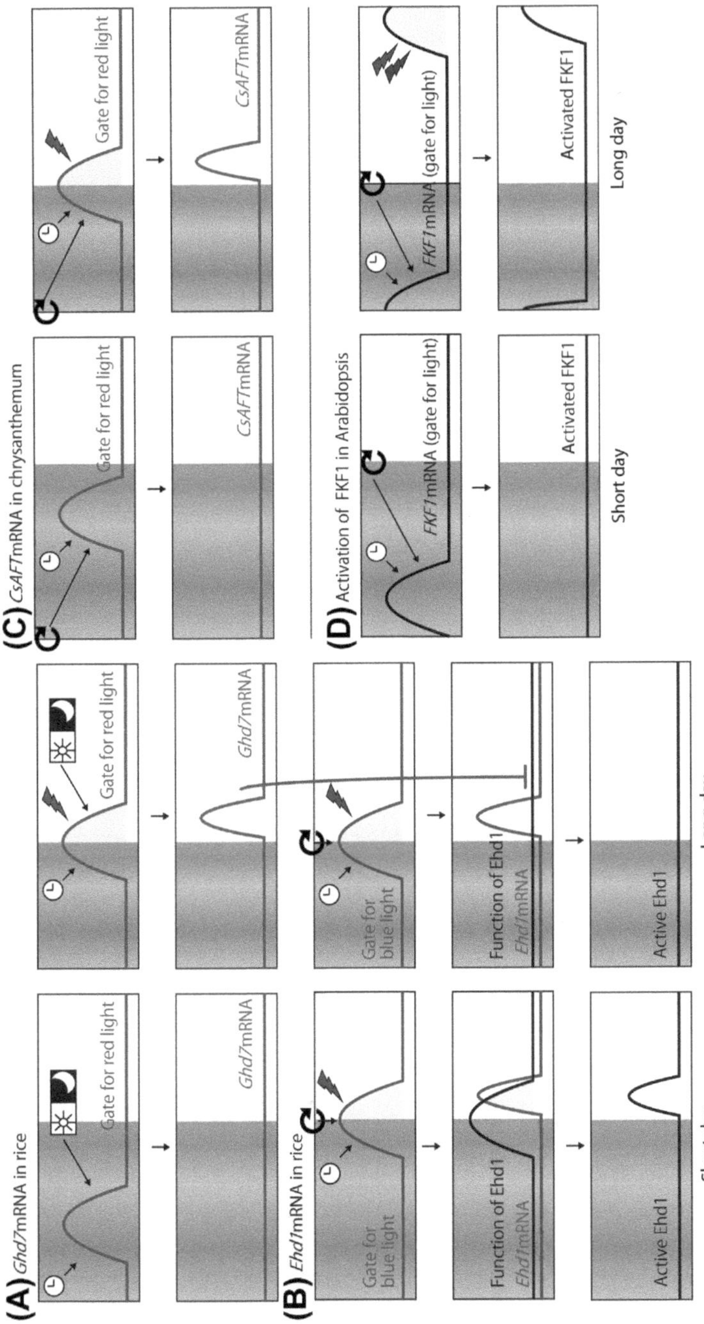
(A) Ghd7mRNA in rice
Gate for red light
Ghd7mRNA
(B) Ehd1mRNA in rice
Gate for blue light
Function of Ehd1
Ehd1mRNA
Active Ehd1
Short day
Long day
(C) CsAFTmRNA in chrysanthemum
Gate for red light
CsAFTmRNA
(D) Activation of FKF1 in Arabidopsis
FKF1mRNA (gate for light)
Activated FKF1
Short day
Long day

Figure 4.2 Gating mechanisms for day-length recognition. (A) The gate for red light to induce *Ghd7* mRNA in rice (Itoh et al., 2010). The *Ghd7* gate is entrained by day length and is open around midnight under short-day conditions and around dawn under long-day conditions. Red light perception by phytochromes coincident with the gate opening time under long-day conditions results in *Ghd7* mRNA induction. (B) The gate for blue light to induce *Ehd1* mRNA in rice (Itoh et al., 2010). The *Ehd1* gate is entrained by the circadian clock, and is open around dawn regardless of day length. As a result, *Ehd1* mRNA is induced at dawn, but is repressed by Ghd7 only under long-day conditions. Moreover, Ehd1 activity promotes *Hd3a* expression only in the morning under short-day conditions. Thus, when *Ehd1* mRNA and the Ehd1 activity are coincident at dawn under short-day conditions, *Hd3a* mRNA is induced. (C) The gate for red light to induce *CsAFT* mRNA in *Chrysanthemum* (Higuchi et al., 2013). The *CsAFT* gate is entrained by the circadian clock or light-to-dark transition, and is open around midnight under short-day conditions and around dawn under long-day conditions. Under long-day conditions, red light signals are perceived when the gate for *CsAFT* expressions opens, and *CsAFT* mRNA is induced. (D) The gate for blue light to photoactivate *FKF1* in *Arabidopsis* (Imaizumi et al., 2003). The peak of the diurnal rhythm of *FKF1* mRNA is around dusk, and is similar under long- and short-day conditions. Under long-day conditions, the *FKF1* mRNA level starts to increase before dusk, whereas under short-day conditions it increases after dusk. Therefore, the FKF1 protein synthesis can be coincident with the blue light signal under long-day conditions, and is followed by FKF1 photoactivation. The clock symbols indicate entrainment by the circadian clock, the circular arrows indicate light transitions, the sun/moon symbol indicates day-length information and the red and blue lightning symbols indicate red or blue light signalling, respectively. (See the colour plate.)

long-day conditions, when the red light signal coincides with the 'open gate' state, *Ghd7* mRNA is induced. The role of three rice phytochromes in the *Ghd7* mRNA induction after a red light pulse was examined by an extensive expression analysis in seedlings of all single and double phytochrome mutants (Osugi et al., 2011). Among six mutants tested, *Ghd7* was absent in *phyAphyB* and *phyAphyC*, indicating that *Ghd7* transcription is mediated by *PHYA* alone or any combination of *PHYB* or *PHYC* with another phytochrome, but not *PHYB* or *PHYC* alone. *phyAphyB* and *phyAphyC* flowered much earlier than other mutants, especially under long-day conditions (Takano et al., 2005).

According to the external coincidence model, the gate is directly set by the circadian clock. Thus, the gate would open at an almost fixed time relative to either light-to-dark or dark-to-light transitions independently of day length, as the plant circadian clock is entrained mainly by these light transitions (Kim et al., 2007; Kim, Song, Taylor, & Carré, 2003; Más, Kim, Somers, & Kay, 2003; McWatters, Bastow, Hall, & Millar, 2000; Salomé & McClung, 2005; Yakir et al., 2009). The *Ghd7* gate for red light is, however, day-length dependent; the *Ghd7* gate opens at midnight under short-day conditions whereas it is open in the morning under long-day conditions (Itoh et al. 2010; Osugi et al., 2011). These observations suggested that the *Ghd7* gate for red light is not set simply by light transitions or the circadian clock, and implies a more complex regulation.

4.2.3 Circadian Gating Mechanism Functioning as a Morning Sensor to Confer Photoperiodic Flowering in Rice

The *Ehd1* gate is set up in response to blue light (Figure 4.2(B); Itoh et al., 2010). In contrast to the *Ghd7* gate for red light, the *Ehd1* gate for blue light is not shifted by the day length. Thus, independently of day-length, plants perceive blue light to induce *Ehd1* mRNA when the *Ehd1* gate is open in the subjective morning, indicating that *Ehd1* is induced by blue light even under long-day conditions. It is likely that the *Ehd1* gate for blue light is tightly regulated by the circadian clock, since it is open at an almost constant timing after the light-on transition (Itoh et al., 2010). The *Ehd1* gate for blue light is 'not open' in *osgi*, a rice mutant of a core component of the circadian clock (Itoh et al., 2010).

In the presence of the functional allele of *Ghd7*, *Ehd1* mRNA exhibits clear photoperiodic responses to specific day lengths. However, in the absence of a functional allele of *Ghd7*, the *Ehd1* gate for blue light alone cannot produce such photoperiodic responses, since the *Ehd1* gate functions only as a morning sensor.

As mentioned above, *Ehd1* may also respond to day length at the posttranscriptional level. Overexpression of *Ehd1* (Actin::*Ehd1*) resulted in a clear acceleration of flowering under short-day conditions but not under long-day conditions (Osugi et al., 2011). The role of *Hd1*, a floral repressor under long-day conditions, could be excluded in this experiment because the Actin::*Ehd1* transgene was introduced in Taichung 65, a cultivar that bears a loss-of-function allele of *hd1* (Doi et al., 2004). Therefore, the floral promoter Ehd1 protein may be inactivated under long-day conditions. Expression analyses in leaves collected every 3 h during a 24 h cycle showed that under long-day conditions, *Hd3a* expression was below a detectable level in the Actin::*Ehd1* transgenic line and wild-type plants (Osugi and Izawa, unpublished data). Under short-day conditions, *Hd3a* was upregulated around dawn in the Actin::*Ehd1* line, indicating that Ehd1 may induce *Hd3a* only during a certain time. Under long-day conditions, *Hd3a* is induced in *phyAphyC* and *phyAphyB* but not in *phyB* and *phyBphyC*, although *Ehd1* is derepressed in all four mutants (Osugi et al., 2011). Thus, *PHYA* is a possible repressor of Ehd1 function under long-day conditions. Since *Ehd1* encodes a B-type response regulator and phytochromes have a kinase activity (Doi et al., 2004; Li, Li, Wang, & Wang, 2011; Phee et al., 2008; Yeh & Lagarias, 1998), PHYA might regulate Ehd1 by phosphorylation.

Considering the posttranscriptional regulation of *Ehd1*, the *Ehd1* gate for blue light could be an important part of the photoperiodic response of *Ehd1* activity to specific day lengths, regardless of the critical repressing activity of Ghd7. *Ehd1* would be induced in the morning to promote floral induction because only *Ehd1* expressed at that time could induce *Hd3a* mRNA Since PHYA alone could induce *Ghd7* mRNA expression under long-day conditions, it is possible that Ghd7 protein could control Ehd1 activity under long-day conditions. In addition, the posttranscriptional regulation of *Ehd1* to control *Hd3a* could be also regulated by the circadian clock, since it seems to show a diurnal rhythm under short-day conditions. Thus, *Ehd1* can induce *Hd3a* mRNA through two distinct mechanisms: the *Ehd1* gate for blue light and photoperiodic posttranscriptional regulation of *Ehd1*. The coincidence of these two regulations may help to set up an acute threshold for the critical day-length recognition of *Hd3a* expression in rice seedlings.

4.2.4 Gating Mechanisms in Other Plant Species

The response of the *Ghd7* mRNA level to day length is formed by a complex gating mechanism, which ensures that rice plants can perceive red light stimuli at dawn only under noninductive long-day conditions. A similar gating mechanism has recently been reported in *chrysanthemum* (*Chrysanthemum seticuspe*), another short-day plant (Figure 4.2(C); Higuchi et al., 2013). Similar to florigen genes, *chrysanthemum Anti-florigenic FT/TFL1 family protein* (*CsAFT*) is induced in leaves, but it represses floral transition at the shoot apex antagonistically to florigen genes. *CsAFT* is regulated by a gating mechanism: red light, mediated by phytochrome(s), induces *CsAFT* mRNA only when the *CsAFT* gate is open (around midnight under short-day conditions and at dawn under long-day conditions). Thus, red light signals repress flowering only under long-day conditions, when the *CsAFT* gate is open. As a result, the flowering repressor *CsAFT* is highly induced under noninductive long-day conditions.

Both *Ghd7* and *CsAFT* are flowering repressors, and the gating mechanisms involved in their expression share several similarities but also some differences: (1) they are induced by phytochrome-mediated red light signalling and (2) opening times of their gates coincide with red light under non-inductive long-day conditions (Figure 4.2(A) and (B); Higuchi et al., 2013; Itoh et al., 2010). However, the diurnal rhythms of these gates are entrained differently. The opening time of the *Ghd7* gate is not determined by the time from dawn or dusk, whereas that of the *CsAFT* gate is an almost fixed time after dusk (Higuchi et al., 2013; Itoh et al., 2010). This suggests that the *Ghd7* gate can respond to day length, whereas the *CsAFT* gate might be simply regulated by the circadian clock or dusk signals. Although it remains to be investigated how the circadian clock and light signalling are involved in regulation of the gates, the mechanisms might be distinct for *Ghd7* and *CsAFT*.

Similar gating mechanisms stimulated under specific day lengths are reported not only in short-day plants but also in long-day plants. In *Arabidopsis*, the *FKF1* gene encodes an F-box protein that is activated by blue light and promotes flowering through multiple pathways (Figure 4.2(D); Imaizumi et al., 2005, 2003; Sawa et al., 2007; Song et al., 2012). The circadian clock determines the peak of *FKF1* mRNA before the end of daytime under long-day conditions and after the end of daytime (or at the beginning of the night) under short-day conditions. Therefore, FKF1 protein, that is immediately translated, can perceive blue light around dusk (but not around dawn) and promotes flowering only under long-day conditions. This behaviour of FKF1 protein can form a gate which is day-length dependent. This mechanism dependent on FKF1 is described in Chapter One and is not discussed in detail here. As suggested by these data for rice, *chrysanthemum*, and *Arabidopsis*, the day-length-dependent gating mechanisms for light stimuli are likely widely conserved among plant species.

4.3 REGULATION OF FLOWERING TIME THROUGH NONGATING MECHANISMS IN RESPONSE TO DAY LENGTH

Although the gating mechanisms for photoperiodic flowering are critical for day-length recognition and are conserved among many species, such mechanisms have not been reported in some light signalling pathways involved in day-length recognition.

4.3.1 Functional Regulation of Ghd7

Ghd7 is a strong flowering repressor in rice, especially under long-day conditions (Xue et al., 2008). Although the expression level of *Ghd7* correlates well with its long-day-specific activity, some reports indicate that posttranscriptional regulation of Ghd7 activity also contributes to the control of flowering time in rice.

4.3.1.1 Phytochrome B

PHYB may modulate Ghd7 protein activity to repress *Ehd1*. In the *phyB* mutant, *Ghd7* expression responds to day length, and is higher under long-day conditions than in wild type (Osugi et al., 2011). Yet, *Ehd1* is derepressed regardless of day length in both the *phyB* mutant and *ghd7* mutant (Xue et al., 2008; Itoh et al., 2010). *Ghd7* expressed at dawn represses *Ehd1* on the following morning; therefore, Ghd7 activity could be modulated by *PHYB* at later times of the day, perhaps around dusk or during the night when the Pr/Pfr proportion changes in response to day length.

4.3.1.2 Heading date 16

Heading date 16 (*Hd16*) encodes casein kinase I and delays flowering, especially under long-day conditions (Matsubara et al., 2008, 2012; Kwon et al., 2014). Hd16 phosphorylates Ghd7 in vitro. *Hd16* does not alter *Ghd7* expression, but their protein products physically interact (Hori et al., 2013). The strong repressive activity of Ghd7 to delay flowering time is reduced in the *hd16* mutant background, whereas *Hd16* only slightly affects flowering time in the *ghd7* background. *Hd16* expression is slightly higher under long-day conditions than under short-day conditions, indicating it might modulate the function of *Ghd7*, especially under long-day conditions. Whether the long-day-specific function of *Hd16* is due to the day-length-responsive activity of Hd16 itself or to long-day-specific upregulation of *Ghd7* mRNA remains to be established.

4.3.2 Functional Regulation of Hd1

The transcription factor Hd1 upregulates the expression of the *Hd3a* florigen under short-day conditions, but represses it under long-day conditions (Kojima et al., 2002; Yano et al., 2000). Neither the quantity nor the diurnal rhythm of *Hd1* mRNA varies considerably between short- and long-day conditions. Therefore, the Hd1 activity may be posttranscriptionally modulated in response to day length. Several candidate modulators of Hd1 are described below.

4.3.2.1 Phytochromes

Phytochromes are necessary for functional modulation of Hd1 (Izawa et al., 2002). In the genetic background of functional phytochromes, a defect in *Hd1* causes upregulation of *Hd3a* mRNA under long-day conditions. Combined with the *se5* mutation, the *hd1* mutation downregulates *Hd3a*. Thus, phytochromes switch the *Hd1* function from a promoter to a repressor of *Hd3a* expression under long-day conditions. In short-day plants, light pulses at midnight under short-day conditions delay flowering time. This treatment is called 'night break', and red light is the most effective (Borthwick, et al., 1952). In rice, the night break delays flowering time through *Hd3a* repression (Ishikawa et al., 2005). This effect requires the functional alleles of not only *Hd1*, but also *PHYB* (Ishikawa et al., 2005).

4.3.2.2 Ghd7 and More

Ghd7 is required for *Hd1* to delay flowering under long-day conditions. Under natural long-day conditions, a defect in *Hd1* causes early flowering in the wild type (Nonoue et al., 2008; Yano et al., 2000), but has only a slight effect in the *ghd7* background (Nonoue et al., 2008). Ghd7 represses *Ehd1* in response to the critical day length (Itoh et al., 2010). In two-week-old rice seedlings, Hd1 may not contribute to the induction of the *Hd3a* florigen upon critical day-length recognition. Examination of whether the functional modulation of Hd1 is also an acute response to day length at more mature rice stages is needed to fully reveal the importance of *Ghd7* in the functional modulation of Hd1.

Phytochromes can modulate the function of Hd1 (Izawa et al., 2002) through *Ghd7* induction, as *Ghd7* is not induced in *se5* (Itoh et al., 2010).

In this scenario, *Ghd7* would be induced by phytochromes only under long-day conditions through the gating mechanism, and would modulate the function of Hd1. An analysis of a double mutant of *se5 ghd7* could reveal whether *Ghd7* is the sole pathway for phytochromes to alter the function of Hd1.

Heading date 17, Ef7, Os Early Flowering 3-1 (*Hd17/Ef7/OsELF3-1*) may also modulate Hd1 function through *Ghd7*. *Hd17* encodes a homologue of *Arabidopsis ELF3* (Matsubara et al., 2012; Murakami, Tago, Yamashino, & Mizuno, 2007; Saito et al., 2012; Zhao et al., 2012). *ELF3* is a component of the circadian clock, and modulates light input to the core oscillator, especially for phytochrome-mediated signals (Kolmos et al., 2011; McWatters et al., 2000). *Hd17* promotes the repressor activity of Hd1 depending on the genetic background, because the effect of *Hd1* on flowering time is weakened in the *hd17* mutant under natural long-day conditions in rice cultivar 'Nipponbare', and to a lesser extent in cultivar 'Ginbozu' (Matsubara et al., 2012; Saito et al., 2012). *Hd17* also upregulates *Ghd7* mRNA (Matsubara et al., 2012; Saito et al., 2012; Zhao et al., 2012) and thus may modulate the function of Hd1 through *Ghd7* expression. As *ELF3* modifies the phytochrome-mediated signals to the circadian clock, *Hd17* could be involved in gating phytochrome-mediated red light signalling to induce *Ghd7*.

Under long-day conditions, *Hd1* may repress *Hd3a* and *RFT1*, possibly throughout development, since *Hd1/ehd1* flowers later than *hd1/ehd1*,by more than 100 days (Doi et al., 2004). Although *Ghd7* seems to be required for Hd1 to repress *Hd3a* under long-day conditions, mRNA expression of *Ghd7* is strongly reduced 4–5 weeks after imbibition, when floral induction has not yet occurred (Itoh & Izawa, 2013). As a result, *Ghd7* expression is very low in adult plants under long day, when Hd1 can repress flowering. There are two possible explanations for the functional modulation of Hd1 at the mature stages. (1) Unidentified factors may modulate Hd1 function (Itoh & Izawa, 2013). In this case, *Ghd7* would modulate Hd1 function until its own expression is reduced. Afterwards, unidentified factors could regulate Hd1 function similarly to *Ghd7* during the first weeks after germination. (2) As *Ghd7* is posttranscriptionally regulated by *PHYB* and *Hd16* (Osugi et al., 2011; Hori et al., 2013), its low expression level may be still sufficient to modulate Hd1 function in response to long-day conditions. Both *PHYB* and *Hd16* are constitutively expressed in leaf blades throughout various developmental stages (Hori et al., 2013; the Rice Expression Profile Database, http://ricexpro.dna.affrc.go.jp).

4.3.2.3 Heading date 6

Heading date 6 (*Hd6*) encodes the α subunit of casein kinase II, and contributes to flowering time control, but not to regulation of circadian clock function (Ogiso, Takahashi, Sasaki, Yano, & Izawa, 2010; Takahashi, Shomura, Sasaki, & Yano, 2001). *Hd6* requires a functional allele of *Hd1* to regulate flowering time, and promotes the Hd1 repressive activity under long-day conditions. Hd6 activity may be post-transcriptionally regulated by day length, as *Hd6* mRNA expression shows neither diurnal rhythm nor response to day length. Hd1 is not phosphorylated by Hd6, and it is currently unclear what the target of the Hd6 kinase is.

4.4 DEVELOPMENTAL RESPONSES TO DAY LENGTH

Many developmental processes other than flowering time respond to day length. In *Arabidopsis*, the hypocotyl elongates more under short-day conditions than under long-day conditions (Niwa, Yamashino, & Mizuno, 2009). Phytochrome-interacting factors (*PIFs*), especially *PIF3–PIF5*, play pivotal roles in this process (Nozue et al., 2007; Soy et al., 2012). *PIFs* encode basic helix-loop-helix transcription factors and positively regulate hypocotyl growth (reviewed in Leivar and Quail (2011)). Light exposure induces degradation of PIFs upon direct interaction with phytochromes and their subsequent phosphorylation (Al-Sady, Ni, Kircher, Schäfer, & Quail, 2006; Park et al., 2004). In addition, the DNA-binding activity of residual PIF proteins is abolished by the light-induced DELLA family proteins (Achard et al., 2007; Feng et al., 2008; de Locus et al., 2008; Yamashino et al., 2013). Thus, the PIF proteins are abundant and functional in the dark, when promotion of hypocotyl elongation is highest. *PIF*-mediated hypocotyl elongation in *Arabidopsis* also uses gating mechanisms to recognise the day length. In this section, gating and nongating mechanisms underlying photoperiodic hypocotyl elongation are described.

4.4.1 Hypocotyl Growth Regulation by Gating Mechanisms: PIF4 and PIF5

Transcription of *PIF4* and *PIF5* is regulated by the circadian clock (Nozue et al. 2007). Their transcripts peak at an almost fixed time after dusk regardless of day length (Niwa et al., 2009; Nozue et al. 2007). Under short-day conditions, *PIF4* and *PIF5* transcripts start to increase at the end of night, and the proteins accumulate until dawn. Under long-day conditions, *PIF4* and *PIF5* mRNAs are expressed after dawn; thus, proteins are degraded and inactivated as described above. As a result, hypocotyl growth is more pronounced under short-day conditions. This control may represent a gating mechanism, where the levels of *PIF4* and *PIF5* mRNAs are diurnally controlled by the circadian clock and protein degradation and inactivation occur when the proteins are present during the light phase, resulting in some day-length-dependent gating effects.

4.4.2 Hypocotyl Growth Regulation by a Nongating Mechanism: PIF3

Unlike *PIF4* and *PIF5*, the expression of *PIF3* is constant during the whole day (Soy et al., 2012). However, the PIF3 protein accumulates at night, and its levels exhibit a diurnal rhythm with a peak at the end of the night. PIF3 is degraded by light, and its levels are low during the day. In a 12/12 light/dark cycle, PIF3 protein levels do not reach a peak during the night, whereas longer nights lead to higher PIF3 accumulation. In addition, light-induced DELLA proteins inactivate residual PIF3 during the daytime (Achard et al., 2007; Feng et al., 2008; de Locus et al., 2008). As a result, *PIF3* promotes hypocotyl growth in response to day length through its dosage or through the duration of its phase of expression. Although the expression of phytochromes is partially regulated by the circadian clock, *PIF3* may not require the circadian clock to recognise day length (Sharrock & Clack, 2002). This excellent work demonstrates that plants can recognise day length without any gating mechanism controlled by the circadian clock, by simply monitoring night duration on the basis of protein degradation and inactivation. Such time-keeping molecular mechanism independent of the gating mechanism controlled by the circadian clock has not been reported in the flowering time control in rice.

4.5 DAY-LENGTH RECOGNITION IN DIFFERENT PLANT SPECIES

4.5.1 Day-Length Recognition in the Flowering Time Control in Rice

To control flowering time, rice has at least four distinct molecular mechanisms of day-length recognition, some components of which have been described above (Figure 4.3(A)).

1. The expression of *Ghd7*, a flowering repressor, is induced by red light under noninductive long-day conditions (Figure 4.2(A); Itoh et al., 2010). This induction is regulated by a photoperiodic gating mechanism,

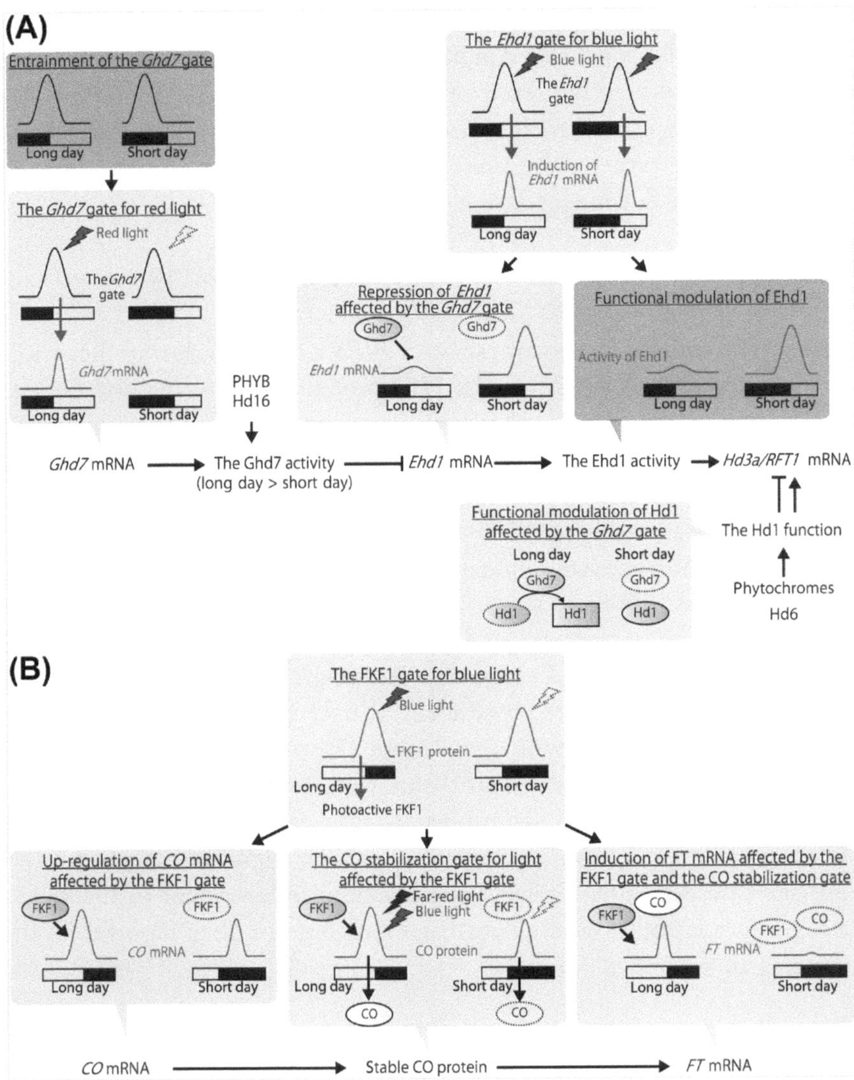

Figure 4.3 Multiple molecular mechanisms for day-length recognition in (A) rice and (B) *Arabidopsis*. Green rectangles show responses to day length regulated by gating mechanisms or their direct effects. Blue rectangles show mechanisms whose regulation is currently unknown. Black waves represent diurnal rhythms of gates. Green waves represent diurnal rhythms of mRNA expression or protein activity. Circles show proteins. Dashed circles indicate proteins showing reduced activity or abundance. Circled and squared *Hd1* indicate that the protein promotes or represses *Hd3a* expression, respectively. (See the colour plate.)

and red light usually coincides with the opening time of the gate at dawn only under long-day conditions.

2. Ghd7 activity to repress *Ehd1* is enhanced by *PHYB* and *Hd16* under long-day conditions (Osugi et al., 2011; Hori et al., 2013). As the latter genes are expressed regardless of day length, it would be interesting to investigate their long-day-specific functions.
3. The combination of morning-specific induction of *Ehd1* by blue light and posttranscriptional regulation of *Ehd1* confers a short-day-specific floral promoter activity to Ehd1 (Figure 4.2(B); Itoh et al., 2010; Osugi et al., 2011). It is noteworthy that *Ehd1* induction by blue light is regulated by gating mechanisms and requires proper circadian clock function (especially a functional allele of *OsGI*). This morning-specific *Ehd1* induction is important for the response to day length, since the Ehd1 activity to induce *Hd3a* can be permitted only around dawn even under short-day conditions.
4. *Hd1* increases *Hd3a* expression under short-day conditions and represses it under long-day conditions. *Hd1* is posttranscriptionally modulated in response to day length to regulate the *Hd3a* mRNA level (Izawa et al., 2002; Yano et al., 2000). Many proteins, including phytochromes and *Ghd7*, have been identified as potential regulators of *Hd1* (Izawa et al., 2002; Nonoue et al., 2008). *Ghd7* can convert Hd1 into a repressor of *Hd3a*. Phytochromes also promote the repressor activity of Hd1, possibly through the induction of *Ghd7* mRNA. Interestingly, both *Ghd7* and phytochromes are also involved in the regulation of *Ehd1* expression (Xue et al., 2008; Itoh et al., 2010). Therefore, *Hd1* and *Ehd1*, the two predominant regulators of *Hd3a*, might be similarly regulated, and some of the four mechanisms described above could function in both pathways coordinately.

4.5.2 Day-Length Recognition in the Flowering Time Control in *Arabidopsis*

Similarly to rice, multiple mechanisms recognising day length regulate flowering time in *Arabidopsis* (Figure 4.3(B), Chapter One and Chapter Three of this Volume).

1. FKF1 is activated by blue light under long-day conditions (Figure 4.2(D); Imaizumi et al., 2003; Sawa et al., 2007). Photoactivated FKF1 promotes *FT* mRNA expression through a complex pathway where the protein affects multiple steps of the signalling cascade by upregulating *CO* mRNA expression, promoting CO protein

stabilisation and inducting *FT* mRNA (Imaizumi et al., 2005; Song et al., 2012). Upregulation of *CO* and *FT* is enhanced by the degradation of CYCLING DOF FACTORs transcriptional repressors, through their direct interaction with FKF1 (Fornara et al., 2009; Imaizumi et al., 2005). These pathways are regulated by a gating mechanism for blue light, and FKF1 serves as a single regulator of the gate.

2. In addition to FKF1, PHYA- and CRY2-mediated light signalling stabilises CO protein under long-day conditions (Valverde et al., 2004; Zuo et al., 2011). CO is degraded by the ubiquitin-proteasome system in the dark (Valverde et al., 2004; Jang et al., 2008). *CO* mRNA peaks at dusk almost independently of day length (Suárez-López et al., 2001; Yanovsky & Kay, 2002). Therefore, under long-day conditions but not under short-day conditions, the coincidence of CO protein with light at the end of the day results in its protection from degradation and activation of flowering (Jang et al., 2008; Liu et al., 2008; Valverde et al., 2004; Yanovsky & Kay, 2002). This corresponds to a gating mechanism of PHYA- and CRY2-mediated light signalling, in which *CO* mRNA and CO protein are the targets of gate regulation.
3. *PHYB* also contributes to day-length recognition, since it delays flowering, especially under noninductive short-day conditions (Goto, Kumagai, & Koornneef, 1991). In contrast to PHYA, which stabilises CO protein around dusk, PHYB signalling degrades it in the morning (Valverde et al., 2004). The morning-specific function of PHYB requires the *PHYTOCHROME-DEPENDENT LATE-FLOWERING* (*PHL*) gene (Endo, Tanigawa, Murakami, Araki, & Nagatani, 2013). The PHL protein localises to the nucleus from the middle to the end of the daytime, and represses PHYB activity. However, PHL might respond not to day length but to circadian time, since *PHL* mRNA is expressed and localised to the nucleus regardless of day length. Therefore, in addition to *PHL*, *PHYB* uses other pathways to repress *FT* mRNA in response to day length. Possible candidates are *VASCULAR PLANT ONE–ZINC FINGER 1 and 2*, which increase *FT* mRNA level downstream of *PHYB* (Yasui et al., 2012). *HIGH EXPRESSION OF OSMOTICALLY RESPONSIVE GENES 1* (*HOS1*) could also act downstream of *PHYB*, since *HOS1* modulates CO stabilisation at a similar circadian time as *PHYB* (Lazaro, Valverde, Piñeiro, & Jarillo, 2012).

4.5.3 Comparison of the Responses to Day Length between Rice and *Arabidopsis*

Both rice and *Arabidopsis* adopt gating mechanisms for day-length recognition, and share several conserved genes. However, gating mechanisms in these species differ in several respects.

1. They respond differently to inductive and noninductive day lengths. In rice, gating mechanisms respond to noninductive long-day conditions mainly by inducing a flowering repressor. Although *Ehd1* is induced by blue light under inductive short-day conditions, this is due to the absence of Ghd7, which is produced and functions only under noninductive long-day conditions. In *Arabidopsis*, expression of a floral promoter, CO protein, is produced under inductive long-day conditions.
2. Based on computational modelling, *FT* mRNA does not show any critical day-length response in *Arabidopsis*, rather gradually increases according to light duration (Salazar et al., 2009). Thus, gate responses under various photoperiods may be different in the two species. In rice, length of light duration may not be important to confer the critical response in day-length recognition, since a short light pulse is sufficient to induce *Ghd7* expression although *Ehd1* induction required strong and relatively long blue light signals (Itoh et al., 2010; Itoh & Izawa, 2013). These are confirmed by light pulse treatment experiments (Itoh et al., 2010; Osugi et al., 2011). It is likely that acute responses for Ghd7 expression may contribute to recognition of critical day lengths for *Ehd1* and *Hd3a* expression. In *Arabidopsis*, light signals at dusk after longer light duration (when the *CO* mRNA peaks) may allow translation and stabilisation of CO protein more effectively (Valverde et al., 2004). However, how acutely the gates in *Arabidopsis* respond to light signals remains to be reported.
3. The gate opening time is different. In rice, at least three of the four photoperiodic mechanisms recognise light signals at dawn. Around dawn, red light signalling induces *Ghd7*, whereas blue light signalling induces *Ehd1* through gating mechanisms (Itoh et al., 2010). Post-transcriptional modulation of Hd1 may also use light signals at dawn, since *Ghd7* induced by red light at dawn modulates the Hd1 function on the following morning (Itoh et al., 2010; Nonoue et al., 2008). In *Arabidopsis*, the two gating mechanisms that photoactivate FKF1 protein and stabilise CO protein (Salazar et al., 2009; Song et al., 2012) use light signals around dusk. Under natural field conditions, but not

under artificial experimental conditions, dawn light and dusk light have different spectra. Thus, the gate opening time could be a consequence of the wavelength required for promoting the flowering process.

4.5.4 Day-Length Recognition in Other Plant Species

Although the gating mechanisms for day-length recognition reported in rice and *Arabidopsis* are clearly different, the information available for these two species is not sufficient to explain all photoperiodic responses in other plants. For instance, assuming that the expression of *FT*-like genes is induced at the light-sensitive phase, there are dawn-responsive long-day plants, such as *Gentiana capitata*, and dusk-responsive short-day plants, such as *Cucurbita maxima* (Lin et al., 2007; Imamura, Nakatsuka, Higuchi, Nishihara, & Takahashi, 2011). *Gentiana capitata* should respond to inductive long-day or noninductive short-day conditions at dawn, whereas *C. maxima* should respond to inductive short-day or non-inductive long-day conditions at dusk. Thus, many gating mechanisms might be species specific, consistently with previous studies suggesting that plant responsiveness to day length has evolved independently in different species (Ballerini & Kramer, 2011).

A molecular mechanism quite different from those in rice and *Arabidopsis* has been reported in the short-day plant *Pharbitis nil* (Hayama, Agashe, Luley, King, & Coupland, 2007). *PnFT1* and *PnFT2 (PnFT* stands for *Pharbitis nil FT)*, the *FT*-like genes in this species, are upregulated around dawn under inductive short-day conditions. In contrast to rice, the induction of *PnFT1* and *PnFT2* is not stimulated by light. *PnFT1* and *PnFT2* are induced at a fixed time after light-off regardless of the day length, and their accumulation requires sufficient night duration. Since the expression of circadian clock-regulated genes is also reset at the light-to-dark transition, the authors proposed that the circadian clock is entrained at this time, and *PnFT1* and *PnFT2* are induced at a certain circadian time. Interestingly, the night duration required to promote flowering depends on the ambient temperature, although it is widely accepted that the circadian clock compensates for the differences in the ambient temperature (Takimoto & Hamner, 1964). Therefore, it is controversial whether the circadian clock directly regulates *PnFT1* and *PnFT2* mRNA without any stimulation, or whether there is an unknown gating mechanism, which is reset at dusk and is sensitive to temperature. Distinct molecular mechanisms in rice, *Arabidopsis*, *P. nil*, and other plants suggest that research in *Arabidopsis* cannot cover the full spectrum of photoperiodic responses.

4.6 PERSPECTIVES

Rice and *Arabidopsis* have developed multiple gating mechanisms to recognise day length (Figure 4.3). In rice, the gating mechanisms for red light that induce *Ghd7* fit Pittendrigh's external coincidence model, though the gate opening time is regulated in a more complex manner. In *Arabidopsis*, FKF1 photoactivation and stabilisation of CO protein are also well explained by this model (Imaizumi et al., 2003; Valverde et al., 2004; Sawa et al., 2007; Zuo et al., 2011). The use of external light signals at the gate opening time for day-length recognition may be conserved among different species. Although individual gating mechanisms are well explained by the external coincidence model, many molecular mechanisms are integrated in rice and *Arabidopsis*. Thus, we now realise that day length is recognised not by one simple molecular mechanism (reflected in the external coincidence model), but by several mechanisms. In this chapter, we mainly discussed the importance of the gating mechanisms and light at a specific circadian time, but some mechanisms may behave like an hourglass and measure the elapsed time, such as night duration (such as the *PIF3*-dependent mechanism for *Arabidopsis* hypocotyl elongation). In *P. nil*, the induction of florigen genes occurs at an almost fixed time from dusk, and flowering time is temperature sensitive (Hayama et al., 2007; Takimoto & Hamner, 1964). It is interesting that the temperature-sensitive molecules, like the Pfr form of phytochrome accumulating in the dark, participate in responses to day length.

Rice is one of the most important crops in the world. Genetic modification of its flowering time during complex domestication processes and breeding by artificial crossing have critically helped its local adaptation to various environmental conditions. Thus, studies of day-length recognition in rice have not only fundamental scientific importance, but also a considerable agricultural impact. In this paper, we argue that, despite certain evolutionary conservation, day-length recognition in rice and *Arabidopsis* differs more than researchers in this field anticipated. Although rice research may be time consuming (mainly because of the relatively long time needed for material generation and the requirement for large growth facilities), various rice genetic resources became available thanks to a number of genome projects worldwide. Recently developed computational models now allow us to predict gene expression in response to changing environments, to disregard the effects of temporal environmental fluctuations, and to obtain reproducible gene expression data from rice plants grown in the field (Nagano et al., 2012). Studies of day-length recognition will improve future breeding and cultivation of rice and other crops, such as wheat and corn.

REFERENCES

Abe, M., Kobayashi, Y., Yamamoto, S., Daimon, Y., Yamaguchi, A., Ikeda, Y., et al. (2005). FD, a bZIP protein mediating signals from the floral pathway integrator FT at the shoot apex. *Science, 309*, 1052–1056.

Achard, P., Liao, L., Jiang, C., Desnos, T., Bartlett, J., Fu, X., et al. (2007). DELLAs contribute to plant photomorphogenesis. *Plant Physiology, 143*, 1163–1172.

Al-Sady, B., Ni, W., Kircher, S., Schäfer, E., & Quail, P. H. (2006). Photoactivated phytochrome induces rapid PIF3 phosphorylation prior to proteasome-mediated degradation. *Molecular Cell, 23*, 439–446.

Ballerini, E. S., & Kramer, E. M. (2011). In the light of evolution: a reevaluation of conservation in the CO-FT regulon and its role in photoperiodic regulation of flowering time. *Frontiers in Plant Science, 2*, 18.

Borthwick, H. A., Hendricks, S. B., & Parker, M. W. (1952). The reaction controlling floral initiation. *Proceedings of the National Academy of Sciences of the United States of America, 38*, 929–934.

Bünning, E. (1936). Die endogene Tagesrhythmik als Grundlage der photoperiodischen Reaktion. *Ber Deutsch Botan Gesellschaft, 54*, 590–601.

Bünning, E. (1950). Endogenous rhythms in plants. *Annual Review of Plant Physiology, 7*, 71–90.

Chailakhyan, M. K. (1936). New facts in support of the hormonal theory of plant development. *Comptes Rendus de l Academie des Sciences de l URSS, 4*, 79–83.

Chen, M., & Chory, J. (2011). Phytochrome signaling mechanisms and the control of plant development. *Trends in Cell Biology, 21*, 664–671.

Corbesier, L., Vincent, C., Jang, S., Fornara, F., Fan, Q., Searle, I., et al. (2007). FT protein movement contributes to long-distance signaling in floral induction of *Arabidopsis*. *Science, 316*, 1030–1033.

Doi, K., Izawa, T., Fuse, T., Yamanouchi, U., Kubo, T., Shimatani, Z., et al. (2004). Ehd1, a B-type response regulator in rice, confers short-day promotion of flowering and controls FT-like gene expression independently of Hd1. *Genes and Development, 18*, 926–936.

Endo, M., Tanigawa, Y., Murakami, T., Araki, T., & Nagatani, A. (2013). PHYTOCHROME-DEPENDENT LATE-FLOWERING accelerates flowering through physical interactions with phytochrome B and CONSTANS. *Proceedings of the National Academy of Sciences of the United States of America, 110*, 18017–18022.

Feng, S., Martinez, C., Gusmaroli, G., Wang, Y., Zhou, J., Wang, F., et al. (2008). Coordinated regulation of *Arabidopsis thaliana* development by light and gibberellins. *Nature, 451*, 475–479.

Fornara, F., Panigrahi, K. C., Gissot, L., Sauerbrunn, N., Rühl, M., Jarillo, J. A., et al. (2009). *Arabidopsis* DOF transcription factors act redundantly to reduce CONSTANS expression and are essential for a photoperiodic flowering response. *Developmental Cell, 17*, 75–86.

Gao, H., Zheng, X. M., Fei, G., Chen, J., Jin, M., Ren, Y., et al. (2013). Ehd4 encodes a novel and *Oryza*-genus-specific regulator of photoperiodic flowering in rice. *PLoS Genetics, 9*, e1003281.

Garner, W. W., & Allard, H. A. (1920). Effect of the relative length of day and night and other factors of the environment on growth and reproduction in plants. *Journal of Agriculture, 18*, 553–606.

Goto, N., Kumagai, T., & Koornneef, M. (1991). Flowering responses to light-breaks in photomorphogenic mutants of *Arabidopsis thaliana*, a long-day plant. *Plant Physiology, 83*, 209–215.

Griffiths, S., Dunford, R. P., Coupland, G., & Laurie, D. A. (2003). The evolution of CONSTANS-like gene families in barley, rice, and *Arabidopsis*. *Plant Physiology, 131*, 1855–1867.

Hayama, R., Agashe, B., Luley, E., King, R., & Coupland, G. (2007). A circadian rhythm set by dusk determines the expression of FT homologs and the short-day photoperiodic flowering response in *Pharbitis*. *Plant Cell*, *19*, 2988–3000.

Higuchi, Y., Narumi, T., Oda, A., Nakano, Y., Sumitomo, K., Fukai, S., et al. (2013). The gated induction system of a systemic floral inhibitor, antiflorigen, determines obligate short-day flowering in chrysanthemums. *Proceedings of the National Academy of Sciences of the United States of America*, *110*, 17137–17142.

Hirose, F., Shinomura, T., Tanabata, T., Shimada, H., & Takano, M. (2006). Involvement of rice cryptochromes in de-etiolation responses and flowering. *Plant Cell Physiology*, *47*, 915–925.

Hori, K., Ogiso-Tanaka, E., Matsubara, K., Yamanouchi, U., Ebana, K., & Yano, M. (2013). Hd16, a gene for casein kinase I, is involved in the control of rice flowering time by modulating the day-length response. *The Plant Journal*, *76*, 36–46.

Imamura, T., Nakatsuka, T., Higuchi, A., Nishihara, M., & Takahashi, H. (2011). The gentian orthologs of the FT/TFL1 gene family control floral initiation in *Gentiana*. *Plant Cell Physiology*, *52*, 1031–1041.

Imaizumi, T., Schultz, T. F., Harmon, F. G., Ho, L. A., & Kay, S. A. (2005). FKF1 F-box protein mediates cyclic degradation of a repressor of CONSTANS in *Arabidopsis*. *Science*, *309*, 293–297.

Imaizumi, T., Tran, H. G., Swartz, T. E., Briggs, W. R., & Kay, S. A. (2003). FKF1 is essential for photoperiodic-specific light signalling in *Arabidopsis*. *Nature*, *426*, 302–306.

Ishikawa, R., Tamaki, S., Yokoi, S., Inagaki, N., Shinomura, T., Takano, M., et al. (2005). Suppression of the floral activator Hd3a is the principal cause of the night break effect in rice. *Plant Cell*, *17*, 3326–3336.

Itoh, H., & Izawa, T. (2013). The coincidence of critical day length recognition for florigen gene expression and floral transition under long-day conditions in rice. *Molecular Plant*, *6*, 635–649.

Itoh, H., Nonoue, Y., Yano, M., & Izawa, T. (2010). A pair of floral regulators sets critical day length for Hd3a florigen expression in rice. *Nature Genetics*, *42*, 635–638.

Izawa, T. (2007). Adaptation of flowering-time by natural and artificial selection in *Arabidopsis* and rice. *The Journal of Experimental Botany*, *58*, 3091–3097.

Izawa, T., Oikawa, T., Sugiyama, N., Tanisaka, T., Yano, M., & Shimamoto, K. (2002). Phytochrome mediates the external light signal to repress FT orthologs in photoperiodic flowering of rice. *Genes and Development*, *16*, 2006–2020.

Izawa, T., Mihara, M., Suzuki, Y., Gupta, M., Itoh, H., Nagano, AJ,. et al. (2011). Os-GIGANTEA confers robust diurnal rhythms on the global transcriptome of rice in the field. *Plant Cell*, *23*(5):1741–1755. http://dx.doi.org/10.1105/tpc.111.083238.

Izawa, T., Oikawa, T., Tokutomi, S., Okuno, K., & Shimamoto, K. (2000). Phytochromes confer the photoperiodic control of flowering in rice (a short-day plant). *The Plant Journal*, *22*, 391–399.

Jang, S., Marchal, V., Panigrahi, K. C., Wenkel, S., Soppe, W., Deng, X. W., et al. (2008). *Arabidopsis* COP1 shapes the temporal pattern of CO accumulation conferring a photoperiodic flowering response. *The EMBO Journal*, *27*, 1277–1288.

Kami, C., Lorrain, S., Hornitschek, P., & Fankhauser, C. (2010). Light-regulated plant growth and development. *Current Topics in Developmental Biology*, *91*, 29–66.

Kardailsky, I., Shukla, V. K., Ahn, J. H., Dagenais, N., Christensen, S. K., Nguyen, J. T., et al. (1999). Activation tagging of the floral inducer FT. *Science*, *286*, 1962–1965.

Kim, W. Y., Fujiwara, S., Suh, S. S., Kim, J., Kim, Y., Han, L., et al. (2007). ZEITLUPE is a circadian photoreceptor stabilized by GIGANTEA in blue light. *Nature*, *449*, 356–360.

Kim, J. Y., Song, H. R., Taylor, B. L., & Carré, I. A. (2003). Light-regulated translation mediates gated induction of the *Arabidopsis* clock protein LHY. *The EMBO Journal*, *22*, 935–944.

Kobayashi, Y., Kaya, H., Goto, K., Iwabuchi, M., & Araki, T. (1999). A pair of related genes with antagonistic roles in mediating flowering signals. *Science*, *286*, 1960–1962.

Kojima, S., Takahashi, Y., Kobayashi, Y., Monna, L., Sasaki, T., Araki, T., et al. (2002). Hd3a, a rice ortholog of the *Arabidopsis* FT gene, promotes transition to flowering downstream of Hd1 under short-day conditions. *Plant Cell Physiology, 43*, 1096–1105.

Kolmos, E., Herrero, E., Bujdoso, N., Millar, A. J., Tóth, R., Gyula, P., et al. (2011). A reduced-function allele reveals that EARLY FLOWERING3 repressive action on the circadian clock is modulated by phytochrome signals in *Arabidopsis*. *Plant Cell, 23*, 3230–3246.

Komiya, R., Ikegami, A., Tamaki, S., Yokoi, S., & Shimamoto, K. (2008). Hd3a and RFT1 are essential for flowering in rice. *Development, 135*, 767–774.

Komiya, R., Yokoi, S., & Shimamoto, K. (2009). A gene network for long-day flowering activates RFT1 encoding a mobile flowering signal in rice. *Development, 136*, 3443–3450.

Kwon, C. T., Yoo, S. C., Koo, B. H., Cho, S. H., Park, J. W., Zhang, Z., et al. (2014). Natural variation in Early flowering1 contributes to early flowering in japonica rice under long days. *Plant Cell Environment, 37*, 101–112.

Lazaro, A., Valverde, F., Piñeiro, M., & Jarillo, J. A. (2012). The *Arabidopsis* E3 ubiquitin ligase HOS1 negatively regulates CONSTANS abundance in the photoperiodic control of flowering. *Plant Cell, 24*, 982–999.

Leivar, P., & Quail, P. H. (2011). PIFs: pivotal components in a cellular signaling hub. *Trends in Plant Science, 16*, 19–28.

Li, J., Li, G., Wang, H., & Wang, D. X. (2011). *Phytochrome signaling mechanisms. Arabidopsis book.*

Lin, M. K., Belanger, H., Lee, Y. J., Varkonyi-Gasic, E., Taoka, K., Miura, E., et al. (2007). FLOWERING LOCUS T protein may act as the long-distance florigenic signal in the cucurbits. *Plant Cell, 19*, 1488–1506.

Liu, L. J., Zhang, Y. C., Li, Q. H., Sang, Y., Mao, J., Lian, H. L., et al. (2008). COP1-mediated ubiquitination of CONSTANS is implicated in cryptochrome regulation of flowering in *Arabidopsis*. *Plant Cell, 20*, 292–306.

de Lucas, M., Davière, J. M., Rodríguez-Falcón, M., Pontin, M., Iglesias-Pedraz, J. M., Lorrain, S., et al. (2008). A molecular framework for light and gibberellin control of cell elongation. *Nature, 451*, 480–484.

Más, P., Kim, W.Y., Somers, D. E., & Kay, S. A. (2003). Targeted degradation ofTOC1 by ZTL modulates circadian function in *Arabidopsis thaliana*. *Nature, 426*, 567–570.

Matsubara, K., Kono, I., Hori, K., Nonoue, Y., Ono, N., Shomura, A., et al. (2008). Novel QTLs for photoperiodic flowering revealed by using reciprocal backcross inbred lines from crosses between japonica rice cultivars. *Theoretical and Applied Genetics, 117*, 935–945.

Matsubara, K., Ogiso-Tanaka, E., Hori, K., Ebana, K., Ando, T., & Yano, M. (2012). Natural variation in Hd17, a homolog of *Arabidopsis* ELF3 that is involved in rice photoperiodic flowering. *Plant Cell Physiology, 53*, 709–716.

Matsubara, K., Yamanouchi, U., Nonoue, Y., Sugimoto, K., Wang, Z. X., et al. (2011). Ehd3, encoding a plant homeodomain finger-containing protein, is a critical promoter of rice flowering. *The Plant Journal, 66*, 603–612.

Matsubara, K., Yamanouchi, U., Wang, Z. X., Minobe, Y., Izawa, T., & Yano, M. (2008). Ehd2, a rice ortholog of the maize INDETERMINATE1 gene, promotes flowering by up-regulating Ehd1. *Plant Physiology, 148*, 1425–1435.

McWatters, H. G., Bastow, R. M., Hall, A., & Millar, A. J. (2000). The ELF3 zeitnehmer regulates light signalling to the circadian clock. *Nature, 408*, 716–720.

Murakami, M., Tago, Y., Yamashino, T., & Mizuno, T. (2007). Comparative overviews of clock-associated genes of *Arabidopsis thaliana* and *Oryza sativa*. *Plant Cell Physiology, 48*, 110–121.

Nagano, A. J., Sato, Y., Mihara, M., Antonio, B. A., Motoyama, R., Itoh, H., et al. (2012). Deciphering and prediction of transcriptome dynamics under fluctuating field conditions. *Cell, 151*, 1358–1369.

Nagatani, A. (2010). Phytochrome: structural basis for its functions. *Current Opinion in Plant Biology, 13*, 565–570.

Nemoto, Y., Kisaka, M., Fuse, T., Yano, M., & Ogihara, Y. (2003). Characterization and functional analysis of three wheat genes with homology to the CONSTANS flowering time gene in transgenic rice. *The Plant Journal, 36*, 82–93.

Niwa, Y., Yamashino, T., & Mizuno, T. (2009). The circadian clock regulates the photoperiodic response of hypocotyl elongation through a coincidence mechanism in *Arabidopsis thaliana*. *Plant Cell Physiology, 50*, 838–854.

Nonoue, Y., Fujino, K., Hirayama, Y., Yamanouchi, U., Lin, S. Y., & Yano, M. (2008). Detection of quantitative trait loci controlling extremely early heading in rice. *Theoretical and Applied Genetics, 116*, 715–722.

Nozue, K., Covington, M. F., Duek, P. D., Lorrain, S., Fankhauser, C., Harmer, S. L., et al. (2007). Rhythmic growth explained by coincidence between internal and external cues. *Nature, 448*, 358–361.

Osugi, A., Itoh, H., Ikeda-Kawakatsu, K., Takano, M., & Izawa, T. (2011). Molecular dissection of the roles of phytochrome in photoperiodic flowering in rice. *Plant Physiology, 157*, 1128–1137.

Ogiso, E., Takahashi, Y., Sasaki, T., Yano, M., & Izawa, T. (2010). The role of casein kinase II in flowering time regulation has diversified during evolution. *Plant Physiology, 152*, 808–820.

Park, S. J., Kim, S. L., Lee, S., Je, B. I., Piao, H. L., Park, S. H., et al. (2008). Rice Indeterminate 1 (OsId1) is necessary for the expression of Ehd1 (Early heading date 1) regardless of photoperiod. *The Plant Journal, 56*, 1018–1029.

Park, E., Kim, J., Lee, Y., Shin, J., Oh, E., Chung, W. I., et al. (2004). Degradation of phytochrome interacting factor 3 in phytochrome-mediated light signaling. *Plant Cell Physiology, 45*, 968–975.

Phee, B. K., Kim, J. I., Shin, D. H., Yoo, J., Park, K. J., Han, Y. J., et al. (2008). A novel protein phosphatase indirectly regulates phytochrome-interacting factor 3 via phytochrome. *Biochemical Journal, 415*, 247–255.

Pittendrigh, C. S. (1966). The circadian oscillation in *Drosophila pseudoobscura* pupae: a model for the photoperiodic clock. *Zeitschrift fuer Pflanzenphysiologie, 54*, 275–307.

Pittendrigh, C. S., & Minis, D. H. (1964). The entrainment of circadian oscillations by light and their role as photoperiodic clocks. *The American Naturalist, 98*, 261–294.

Putterill, J., Robson, F., Lee, K., Simon, R., & Coupland, G. (1995). The CONSTANS gene of *Arabidopsis* promotes flowering and encodes a protein showing similarities to zinc finger transcription factors. *Cell, 80*(6), 847–857.

Quail, P. H. (2002). Phytochrome photosensory signalling networks. *Nature Reviews Molecular Cell Biology, 3*, 85–93.

Saito, H., Ogiso-Tanaka, E., Okumoto, Y., Yoshitake, Y., Izumi, H., Yokoo, T., et al. (2012). Ef7 encodes an ELF3-like protein and promotes rice flowering by negatively regulating the floral repressor gene Ghd7 under both short- and long-day conditions. *Plant Cell Physiology, 53*, 717–728.

Samach, A., Onouchi, H., Gold, S. E., Ditta, G. S., Schwarz-Sommer, Z., Yanofsky, M. F., et al. (2000). Distinct roles of CONSTANS target genes in reproductive development of *Arabidopsis*. *Science, 288*, 1613–1616.

Salazar, J. D., Saithong, T., Brown, P. E., Foreman, J., Locke, J. C., Halliday, K. J., et al. (2009). Prediction of photoperiodic regulators from quantitative gene circuit models. *Cell, 139*, 1170–1179.

Salomé, P. A., & McClung, C. R. (2005). What makes the *Arabidopsis* clock tick on time? A review on entrainment. *Plant, Cell, and Environment, 28*, 21–38.

Sawa, M., Nusinow, D. A., Kay, S. A., & Imaizumi, T. (2007). FKF1 and GIGANTEA complex formation is required for day-length measurement in *Arabidopsis*. *Science, 318*, 261–265.

Sharrock, R. A., & Clack, T. (2002). Patterns of expression and normalized levels of the five *Arabidopsis* phytochromes. *Plant Physiology, 130*, 442–456.

Song, Y. H., Smith, R. W., To, B. J., Millar, A. J., & Imaizumi, T. (2012). FKF1 conveys timing information for CONSTANS stabilization in photoperiodic flowering. *Science, 336*, 1045–1049.
Soy, J., Leivar, P., González-Schain, N., Sentandreu, M., Prat, S., Quail, P. H., et al. (2012). Phytochrome-imposed oscillations in PIF3 protein abundance regulate hypocotyl growth under diurnal light/dark conditions in *Arabidopsis. The Plant Journal, 71*, 390–401.
Suárez-López, P., Wheatley, K., Robson, F., Onouchi, H., Valverde, F., & Coupland, G. (2001). CONSTANS mediates between the circadian clock and the control of flowering in *Arabidopsis. Nature, 410*, 1116–1120.
Takahashi, Y., Shomura, A., Sasaki, T., & Yano, M. (2001). Hd6, a rice quantitative trait locus involved in photoperiod sensitivity, encodes the alpha subunit of protein kinase CK2. *Proceedings of the National Academy of Sciences of the United States of America, 98*, 7922–7927.
Takano, M., Inagaki, N., Xie, X., Yuzurihara, N., Hihara, F., Ishizuka, T., et al. (2005). Distinct and cooperative functions of phytochromes A, B, and C in the control of deetiolation and flowering in rice. *Plant Cell, 17*, 3311–3325.
Takimoto, A., & Hamner, K. C. (1964). Effect of Temperature and Preconditioning on Photoperiodic Response of *Pharbitis nil. Plant Physiology, 39*, 0024–1030.
Tamaki, S., Matsuo, S., Wong, H. L., Yokoi, S., & Shimamoto, K. (2007). Hd3a protein is a mobile flowering signal in rice. *Science, 316*, 1033–1036.
Valverde, F., Mouradov, A., Soppe, W., Ravenscroft, D., Samach, A., & Coupland, G. (2004). Photoreceptor regulation of CONSTANS protein in photoperiodic flowering. *Science, 303*, 1003–1006.
Wang, H. (2005). Signaling mechanisms of higher plant photoreceptors: a structure-function perspective. *Current Topics in Developmental Biology, 68*, 227–261.
Wu, C., You, C., Li, C., Long, T., Chen, G., Byrne, M. E., et al. (2008). RID1, encoding a Cys2/His2-type zinc finger transcription factor, acts as a master switch from vegetative to floral development in rice. *Proceedings of the National Academy of Sciences of the United States of America, 105*, 12915–12920.
Xue, W., Xing, Y., Weng, X., Zhao, Y., Tang, W., Wang, L., et al. (2008). Natural variation in Ghd7 is an important regulator of heading date and yield potential in rice. *Nature Genetics, 40*, 761–767.
Yakir, E., Hilman, D., Kron, I., Hassidim, M., Melamed-Book, N., & Green, R. M. (2009). Posttranslational regulation of CIRCADIAN CLOCK ASSOCIATED1 in the circadian oscillator of *Arabidopsis. Plant Physiology, 150*, 844–857.
Yamashino, T., Nomoto, Y., Lorrain, S., Miyachi, M., Ito, S., Nakamichi, N., et al. (2013). Verification at the protein level of the PIF4-mediated external coincidence model for the temperature-adaptive photoperiodic control of plant growth in *Arabidopsis thaliana. Plant Signaling Behavior, 8*, e23390.
Yano, M., Katayose, Y., Ashikari, M., Yamanouchi, U., Monna, L., Fuse, T., et al. (2000). Hd1, a major photoperiod sensitivity quantitative trait locus in rice, is closely related to the *Arabidopsis* flowering time gene CONSTANS. *Plant Cell, 12*, 2473–2484.
Yanovsky, M. J., & Kay, S. A. (2002). Molecular basis of seasonal time measurement in *Arabidopsis. Nature, 419*, 308–312.
Yasui, Y., Mukougawa, K., Uemoto, M., Yokofuji, A., Suzuri, R., Nishitani, A., et al. (2012). The phytochrome-interacting vascular plant one-zinc finger1 and VOZ2 redundantly regulate flowering in *Arabidopsis. Plant Cell, 24*, 3248–3263.
Yeh, K. C., & Lagarias, J. C. (1998). Eukaryotic phytochromes: light-regulated serine/threonine protein kinases with histidine kinase ancestry. *Proceedings of the National Academy of Sciences of the United States of America, 95*, 13976–13981.
Zhao, J., Huang, X., Ouyang, X., Chen, W., Du, A., Zhu, L., et al. (2012). OsELF3-1, an ortholog of *Arabidopsis* early flowering 3, regulates rice circadian rhythm and photoperiodic flowering. *PLoS One, 7*, e43705.
Zuo, Z., Liu, H., Liu, B., Liu, X., & Lin, C. (2011). Blue light-dependent interaction of CRY2 with SPA1 regulates COP1 activity and floral initiation in *Arabidopsis. Current Biology, 21*, 841–847.

CHAPTER FIVE

Genetic Control of Reproductive Development in Temperate Cereals

Chiara Campoli* and Maria von Korff*,†,‡,1
*Max Planck Institute for Plant Breeding Research, Cologne, Germany
†Institute of Plant Genetics, Heinrich Heine University, Düsseldorf, Germany
‡Cluster of Excellence on Plant Sciences "From Complex Traits towards Synthetic Modules", Düsseldorf, Germany
[1]Corresponding author: e-mail address: korff@mpipz.mpg.de

Contents

Abstract

Flowering is a central developmental process in the life cycle of a plant. Consequently, the decision to flower has to be taken at the right moment, when internal factors and external cues are at optimum to ensure reproductive success. This is crucial to every plant, and is of particular interest in crop species, where reproductive success has a major impact on yield. This chapter gives an overview of the genetics of flowering time in temperate cereals such as barley and wheat. The major flowering time genes are presented and their interaction is discussed in the light of the current knowledge coming from the model species *Arabidopsis thaliana* and rice. The importance of flowering time genes for adaptation to different environments is discussed. Finally, the impact of flowering time on yield and pleiotropic effects of flowering time genes are presented.

Advances in Botanical Research, Volume 72
ISSN 0065-2296
http://dx.doi.org/10.1016/B978-0-12-417162-6.00005-5

5.1 INTRODUCTION

The onset of flowering is crucial for reproductive success and has a major impact on yield in crops. Temperate cereals, which include economically important species such as wheat and barley, are a significant source of food and animal feed. Improving their yield will be crucial in future years to meet the increasing demands of a growing world population. In temperate climates, plants have to respond to variation in day length and temperature to coordinate flowering time with seasonal changes. Flowering time is a complex trait, controlled by many genes. Variation in flowering time was the basis for the adaptation of wheat and barley to a wide range of environments, different from those typical of the Fertile Crescent, where these cereal crops were first domesticated. In recent years, an increasing number of flowering time genes have been identified and placed into floral pathways. Information from the model species *Arabidopsis thaliana* (hereafter *Arabidopsis*) and rice (*Oryza sativa* L.) has been used to infer possible interactions or suggest orthologous genes. However, the functions of many flowering time orthologues are modified in wheat and barley. Gene duplications in the complex genomes of barley and wheat may have contributed to the diversification of flowering time networks. This chapter reviews the current knowledge on flowering time genes in wheat and barley, their role in adaptation and their impact on yield. Flowering genes are presented and their allelic variation and interactions in the different floral pathways are discussed. A better understanding of the physiological and genetic basis of flowering time will be the key to breed cereal crops adapted to different environments, affected by climate change.

5.2 FLOWERING TIME AND ADAPTATION TO DIFFERENT ENVIRONMENTS

The transition from the vegetative to the reproductive stage is a key adaptive trait that ensures that plants set flowers at the optimum time. Barley and wheat have originated in the Fertile Crescent, but modern varieties are cultivated in a wide range of environments. Allelic diversity at genes regulating response to photoperiod and vernalisation favoured the adaptation of temperate cereals to different environments. Early flowering, for example, has been selected as an adaptation to short-growing seasons, to avoid hot and dry summers in Mediterranean areas. In temperate climates, instead, humid and cool summers allow plants to grow longer and late flowering improves yield. Vernalisation requirement has evolved as a mechanism to prevent plants from flowering before winter. In fact, flowering occurs only after a prolonged exposure to cold, when plants are able to respond to increasing day length and flower. Wheat and barley can be classified into two growth types: winter and spring. Winter growth types include those genotypes that require a prolonged exposure to cold temperature to flower, whereas spring growth types do not respond to vernalisation. However, there is a continuous gradation regarding the vernalisation requirement in different genotypes ranging from spring to extreme winter types (Enomoto, 1929). The progenitor of cultivated barley, *Hordeum vulgare* ssp. *spontaneum* and the three progenitor species of hexaploid wheat, *Triticum urartu*, *Aegilops speltoides* and *Aegilops tauschii*, originated in the Fertile Crescent (Badr et al., 2000; Dubcovsky & Dvorak, 2007). Hexaploid wheat resulted from the hybridization between tetraploid domesticated emmer wheat (*Triticum dicoccum*) and the diploid grass *A. tauschii* (Asplunda, Hagenblada, Matti, & Leinob, 2010). Wild wheat species comprise spring and winter types (Goncharov, 1998; Goncharov & Chikida, 1995). The distribution of the spring-type *A. tauschii* in the eastern part of Iran and in Afghanistan suggests that hexaploid wheat derived from the hybridisation between winter-type emmer wheat and spring-type *A. tauschii*. The derived partial vernalisation requirement was an advantage to successfully adapt to the relatively mild winters of these areas (Iwaki, Haruna, Niwa, & Kato, 2001). Bread wheat is widely cultivated in various parts of the world. The distribution pattern of bread wheat is closely related to the degree of winter coldness and depends on vernalisation requirement and frost resistance (Iwaki et al., 2001). Spring types are mainly cultivated in high latitudes with very cold winters or as winter wheat in low latitudes with relatively warm winters. Winter types, which are generally more frost resistant than spring types, are predominant in medium latitudes with cold winters (Fujita, Kawada, & Tahir, 1992; Iwaki et al., 2001). Wild barley requires a prolonged cold exposure, indicating that the ancestral form in barley is the winter growth habit (Saisho, Ishii, Hori, & Sato, 2011). Barley is nowadays cultivated in a wide range of environments. Varieties with a mild vernalisation requirement are grown in Mediterranean areas and the Middle East. These genotypes are sown in autumn and respond to vernalisation, but they can eventually flower in the absence of vernalisation (Weltzien, 1988, 1989). Winter growth types, which have been selected for cultivation in Northern latitudes, show an improved resistance to low temperature (Cockram et al., 2007). Spring growth types, instead, have been selected to allow spring sowing and late flowering in order to exploit most of the spring and summer months in Central Europe. A further expansion to northern environments, characterised by cold winters and short summers, required the selection of early flowering spring genotypes. These genotypes do not respond to photoperiod or vernalisation and are characterized by the so called *early maturity* (*eam*) or *earliness* per se (*eps*) loci (Faure et al., 2012).

5.3 IMPACT OF FLOWERING TIME ON YIELD IN TEMPERATE CEREALS

Time to flowering is the result of the duration of pre-anthesis phases and depends on the coordination of changes at the shoot apex, spike formation and plant growth. The impact of different developmental phases on yield has been thoroughly studied in wheat. In particular the final number of grains, which is the most important component of cereal yield, is correlated to the spike dry weight and the final number of fertile florets (Reynolds et al., 2009). Some floret primordia are aborted during the maximum stem and spike growth phase, which has been explained with the competition for limited assimilates between spike and stem during that phase (Ghiglione et al., 2008; Gonzalez, Miralles, & Slafer, 2011; Gonzalez, Slafer, & Miralles, 2003). The duration of stem elongation is thus correlated to the number of fertile florets (González et al., 2003; Miralles & Richards, 2000; Slafer, 2003) and has therefore been proposed as a target trait to improve wheat yield potential. Slafer, Abeledo, Miralles, Gonzalez, and Whitechurch (2001) suggested that a longer stem elongation phase would result in a higher number of fertile florets. Increasing the final grain number can thus be achieved by manipulating the length of different developmental phases, while keeping the overall time to flowering unchanged. A better understanding of the physiological and genetic basis of flowering time, including possible signalling in response to different environmental cues, such as photoperiod and temperature may help minimizing floret abortion for a more optimal source-sink balance.

5.4 FLOWERING TIME GENES AND FLORAL PATHWAYS IN TEMPERATE CEREALS

Genes controlling flowering time have been extensively studied in the model plant *Arabidopsis* and placed into genetic networks (Chapters 1, 2 and 3). *Arabidopsis*, like the temperate cereals, is a facultative long day plant comprising many accessions displaying both summer and winter annual growth habits. Likewise, numerous studies in rice have identified the major genes affecting flowering time, deciphering pathways of gene interaction in a model grass species (Chapter 4). Flowering time control has also been a target of numerous studies in temperate cereals, which include economically important species such as barley (*H. vulgare* L.) and wheat (*Triticum aestivum* L.). It has been shown that there is a high degree of conservation for flowering time genes across species, and *Arabidopsis* orthologous genes, in particular of those belonging to the photoperiodic flowering pathway, have been identified in cereals (Cockram et al., 2007; Distelfeld, Li, & Dubcovsky, 2009; Higgins, Bailey, & Laurie, 2010). Information from the model species *Arabidopsis* or from the closely related rice can thus be successfully used to identify genes and suggest possible interactions in temperate cereals. However, because of their different growth strategies, the use of rice as a model for flowering in temperate cereals has some limitations. Rice is a short day tropical plant, with no requirement for vernalisation, while barley and wheat are facultative long day plants, which may require, depending on the genotype, a prolonged exposure to cold before the onset of flowering.

The major flowering genes in barley and wheat have initially been identified by exploiting natural genetic diversity and quantitative trait loci (QTL) studies (Dubcovsky, Lijavetzky, Appendino, Tranquilli, & Dvorak, 1998; Karsai et al., 2005; Laurie, Pratchett, Snape, & Bezant, 1995). Major regulators for flowering time in temperate cereals are encoded by the *Photoperiod 1* gene (*Ppd-H1*, *Ppd-A1*, *Ppd-B1*, *Ppd-D1*; Beales, Turner, Griffiths, Snape, & Laurie, 2007; Díaz, Zikhali, Turner, Isaac, & Laurie, 2012; Turner, Beales, Faure, Dunford, & Laurie, 2005; Wilhelm, Turner, & Laurie, 2009) and the vernalisation genes *Vrn1*, *Vrn2* and *Vrn3* (Yan et al., 2003, 2004, 2006). Additional components of flowering time pathways are the *eps* or *eam* genes. *Eam* loci have been characterised in classical QTL mapping studies (Hanocq, Laperche, Jaminon, Laine, & Le Gouis, 2007; Hanocq, Niarquin, Heumez, Rousset, & Le Gouis, 2004; Kamran et al., 2013; Laurie et al., 1995; Shindo, Tsujimoto, & Sasakuma, 2003) and recently the *eam* loci, *eam6*, *eam8* and *eam10* have been cloned (Campoli et al., 2013; Comadran et al., 2012; Faure et al., 2012; Zakhrabekova et al., 2012). The next paragraphs describe our most recent understanding of gene networks controlling photoperiodic and vernalisation responses in temperate cereals. Figure 5.1 provides an overview of temperate cereal flowering genes and their interactions in the flowering pathways. The map position of the major flowering genes and loci in barley is shown in Figure 5.2.

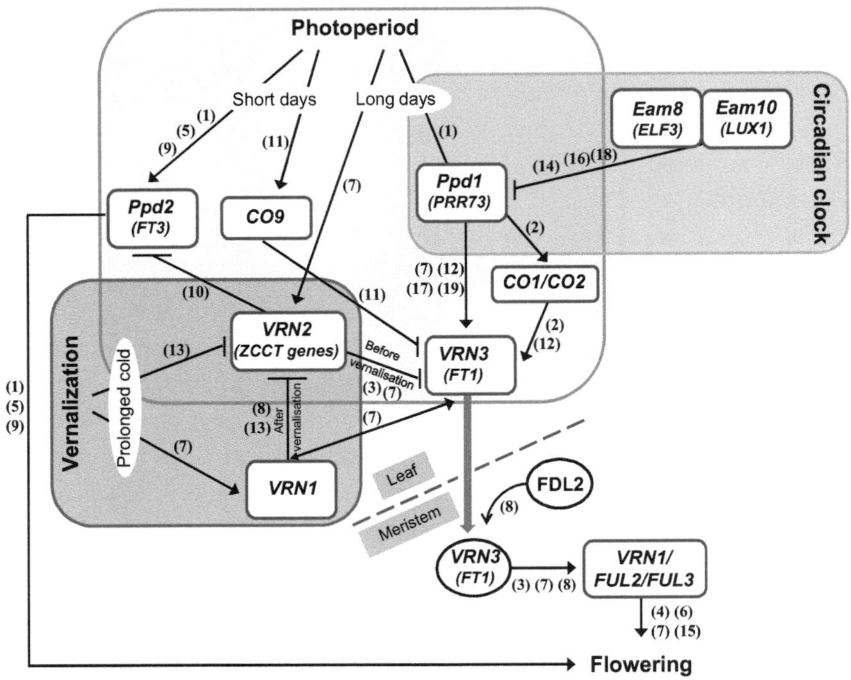

Figure 5.1 Model of flowering time control pathways in wheat and barley. The different external and internal cues are highlighted in different colours. Positive and negative regulatory actions are indicated by arrows and lines with bars, respectively. Boxes indicate genes, while circles indicate proteins. The green arrow shows that the FT1 protein moves from the leaf to the meristem. The figure incorporates different aspects of previously published wheat and barley models. Numbers in brackets indicate literature in which experimental evidences support the model: (1) Laurie et al. (1995); (2) Turner et al. (2005); (3) Yan et al. (2006); (4) Adam et al. (2007); (5) Faure et al. (2007); (6) Shitsukawa et al. (2007); (7) Hemming et al. (2008); (8) Li and Dubcovski (2008); (9) Kikuchi et al. (2009); (10) Casao, Iguarta, et al. (2011); (11) Kikuchi et al. (2011); (12) Campoli, Drosse, et al. (2012); (13) Chen and Dubcovski (2012); (14) Faure et al. (2012); (15) Kinjo et al. (2012); (16) Mizuno et al. (2012); (17) Shaw et al. (2012); (18) Campoli et al. (2013); (19) Shaw et al. (2013). (See the colour plate.)

5.4.1 Photoperiod Response

In temperate cereals the response to photoperiod is primarily controlled by the *Photoperiod 1* (*Ppd1*) gene, located in collinear regions on the short arm of group two chromosomes (Laurie et al., 1995; Law, Sutka, & Worland, 1978; Scarth & Law, 1983; Welsh, Keim, Pirasteh, & Richards, 1973). The *Ppd1* genes encode a pseudo-response regulator (PRR) protein, PRR37, homologous to the *Arabidopsis* PRR3/PRR7 proteins, characterized by a

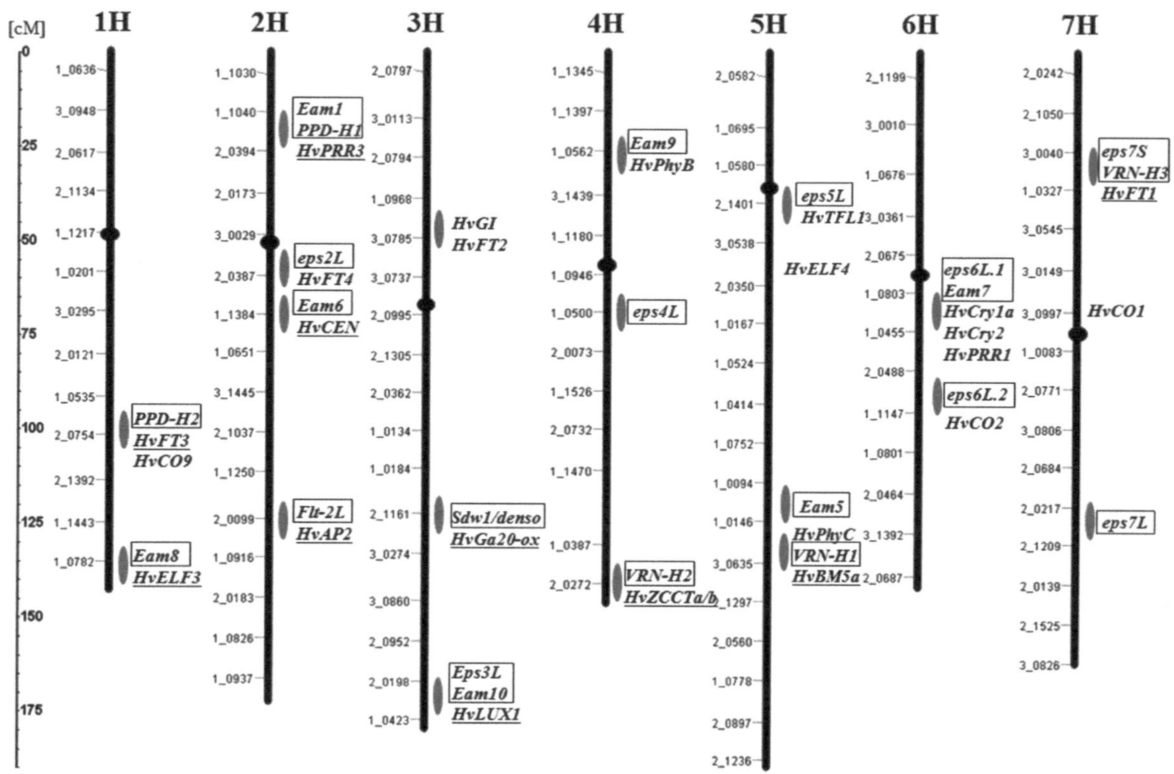

Figure 5.2 Consensus map of quantitative trait loci (QTL) positions for flowering time in barley. Positions of QTL and flowering time candidate genes were projected onto the barley single nucleotide protein (SNP) consensus map of Muñoz-Amatriaín et al. (2011). Markers to the left of the chromosomes represent SNP markers. Black ovals indicate the position of the centromeres. Approximate positions of flowering time QTL are indicated by grey (green in online) ovals to the right of the chromosomes. Names of QTL are boxed. Confirmed genes are underlined, whereas suggested candidate genes for QTL are not. References for candidate genes are reported in the text. The QTL shown are a summary of the following publications: Laurie et al. (1995); Bezant et al. (1996); Marquez-Cedillo et al. (2001); Teulat et al. (2001); Ivandic et al. (2002); Baum et al. (2003); Boyd et al. (2003); Pillen et al. (2003, 2004); Szűcs et al. (2006); von Korff et al. (2006, 2008); Cuesta-Marcos, Casas, et al. (2008); Cuesta-Marcos, Igartua, et al. (2008); Chen et al. (2009a); Borràs-Gelonch et al. (2010); Wang et al. (2010); Rollins et al. (2013).

pseudoreceiver and a CCT (CONSTANS, CONSTANS-like and TOC1) domains. The ancestral, dominant form of *Ppd1* confers an acceleration of flowering under increasing day length. Barley and wheat carry different natural polymorphisms which modify the response to long days. In barley, a recessive mutation in the CCT domain of *ppd-H1* has been selected in spring cultivars grown in Northern agricultural areas. This variant confers delayed flowering and maturity under long days and may thus represent an adaptation to long growing seasons in Central and Northern Europe (Jones et al., 2008; Turner et al., 2005). Recessive *ppd1* alleles have recently been identified in wheat. Similar to barley, loss of function deletions in the wheat *ppd1* homeologous series delay flowering time under long days (Shaw, Turner, Herry, Griffiths, & Laurie, 2013). Nevertheless the effect of a loss of function allele is often masked by the presence of functional alleles on the homeologous chromosomes. This may be the reason why *ppd1* loss of function alleles were not exploited to adapt wheat to long growing seasons (Shaw et al., 2013). Conversely, in wheat, dominant mutations in the *ppd1* genes accelerate flowering under both long and short day conditions and confer day-length neutrality. Early, day-length neutral flowering results in yield benefits in short season agro-environments (Beales et al., 2007; Worland et al., 1998). Deletions in the promoters of *Ppd-A1a* and *Ppd-D1a* cause their constitutive up-regulation and early flowering (Beales et al., 2007; Nishida et al., 2013; Wilhelm et al., 2009). In addition, differences in copy number of *Ppd-B1a* result in higher expression levels of this gene (Díaz et al., 2012). An insertion in the 5′ upstream region of *Ppd-B1*a has also been associated with early flowering (Nishida et al., 2013). Turner et al. (2005) have shown that in barley dominant alleles of *Ppd-H1* are associated with increased expression of *HvFT1*. Similarly, increased expression of the wheat *Ppd1* genes up-regulated *TaFT1* homeologous series in a genome-independent manner. Conversely, a loss of function *ppd1* allele is associated with a reduced *TaFT1* expression, with different alleles having variable and cumulative effects (Shaw et al., 2013; Shaw, Turner, & Laurie, 2012). *HvFT1* and *TaFT1* have been identified as the genes underlying the *Vrn3* locus on the short arm of group seven chromosomes and encode for RAF-kinase inhibitor proteins homologous to *Arabidopsis* Flowering locus T (FT) (Yan et al., 2006). Recently, it has been shown that in certain spring barleys, the presence of multiple copies of *HvFT1* underlying the spring *Vrn-H3* allele is associated with earlier up-regulation of *HvFT1* (Nitcher, Distelfeld, Tan, Yan, & Dubcovsky, 2013). When only one *HvFT1* copy is present, promoter and first intron haplotype differences contribute to smaller effect variation in flowering time, although additional studies are needed to identify the nature of these differences (Casas et al., 2011; Nitcher et al., 2013). In wheat, transposable elements inserted in the intron of a *Vrn3* allele are associated with higher *TaFT* expression level (Yan et al., 2006). In *Arabidopsis*, *FT* encodes the mobile florigen hormone whose cognate protein moves from the leaves through the phloem to the shoot apical meristem. In the apical meristem, the FT protein forms a complex with the bZIP protein Flowering locus D (FD) and binds to the promoter of the meristem identity genes *APETALA1* (*AP1*) and *FRUITFULL* (*FUL*) to induce the switch from vegetative to reproductive growth (Corbesier et al., 2007; Wigge et al., 2005). Li and Dubcovsky (2008) have shown that in wheat VRN3 interacts with an FD-like protein (TaFDL2). TaFDL2 in turn binds to the promoter of *TaVRN1*, which is the wheat homologue of *AP1/FUL*. Currently, it is not known if the VRN3 protein moves from the leaf to the apex also in wheat. However, Tamaki, Matsuo, Wong, Yokoi, and Shimamoto (2007) have shown that in rice the protein encoded by *Hd3a*, orthologous to *VRN3*, moves from the leaf to the shoot apical meristem and induces flowering, suggesting that similar regulatory mechanisms are conserved between dicots and monocots.

In *Arabidopsis*, a central regulator of photoperiodic flowering is *CONSTANS* (*CO*), which encodes a CCT domain protein that triggers *FT* expression upon exposure of plants to long days (Samach et al., 2000). The circadian clock regulates *CO* at the transcriptional level, so that *CO* mRNA abundance is higher at the end of a long day. In addition, CO protein is regulated by the cryptochromes Cry1 and Cry2, the phytochromes PhyA, PhyB, and the ubiquitin ligase Constitutive Photomorphogenic 1 (COP1) that, modify its stability in order to restrict its expression at dusk under long days (Jang et al., 2008; Valverde et al., 2004). Nine orthologues of the *AtCO* gene have been isolated in barley. *HvCO1* and *HvCO2*, located on chromosome 7H and 6H, respectively, show the highest similarity to the *Arabidopsis CO* gene (Griffith, Dunford, Coupland, & Laurie, 2003). In wheat, three genes with a CCT domain have been isolated on the long arm of the homeologous group 6 and named *TaHd1-1*, *TaHd1-2* and *TaHd1-3*, following the nomenclature of the homologue rice gene *Heading date 1* (*Hd1*) (Nemoto, Kisaka, Fuse, Yano, & Ogihara, 2003; Yano et al., 2000). The *TaHd1-2* gene has a deletion in the promoter region containing the GATA-1 box and its expression is not detectable in seedlings of wheat, indicating that in wheat only the other two *CO*-like genes *TaHd1-1* and *TaHd1-3* are functionally active. Of these, *TaHd1-1* could complement the rice *hd1* mutation

(Nemoto et al., 2003). Turner et al. (2005) suggested that the mutation in the CCT domain of *Ppd-H1* of spring barley delayed flowering time by shifting the diurnal expression peaks of *HvCO1* and *HvCO2* mRNA into the dark phase, so that the protein, as in *Arabidopsis*, is not synthesized and *Vrn-H3*/*HvFT1* is not induced. Campoli, Drosse, Searle, Coupland, and von Korff (2012) have shown that transgenic barley lines over-expressing *HvCO1* had increased expression of *HvFT1* and flowered earlier under long and short day conditions. Interestingly, *HvCO1* over-expressing lines maintain a response to photoperiod so that plants growing under short day are still flowering later than plants growing under long days. The analysis of a mapping population segregating for transgenes over-expressing *HvCO1* and the two functional variants of *Ppd-H1* revealed that *Ppd-H1* induced *HvFT1* expression downstream of *HvCO1* transcription (Campoli, Drosse, et al., 2012).

In *Arabidopsis*, *CO* transcription is controlled by GIGANTEA (GI) a plant-specific protein with no known functional domains (Fowler et al., 1999). Sequences with homology to GI have been identified in barley and wheat (Dunford, Griffiths, Christodoulou, & Laurie, 2005; Zhao, Liu, Li, Guan, & Zhang, 2005). However functional conservation between HvGI and TaGI and the *Arabidopsis* orthologue AtGI has not yet been demonstrated. In rice, the over-expression of *OsGI* induced the expression of *Hd1*, which is the rice orthologue of *Arabidopsis CO* (Hayama, Yokoi, Tamaki, Yano, & Shimamoto, 2003). In addition, heterologous expression of the *Brachypodium dystachyon* GI protein in a GI-deficient *Arabidopsis* mutant rescued the late flowering phenotype, suggesting that the role of GI is conserved in grasses (Shin-Young, Sangmin, Pil, Moon-Sik, & Chung-Mo, 2010).

Gene duplications have also occurred at the FT locus in grasses providing an additional source for variation in flowering time control (Higgins et al., 2010). In barley, five different *FT*-like genes were identified, *HvFT1*, *HvFT2*, *HvFT3*, *HvFT4* and *HvFT5* (Faure, Higgins, Turner, & Laurie, 2007); of these *HvFT1* (*Vrn-H3*) has been characterised as a flowering promoter (Kikuchi, Kawahigashi, Ando, Tonooka, & Handa, 2009). However, *HvFT3* has been proposed as a candidate gene for the photoperiod response locus *Ppd-H2*, a major QTL located on the long arm of chromosome 1H, which affects flowering under short days (Faure et al., 2007; Kikuchi et al., 2009). Two major functional variants of *HvFT3* are known (Casao, Iguarta, et al., 2011; Casao, Karsai, et al., 2011; Cuesta-Marcos, Casas, et al., 2008). The dominant functional allele is prevalent in Southern European barley germplasm and promotes flowering under short day conditions when vernalisation requirement is not fully satisfied (Casao, Karsai, et al., 2011). A partial deletion of the gene results in a recessive non-functional allele that is common in winter barley (Faure et al., 2007; Kikuchi et al., 2009). Expression of both, *HvFT1* and *HvFT3*, is repressed by *Vrn-H2* and thus also controlled by the vernalisation pathway (see discussion of the vernalisation pathway below, Casao, Iguarta, et al., 2011;Yan et al., 2006). The key photoperiod response gene *Ppd-H1* is a homologue of the circadian clock *Arabidopsis* genes *PRR3/PRR7*, which suggests that the clock plays an important role in the control of flowering in cereals as discussed in the next chapter.

5.4.2 Circadian Clock

Erwin Bünning (1936) first proposed that photoperiodism is connected to the circadian clock and formulated the external coincidence model. Photoperiodic responses are controlled by the clock-regulated expression of a key component and the effect of light on the activity of this component. Only under inductive conditions, sufficient amounts of the key component are exposed to light, thereby inducing a photoperiodic response. The genetic basis of the circadian clock and the external coincidence model was unravelled in the long day plant *Arabidopsis*, and notable progress has been made in identifying the molecular mechanisms by which *Arabidopsis* recognizes day length and promotes flowering. The circadian clock is an internal timekeeper which synchronises biological processes with the diurnal cycle, using molecular mechanisms that include interlocked transcriptional feedback loops. In *Arabidopsis*, the circadian clock is composed of three negative feedback loops: (1) the inhibition of evening complex (EC) genes *EARLY FLOWERING 3* (*ELF3*), *EARLY FLOWERING 4* (*ELF4*) and *LUX ARRHYTHMO* (*LUX*, also known as *PHYTOCLOCK1*) by the rise of CIRCADIAN CLOCK ASSOCIATED1 (CCA1) and LATE ELONGATED HYPOCOTYL (LHY) late at night, (2) the inhibition of *PRR* genes by the EC early at night, and (3) the inhibition of *LHY/CCA1* by TIMING OF CAB EXPRESSION1 (TOC1) in the morning (Huang et al., 2012; Pokhilko et al., 2012). In addition, the evening-expressed GI protein was modelled as a negative regulator of the EC, which in turn inhibits *TOC1* expression (Pokhilko et al., 2012). In *Arabidopsis*, *CO* expression is induced at the end of long days and controlled by the circadian clock and photoperiod. The interaction between the clock-regulated plant-specific protein GI and the light-regulated ubiquitin ligase FKF1 leads to the degradation of the transcriptional repressors, CYCLING DOF FACTORs, releasing the repression of *CO* mRNA at the end of a long day. Peak expression of *CO* and light need to coincide to stabilise the CO protein. Only under long photoperiods, then, the expression of *CO* occurs within the light phase, which is necessary for the stabilisation of CO protein (Chapter 1). Campoli, Shtaya, Davis, and von Korff (2012) and Higgings et al. (2010) have shown that circadian clock genes are structurally conserved between *Arabidopsis* and barley and that their circadian expression patterns suggest functional conservation. However, phylogenetic analyses revealed that duplications/deletions of clock genes occurred throughout the evolution of eudicots and monocots. For instance, the *PRR* genes duplicated independently in monocots and eudicots, and only one homologue of the two paralogous *Arabidopsis* clock genes *LHY/CCA1* is found in monocots (Campoli, Shtaya, et al., 2012; Takata, Saito, Saito, & Uemura, 2010). It is interesting to note that natural variation at *Ppd1* in barley and wheat are major determinants of photoperiod sensitivity (Beales et al., 2007; Turner et al., 2005), while natural variation at *PRR* genes in *Arabidopsis* did not have a strong effect on flowering time (Ehrenreich et al., 2009). Increased copy number or mutations in the promoter region of *Ppd1* homeologous genes in wheat lead to an increased expression of the gene and early flowering under non-inductive short day conditions also referred as day-length neutrality (Beales et al., 2007). Day-length neutrality has not been widely used in barley breeding programmes, but natural and induced *eam* mutants have been used to breed for early flowering spring barley (Lundqvist, 2009). Two barley *eam* genes, *eam8* and *eam10*, have recently been identified as homologues of the *Arabidopsis* circadian clock regulators *ELF3*, and *LUX/ARRHYTHMO (LUX)*, respectively (Campoli et al., 2013; Faure et al., 2012; Zakhrabekova et al., 2012). Faure et al. (2012) have shown that lines harbouring a non-functional *eam*8 (*hvelf3*) protein had a higher expression of *Ppd-H1*, resulting in an induction of *HvFT1*. Similarly, the presence of a mutation in a highly conserved functional domain of *eam10* (*hvlux1)* leads to a higher expression of *Ppd-H1* and earlier flowering under non-inductive short day conditions (Campoli et al., 2013). Moreover a homologue of *Arabidopsis LUX/PHYTOCLOCK1*, *WPCL1* has been proposed as candidate gene for an early flowering *Triticum monococcum* mutant. As in the *eam10* barley mutant, the mutation in the wheat *LUX*-like sequence leads to an over-expression of *Ppd1* and an activation of *TaFT* expression under non-inductive short day conditions (Mizuno, Nitta, Sato, & Nasuda, 2012). It is interesting to note that mutations in barley genes orthologous to the *Arabidopsis* EC genes have similar effects on downstream photoperiod response genes and flowering time. Genetic studies have shown that *HvELF3* and *HvLUX1* interact with *Ppd-H1* (Figure 5.1; Campoli et al., 2013; Faure et al., 2012). In *Arabidopsis*, ELF3 physically associates with the promoter of *PRR9* to repress its transcription suggesting that transcriptional targets of ELF3 are partly conserved between *Arabidopsis* and barley (Dixon et al., 2011; Herrero et al., 2012). Interestingly, despite the pronounced clock defect, independent *eam8* mutations have been selected as a strategy to adapt to short-growing season and expand to northern latitudes (Faure et al., 2012). In contrast to *Arabidopsis* clock mutants, barley *eam8* mutants are not impaired in growth and carbon metabolism suggesting that clock output traits are different between *Arabidopsis* and barley (Habte, Müller, Shtaya, Davis, & von Korff, 2014).

Nevertheless, *eam* loci may contribute to adaptation to certain agro-environments and represent a new source of variation. Additional studies on circadian clock mutants in temperate cereals are thus of particular interest to understand the effects of this variation on crop productivity and fitness.

5.4.3 Vernalisation Response

In temperate cereals, vernalisation response is mainly controlled by two loci: *Vrn1* and *Vrn2* mapping to collinear regions of chromosomes group 5 and 4, respectively. *Vrn-A^{m}1* was first cloned in *T. monococcum* and encodes for a MADS[1]-box transcription factor with high similarity to the *Arabidopsis* meristem identity genes *APETALA1*, *CAULIFLOWER*, and *FRUITFULL* (Yan et al., 2003). Orthologous genes were identified in collinear regions of barley chromosome 5, *Vrn-H1*, and wheat chromosomes 5A, 5B and 5D, named *Vrn-A1*, *Vrn-B1* and *Vrn-D1*, respectively (Trevaskis, Bagnall, Ellis, Peacock, & Dennis, 2003). Recessive alleles at this locus are associated with the winter growth habit and are expressed only after a prolonged exposure to cold (Danyluk et al., 2003; Trevaskis et al., 2003; Yan et al., 2003). In *T. monococcum* the *maintained vegetative phase* (*mvp*) mutant, which carries a deletion in a region containing the *Vrn1* locus never transitioned from the vegetative to the reproductive phase, suggesting that *Vrn1* is critical for the transition to reproductive growth (Shitsukawa et al., 2007). However, a subsequent study pointed out that the large deletion in the *mvp* mutant contained additional genes, including the red/far red light photoreceptor *TmPhyC* and the MADS-box transcription factor *TaAGLG1* (Distelfeld & Dubcovski, 2010). In addition, *Vrn1* null mutants detected in a TILLING population of tetraploid wheat were able to flower suggesting the existence of redundant flowering time genes with meristem identity functions

[1] The abbreviation MADS comes from the first letters of the founding members of the family: Mini Chromosome Maintenance 1 (MCM1) of yeast, Agamous (AG) of Arabidopsis, Deficiens (DEF) of Antirrhinum and Serum Response Factor (SRF) of humans.

(Chen & Dubcovski, 2012). In barley and wheat, insertions and deletions in the first intron of *Vrn1* cause up-regulation of the gene independently of vernalisation (Cockram et al., 2007; Fu et al., 2005; Szűcs et al., 2007; von Zitzewitz et al., 2005). Hemming, Fieg, Peacock, Dennis, and Trevaskis (2009) compared the first intron sequences of 11 barley *Vrn-H1* alleles and identified specific regions associated with repression in non-vernalised plants. The same regions, however, are not required for cold induction of the gene. In wheat, small insertions and deletions or single nucleotide polymorphisms within the proximal promoter, in particular in a region containing a CArG-box, are associated with higher levels of *Vrn1* expression and reduced vernalisation requirement (Chu et al., 2011; Yan et al., 2003; 2004; Zhang, Wu, Yang, Liu, & Zhou, 2012). It has been shown that a 2-Kb fragment upstream of the *Vrn-H1* starting codon is sufficient to drive expression in the shoot apex and leaves of barley, and to induce a reporter gene after cold exposure (Alonso-Peral, Oliver, Casao, Greenup, & Trevaskis, 2011). The expression of *Vrn1* is quantitative, with longer exposure to cold conferring higher expression. *Vrn1* expression remains at elevated levels when plants return to ambient temperatures after vernalisation, suggesting a possible epigenetic regulation (Danyluk et al., 2003; Hemming, Peacock, Dennis, & Trevaskis, 2008; Sasani et al., 2009; Trevaskis, Hemming, Peacock, & Dennis, 2006; von Zitzewitz et al., 2005; Yan et al., 2003). Oliver, Finnegan, Dennis, Peacock, and Trevaskis (2009) showed that after vernalisation silent histone marks (histone 3 lysine 27 tri-methylation, H3K27me3) decreased in regions located between the promoter and the end of the first intron, while active marks for transcription (histone 3 lysine 4 tri-methylation, H3K4me3) occurred at the promoter and first intron of the *Vrn-H1* locus of barley. *Vrn-H1* is induced rapidly, within the first 24 h after cold exposure. However, after return to ambient temperature, induced expression of *Vrn-H1* is maintained only in case of prolonged cold exposure and this is associated with increased histone acetylation (Oliver, Deng, Casao, & Trevaskis, 2013). Analysis of histone modifications in the *TaVRN1* promoter of wheat revealed no significant changes for H3K27me3 after vernalisation, but an increase of H3K4me3 in the winter genotype (Diallo, Ali-Benali, Badawi, Houde, & Sarhan, 2012). In addition, Khan et al. (2013) observed cold-induced hypermethylation in the first intron of the *Vrn-A1* gene, which is associated with its expression and reset in the next generation. Díaz et al. (2012) have demonstrated that copy number variation of *VRN1* correlated with the expression level and vernalisation requirement. Winter barley and wheat show a higher frost resistance compared to spring ones. In winter genotypes the process of vernalisation, the acquisition of competence to flower after prolonged cold exposure is mirrored by the process of cold acclimation, the increase of frost tolerance after exposure to low non-freezing temperature (Thomashow, 1990, 1999). The ability to cold acclimate decrease during development and *Vrn1* has been indicated to play an important role in this (Limin & Fowler, 2006). *Vrn1* expression in the leaves is thus necessary to initiate the cascade that down-regulates the cold acclimation pathway (Dhillon et al., 2010). Allelic variation at the *Vrn1* locus, in addition, determines differences in frost tolerance, which indicates a pleiotropic effect of *Vrn1* (Dhillon et al., 2010). This suggests a possible double role for *Vrn1*: in the leaves to down-regulate the cold acclimation pathways after winter and in the meristem to induce the switch to reproductive development.

The increase in expression of *Vrn1* down-regulates the flowering repressor *Vrn2* (Chen & Dubcovsky, 2012; Loukoianov, Yan, Blechl, Sanchez, & Dubcovsky, 2005). *Vrn-A^{m}2* was identified on the long arm of chromosome 4 of *T. monococcum*, encoding a ZCCT (Zinc finger and CCT domain) gene with no clear orthologues in *Arabidopsis*. Spring alleles carry a mutation in a highly conserved residue of the CCT domain, which most likely disrupts the functionality of the protein (Yan et al., 2004). In barley, the *Vrn-H2* region on chromosome 4HL includes two complete and one truncated ZCCT genes, *ZCCT-Ha*, *ZCCT-Hb* and *ZCCT-Hc*, respectively. Spring barley carries a complete deletion of the locus (Yan et al., 2004). *Vrn2* is expressed under long days, but not under short days and is down-regulated after vernalisation (Trevaskis et al., 2006). The analysis of *Vrn1* null mutants of tetraploid wheat revealed that *Vrn1* was necessary to keep *Vrn2* repressed after prolonged cold exposure, but that was not necessary for the down-regulation of *Vrn2* during vernalisation (Chen & Dubcovsky, 2012). Kikuchi, Kawahigashi, Oshima, Ando, and Handa (2011) have shown that *HvCO9* delays flowering under non-inductive short day conditions, possibly by down-regulating *HvFT1*. *HvCO9* belongs to the same grass-specific *CO*-like subfamily of the flowering repressors *Vrn-H2* in barley and *Ghd7* in rice (Xue et al., 2008). Cockram, Howells, and O'Sullivan (2010) have shown that the chromosomal region on 4H containing the *Vrn2* locus has originated from a duplication of a chromosomal region on chromosome 1 carrying the *HvCO9* locus. The *Vrn2* locus may thus be derived from a targeted duplication of *HvCO9* to the homologous region after the divergence of *Triticeae* (Kikuchi, Kawahigashi, Oshima, Ando, & Handa, 2011). Interestingly, grasses have developed systems for flowering repression that are different from those of

Arabidopsis. Despite homology between *Arabidopsis* and cereal flowering time genes, gene duplication may have favoured functional diversification of flowering time pathways. The genetic control of vernalisation involves different genes in cereals and *Arabidopsis* suggesting that vernalisation response has evolved independently in monocots and dicots. Interestingly, the mechanism of vernalisation response through epigenetic regulation of key genes may be present in both lineages.

5.4.4 Integration of the Photoperiod and Vernalisation Pathways

Epistatic interactions observed among the *Vrn1*, *Vrn2* and *Vrn3* (*FT1*) genes suggest that they are part of the same regulatory gene network. However, the existence of feedback regulatory loop among them has complicated the interpretation of flowering experiments, so that different models of interaction have been proposed. The most recent model suggests that in a winter-type cereal, Vrn2 represses *FT1* in autumn before vernalisation, to counteract the Ppd1 dependent long day induction of *FT1* and avoid flowering prior winter (Figure 5.1). During winter, the prolonged exposure to cold induces *Vrn1* expression, which in turns down-regulates *Vrn2* (Hemming et al., 2008; Li & Dubcovski, 2008; Yan et al., 2006). Vernalisation can also directly down-regulate *Vrn2*, which is kept in a repressed state after cold exposure by *Vrn1* (Chen & Dubcovsky, 2012). After vernalisation, *Ppd1* and *CO* up-regulate *FT1* under long day conditions (Hemming et al., 2008; Yan et al., 2006). Alternative interactions among genes have also been proposed. For example, Shimada et al. (2009) described that, in wheat, the up-regulation of *Vrn1* under long days was followed by the accumulation of *TaFT* transcripts. *TaFT* was not expressed in the *mvp* mutant of *T. monococcum*, which carries a deletion in a region containing the *Vrn1* locus. Consequently, the authors suggested that *Vrn1* is upstream of *FT*1 and up-regulates *FT1* expression under long day conditions.

The vernalisation and photoperiodic pathways, thus, converge on FT1, which integrates environmental signals and promotes the expression of meristem identity genes. As mentioned before, there are contrasting information in the current literature on the role of the MADS-box transcription factor *Vrn1* in the meristem during the transition to flowering (Chen & Dubcovski, 2012; Distelfeld & Dubcovski, 2010; Shitsukawa et al., 2007). Studies in wheat and barley have identified additional MADS-box genes, including *FUL2* and *FUL3* that share with *Vrn1* a similar spatial and temporal expression pattern and can induce flowering when over-expressed in

Arabidopsis and rice (Adam et al., 2007; Kinjo, Shitsukawa, Takumi, & Murai, 2012; Preston & Kellogg, 2007; Schmitz et al., 2000). Additional studies are needed to identify meristem identity genes which control floral transition and inflorescence development in wheat and barley.

5.5 ADDITIONAL FLOWERING GENES IN TEMPERATE CEREALS: THEIR ROLE IN FLOWERING TIME, ADAPTATION AND PLEIOTROPIC EFFECTS

In addition to the major photoperiod and vernalisation loci, which have been consistently identified in crosses between winter and spring varieties (Hanocq et al., 2004; Kuchel, Hollamby, Langridge, Williams, & Jefferies, 2006; Laurie et al., 1995; Sameri, Pourkheirandish, Chen, Takuji-Tonooka, & Komatsuda, 2011; Shindo et al., 2003), a number of minor effect loci have been detected in wheat and barley. The more simple diploid nature of barley has facilitated the positional cloning of these loci, while to overcome the complication of the large and redundant hexaploid wheat genome, often diploid and tetraploid wheat varieties have been studied (Bullrich, Appendino, Tranquilli, Lewis, & Dubcovsky, 2002). Loci with minor effect have often been detected in crosses between exotic (landraces or wild) barley germplasm (Figure 5.2; Baum et al., 2003; Pillen, Zacharias, & Leon, 2004; Rollins et al., 2013; Teulat, Merah, Souyris, & This, 2001; von Korff et al., 2008; von Korff, Léon, & Pillen, 2010; von Korff, Wang, Leon, & Pillen, 2006; Wang et al., 2010). In wheat QTL meta-analysis has recently indicated the presence of more than 90 QTL for heading date, spread over almost the entire genome (Hanocq et al., 2007; Griffiths et al., 2009). These minor effect loci are often independent from external cues and may have a role in fine tuning flowering time. In addition, they often show pleiotropic effects and are thus valuable targets for breeding programmes. In the last years, the genes underlying some of these loci have been cloned or candidate genes have been proposed, although their position in the cereal flowering pathway is not fully understood. With the aim to provide some interesting examples, this paragraph describes the identification, genetic and pleiotropic effects of some of these loci.

The *Eam6* locus in the centromeric region of chromosome 2H has been detected in crosses involving wild barley or Mediterranean landrace genotypes (Marquez-Cedillo et al., 2001; Pillen et al., 2004; Rollins et al., 2013; von Korff et al., 2008; Wang et al., 2010). This locus has major effects on flowering time in autumn sowing in Mediterranean and Australian environments and has been associated with variation in the duration of the basic vegetative period and yield component traits, such as kernel weight, plant height and peduncle length (Boyd et al., 2003; Cuesta-Marcos, Casas, et al., 2008; Cuesta-Marcos, Igartua, et al., 2008; Rollins et al., 2013). Recently, the *eam6* locus was identified as an orthologue of Antirrhinum *CENTRORADIALIS* (*HvCEN*), homologous to *Arabidopsis TFL1* (Comadran et al., 2012). *TFL1* is a member of the *FT*-like gene family, but in contrast to *FT* encodes a flowering repressor. Comadran et al. (2012) indicated that natural variation at *HvCEN* contributed to the adaptation of barley to higher latitudes with cool and wet summers. On the long arm of chromosome 2H, variation for flowering time control was identified at the *Flowering Time-2L (FLT-2L)* locus (Baum et al., 2003; Borràs-Gelonch, Slafer, Casas, van Eeuwijk, & Romagosa, 2010; Boyd et al., 2003; Eleuch et al., 2008; Ivandic et al., 2002; Pillen, Zacharias, & Leon, 2003; Pillen et al., 2004; Rollins et al., 2013; Teulat et al., 2001; von Korff et al., 2006; 2008; 2010). The locus was also associated to frost resistance at heading and affected plant height and rachis internode length (Reinheimer, Barr, & Eglinton, 2004; Chen, Baumann, Fincher, & Collins, 2009; Chen et al., 2009). *Flt-2L* was fine mapped to a region which included *HvAP2*, a gene encoding an AP2 domain protein, with sequence similarity to the wheat domestication gene *Q* located on chromosome 5A, which confers compact spike, reduced plant height, and delays ear emergence, a phenotype similar to the barley *Flt-2L* mutation (Chen, Baumann, et al., 2009). A QTL for flowering time has also been identified in numerous crosses on the long arm of chromosome 3H (Baum et al., 2003; Bezant, Laurie, Pratchett, Chojecki, & Kearsey, 1996; Boyd et al., 2003; Cuesta-Marcos, Casas, et al., 2008; Laurie et al., 1995; Rollins et al., 2013; Szűcs et al., 2006). The exotic early flowering allele at this locus was correlated with increased plant height and reduced yield under favourable conditions, but increased yield under marginal rain-fed conditions (von Korff et al., 2006, 2008). This QTL coincides with the *sdw1/denso* locus which reduces growth and was selected to reduce lodging and to optimise yield under favourable conditions. Recently, *Ga20-oxidase*, a gene involved in the synthesis of gibberellin has been proposed as a candidate for this locus (Jia et al., 2009). In the centromeric region of chromosome 6H QTL for flowering time have also been identified, which coincided with QTL for plant height and yield, where the wild barley alleles reduced time to flowering, plant height and yield under favourable conditions (Bezant et al., 1996; Cuesta-Marcos, Casas, et al., 2008; Cuesta-Marcos, Igartua, et al., 2008; Ivandic et al., 2002; von Korff et al., 2006; Laurie et al., 1995; Pillen et al., 2004). The blue/UV-A light cryptochrome photoreceptors Cry1a and Cry2 which regulate plant growth and development (Quail, 2002) map to the centromeric region of 6H (Szűcs et al., 2006). Furthermore, the same region of 6H harbours the *eam7* mutation which determines photoperiod insensitivity and early flowering under long day conditions (Stracke & Börner, 1998).

Eps QTLs with minor effect have been reported on several wheat chromosomes (Hanocq et al., 2004, 2007; Kamran et al., 2013; Shindo et al., 2003). On wheat chromosome 2B, two *eps* QTL have been identified, one close to, but distinct from the *Ppd-B1* locus, and a second one close to the centromere in the collinear position of the barley *eps2S/eam6* locus (Scarth & Low, 1983; Shindo et al., 2003). Another major *eps* QTL was reported on chromosome 3A, and was shown to have significant effects on plant height, thousand kernel weight and number of grains per plant (Shah, Gill, Yen, Kaeppler, & Ariyarathne, 1999). An *eps* locus with a very strong effect (up to 49 days of delaying in flowering time) was detected on chromosome 1A in a cross between wild and cultivated diploid wheat (*T. monococcum*) and designed *Eps-*A^{m}*1* (Bullrich et al., 2002). The locus has been mapped to a similar region as the barley *eam8* locus (Zakhrabekova et al., 2012). *Eps-*A^{m}*1* was shown to be affected by temperature and to have pleiotropic effects on duration of different developmental phases, spikelet number and yield (Bullrich et al., 2002; Lewis, Faricelli, Appendino, Valarik, & Dubcovsky, 2008).

5.6 CONCLUSIONS

Plants respond to environmental cues, such as day length and temperature, to coordinate flowering with seasonal changes and to flower at the appropriate time. Developmental plasticity was a key for adaptation and cultivation of wheat and barley in different environments. The genetic and molecular understanding of floral transition in temperate cereals has greatly improved in the last years. The transition to flowering depends on a delicate balance of promoting and repressing factors, which integrate environmental signals and transmit them to the meristem. A deeper knowledge of cereal flowering pathways has indicated similarities with the model plant *Arabidopsis*, but also highlighted differences. Orthologous genes have been found, in particular in the photoperiodic pathway and the circadian clock, but often their connectivity and response to endogenous and environmental factors are different. The genetic control of vernalisation is clearly different between *Arabidopsis* and wheat/barley suggesting that vernalisation response has evolved independently in these two plant lineages. Gene duplications as seen for *FT* and *CO*-like genes may have contributed to the sub-functionalisation of flowering gene paralogues and diversification of signal perception, transmission and integration. However, the roles of these different paralogues in wheat and barley are not yet very well understood. In addition, the molecular nature of a large number of flowering time QTL in wheat and barley has not yet been unravelled. Genetic interactions of the major flowering time genes expressed in the leaf have been described. However, studies on the genetic-molecular networks controlling meristem development in wheat and barley are in their infancy. Deciphering the genetic control of the shoot apical meristem, inflorescence and flower meristem development in wheat and barley will be crucial for improving yield and adaptation to different environment.

REFERENCES

Adam, H., Ouellet, F., Kane, N. A., Agharbaoui, Z., Major, G., Tominaga, Y., et al. (2007). Overexpression of TaVRN1 in *Arabidopsis* promotes early flowering and alters development. *Plant and Cell Physiology, 48*, 1192–1206.

Alonso-Peral, M. M., Oliver, S. N., Casao, M. C., Greenup, A. A., & Trevaskis, B. (2011). The promoter of the cereal VERNALIZATION1 gene is sufficient for transcriptional induction by prolonged cold. *PLoS ONE, 6*(12), e29456.

Asplunda, L., Hagenblada, J., Matti, W., & Leinob, C. (2010). Re-evaluating the history of the wheat domestication gene NAM-B1 using historical plant material. *Journal of Archaeological Science, 37*(9), 2303–2307.

Badr, A., Müller, K., Schäfer-Pregl, R., El Rabey, H., Effgen, S., Ibrahim, H. H., et al. (2000). On the origin and domestication history of barley (*Hordeum vulgare*). *Molecular Biology and Evolution, 17*(4), 499–510.

Baum, M., Grando, S., Backes, G., Jahoor, A., Sabbagh, A., & Ceccarelli, S. (2003). QTLs for agronomic traits in the Mediterranean environment identified in recombinant inbred lines of the cross 'Arta' x *H. spontaneum* 41-1. *Theoretical and Applied Genetics, 107*(7), 1215–1225.

Beales, J., Turner, A., Griffiths, S., Snape, J. W., & Laurie, D. A. (2007). A pseudo-response regulator is misexpressed in the photoperiod insensitive Ppd-D1a mutant of wheat (*Triticum aestivum* L.). *Theoretical and Applied Genetics, 115*, 721–733.

Bezant, J., Laurie, D., Pratchett, N., Chojecki, J., & Kearsey, M. (1996). Marker regression mapping of QTL controlling flowering time and plant height in a spring barley (*Hordeum vulgare* L.) cross. *Heredity, 77*, 64–73.

Borràs-Gelonch, G., Slafer, G. A., Casas, A. M., van Eeuwijk, F., & Romagosa, I. (2010). Genetic control of pre-heading phases and other traits related to development in a double-haploid barley (*Hordeum vulgare L.*) population. *Field Crops Research, 119*(1), 36–47.

Boyd, W. J. R., Li, C. D., Grime, C. R., Cakir, M., Potipibool, S., Kaveeta, L., et al. (2003). Conventional and molecular genetic analysis of factors contributing to variation in the timing of heading among spring barley (*Hordeum vulgare* L.) genotypes grown over a mild winter growing season. *Australian Journal of Agricultural Research, 54*(12), 1277–1301.

Bullrich, L., Appendino, M. L., Tranquilli, G., Lewis, S., & Dubcovsky, J. (2002). Mapping of a thermosensitive *earliness per se* gene on *Triticum monococcum* chromosome $1A^{m}$. *Theoretical and Applied Genetics, 105*, 585–593.

Bünning, E. (1936). Die endogene Tagesrhythmik als Grundlage der photoperiodischen Reaktion. *Berichte der Deutschen botanischen Gesellschaft, 54*, 590–607.

Campoli, C., Drosse, B., Searle, I., Coupland, G., & von Korff, M. (2012). Functional characterisation of *HvCO1*, the barley (*Hordeum vulgare*) flowering time ortholog of CONSTANS. *The Plant Journal, 69*, 868–880.

Campoli, C., Pankin, A., Drosse, B., Casao, C. M., Davis, S. J., & von Korff, M. (2013). *HvLUX1* is a candidate gene underlying the *early maturity 10* locus in barley: phylogeny, diversity, and interactions with the circadian clock and photoperiodic pathways. *New Phytologist, 199*, 1045–1059.

Campoli, C., Shtaya, M., Davis, S. J., & von Korff, M. (2012). Expression conservation within the circadian clock of a monocot: natural variation at barley *Ppd-H1* affects circadian expression of flowering time genes, but not clock orthologs. *BMC Plant Biology, 12*, 97.

Casao, M. C., Iguarta, E., Karsai, I., Lasa, J. M., Gracia, M. P., & Casas, A. M. (2011). Expression analysis of vernalization and day-length response genes in barley (*Hordeum vulgare* L.) indicates that VRNH2 is a repressor of PPDH2 (HvFT3) under long days. *Journal of Experimental Botany, 62*, 1939–1949.

Casao, M. C., Karsai, I., Igartua, E., Gracia, M. P., Veisz, O., & Casas, A. M. (2011). Adaptation of barley to mild winters: a role for PPD-H2. *BMC Plant Biology, 11*, 164.

Casas, A. M., Djemel, A., Ciudad, F. J., Yahiaoui, S., Ponce, L. J., Contreras-Moreira, B., et al. (2011). *HvFT1 (VrnH3)* drives latitudinal adaptation in Spanish barleys. *Theoretical and Applied Genetics, 122*, 1293–1304.

Chen, A., Baumann, U., Fincher, G. B., & Collins, N. C. (2009a). *Flt-2L*, a locus in barley controlling flowering time, spike density, and plant height. *Functional and Integrative Genomics, 9*(2), 243–254.

Chen, A., & Dubcovsky, J. (2012). Wheat TILLING mutants show that the vernalization gene VRN1 down-regulates the flowering repressor VRN2 in leaves but is not essential for flowering. *PLoS Genetics, 8*(12), e1003134.

Chen, A., Reinheimer, J. L., Brûlé-Babel, A., Baumann, U., Fincher, G. B., & Collins, N. C. (2009b). Genes and traits associated with barley 2H and 5H chromosome regions controlling sensitivity of reproductive tissues to frost. *Theoretical and Applied Genetics, 118*(8), 1465–1476.

Chu, C. G., Tan, C. T., Yu, G. T., Zhong, S., Xu, S. S., & Yan, L. (2011). A novel retrotransposon inserted in the dominant *Vrn-B1* allele confers spring growth habit in tetraploid wheat (*Triticum turgidum* L.). *G3: Genes | Genomes | Genetics, 1*, 637–645.

Cockram, J., Howells, R. M., & O'Sullivan, D. M. (2010). Segmental chromosomal duplications harbouring group IV CONSTANS-like genes in cereals. *Genome, 53*, 231–240.

Cockram, J., Jones, H., Leigh, F. J., O'Sullivan, D., Powell, W., Laurie, D. A., et al. (2007). Control of flowering time in temperate cereals: genes, domestication, and sustainable productivity. *Journal of Experimental Botany, 58*, 1234–1244.

Comadran, J., Kilian, B., Russell, J., Ramsay, L., Stein, N., Ganal, M., et al. (2012). Natural variation in a homolog of *Antirrhinum CENTRORADIALIS* contributed to spring growth habit and environmental adaptation in cultivated barley. *Nature Genetics, 44*, 1388–1392.

Corbesier, L., Vincent, C., Jang, S., Fornara, F., Fan, Q., Searle, I., et al. (2007). FT protein movement contributes to long-distance signaling in floral induction of *Arabidopsis*. *Science, 316*, 1030–1033.

Cuesta-Marcos, A., Casas, A., Yahiaoui, S., Gracia, M., Lasa, J., & Igartua, E. (2008). Joint analysis for heading date QTL in small interconnected barley populations. *Molecular Breeding, 21*(3), 383–399.

Cuesta-Marcos, A., Igartua, E., Ciudad, F., Codesal, P., Russell, J., Molina-Cano, J., et al. (2008). Heading date QTL in a spring × winter barley cross evaluated in Mediterranean environments. *Molecular Breeding, 21*(4), 455–471.

Danyluk, J., Kane, N. A., Breton, G., Limin, A. E., Fowler, D. B., & Sarhan, F. (2003). TaVRT-1, a putative transcription factor associated with vegetative to reproductive transition in cereals. *Plant Physiology, 132*, 1849–1860.

Dhillon, T., Pearce, S., Stockinger, E., Distelfeld, A., Li, C., Knox, A. K., et al. (2010). Regulation of freezing tolerance and flowering in temperate cereals: the *VRN-1* connection. *Plant Physiology*, *153*, 1846–1858.
Diallo, A. O., Ali-Benali, M. A., Badawi, M., Houde, M., & Sarhan, F. (2012). Expression of vernalization responsive genes in wheat is associated with histone H3 trimethylation. *Molecular Genetics and Genomics*, *287*, 575–590.
Díaz, A., Zikhali, M., Turner, A. S., Isaac, P., & Laurie, D. A. (2012). Copy number variation affecting the *Photoperiod-B1* and *Vernalization-A1* genes is associated with altered flowering time in wheat (*Triticum aestivum*). *PLoS One*, 7(3), e33234.
Distelfeld, A., & Dubcovsky, J. (2010). Characterization of the maintained vegetative phase deletions from diploid wheat and their effect on VRN2 and FT transcript levels. *Molecular Genetics and Genomics*, *283*, 223–232.
Distelfeld, A., Li, C., & Dubcovsky, J. (2009). Regulation of flowering in temperate cereals. *Current Opinion in Plant Biology*, *12*, 178–184.
Dixon, L. E., Knox, K., Kozma-Bogmar, L., Southern, M. M., Pokhilko, A., & Millar, A. J. (2011). Temporal repression of core circadian genes is mediated through EARLY FLOWERING 3 in *Arabidopsis*. *Current Biology*, *21*, 120–125.
Dubcovsky, J., & Dvorak, J. (2007). Genome plasticity a key factor in the success of polyploid wheat under domestication. *Science*, *316*, 1862–1866.
Dubcovsky, J., Lijavetzky, D., Appendino, L., Tranquilli, G., & Dvorak, J. D. (1998). Comparative RFLP mapping of *Triticum monococcum* genes controlling vernalization requirement. *Theoretical and Applied Genetics*, *97*, 968–975.
Dunford, R. P., Griffiths, S., Christodoulou, V., & Laurie, D. A. (2005). Characterisation of a barley (*Hordeum vulgare* L.) homologue of the *Arabidopsis* flowering time regulator GIGANTEA. *Theoretical and Applied Genetics*, *110*, 925–931.
Ehrenreich, I. M., Hanzawa, Y., Chou, L., Roe, J. L., Kover, P. X., & Purugganan, M. D. (2009). Candidate gene association mapping of *Arabidopsis* flowering time. *Genetics*, *183*, 325–335.
Eleuch, L., Abderrazek, J., Grando, S., Ceccarelli, S., von Korff, M., Hajer, A., et al. (2008). Genetic diversity and association analysis for salinity tolerance, heading date and plant height of barley germplasm using SSR markers. *Journal of Integrative Plant Biology*, *50*, 1004–1014.
Enomoto, N. (1929). On the physiological difference between the spring and winter types in wheat and barley. *Journal of the Imperial Agricultural Experimental Station*, *1*, 107–138 (in Japanese with English summary).
Faure, S., Higgins, J., Turner, A., & Laurie, D. A. (2007). The FLOWERING LOCUS T-like gene family in barley (*Hordeum vulgare*). *Genetics*, *176*, 599–609.
Faure, S., Turner, A. S., Gruszka, D., Christodoulou, V., Davis, S. J., von Korff, M., et al. (2012). Mutation at the circadian clock gene EARLY MATURITY 8 adapts domesticated barley (*Hordeum vulgare*) to short growing seasons. *Proceedings of the National Academy of Sciences of the United States of America*, *109*, 8328–8333.
Fowler, S., Lee, K., Onouchi, H., Samach, A., Richardson, K., Morris, B., et al. (1999). GIGANTEA: a circadian clock-controlled gene that regulates photoperiodic flowering in *Arabidopsis* and encodes a protein with several possible membrane-spanning domains. *The EMBO Journal*, *18*, 4679–4688.
Fu, Y. B., Peterson, G. W., Richards, K. W., Somers, D., DePauw, R. M., & Clarke, J. M. (2005). Allelic reduction and genetic shift in the Canadian hard red spring wheat germplasm released from 1845 to 2004. *Theoretical and Applied Genetics*, *110*, 1505–1516.
Fujita, M., Kawada, N., & Tahir, M. (1992). Relationship between cold resistance, heading traits and ear primordia development of wheat cultivars. *Euphytica*, *64*, 123–130.
Ghiglione, H. O., Gonzalez, F. G., Serrago, R., Maldonado, S. B., Chilcott, C., Cura, J. A., et al. (2008). Autophagy regulated by day length determines the number of fertile florets in wheat. *The Plant Journal*, *55*, 1010–1024.

Goncharov, N. P. (1998). Genetic resources of wheat related species: the *Vrn* genes controlling growth habit (spring *vs.* winter). *Euphytica, 100*, 371–376.
Goncharov, N. P., & Chikida, N. N. (1995). Genetics of growth habit in *Aegilops squarrosa* L. *Russian Journal of Genetics, 31*, 343–346.
González, F. G., Miralles, D. J., & Slafer, G. A. (2011). Wheat floret survival as related to pre-anthesis spike growth. *Journal of Experimental Botany, 62*(14), 4889–4901.
González, F. G., Slafer, G. A., & Miralles, D. J. (2003). Floret development and spike growth as affected by photoperiod during stem elongation in wheat. *Field Crops Research, 81*(1), 29–38.
Griffiths, S., Dunford, R. P., Coupland, G., & Laurie, D. A. (2003). The evolution of *CONSTANS*-like gene families in barley, rice, and *Arabidopsis*. *Plant Physiology, 131*, 1855–1867.
Griffiths, S., Simmonds, J., Leverington, M., Wang, Y., Fish, L., Sayers, L., et al. (2009). Meta-QTL analysis of the genetic control of ear emergence in elite European winter wheat germplasm. *Theoretical and Applied Genetics, 119*, 383–395.
Habte, E., Müller, L. M., Shtaya, M., Davis, S. J., & von Korff, M. (2014). Osmotic stress at the barley root affects expression of circadian clock genes in the shoot. *Plant, Cell and Environment*. http://dx.doi.org/10.1111/pce.12242.
Hanocq, E., Laperche, A., Jaminon, O., Laine, A. L., & Le Gouis, J. (2007). Most significant genome regions involved in the control of earliness traits in bread wheat, as revealed by QTL metaanalysis. *Theoretical and Applied Genetics, 114*, 569–584.
Hanocq, E., Niarquin, M., Heumez, E., Rousset, M., & Le Gouis, J. (2004). Detection and mapping of QTL for earliness components in a bread wheat recombinant inbred lines population. *Theoretical and Applied Genetics, 110*, 106–115.
Hayama, R., Yokoi, A., Tamaki, S., Yano, M., & Shimamoto, K. (2003). Adaptation of photoperiodic control pathways produces short-day flowering in rice. *Nature, 422*, 719–722.
Hemming, M. N., Fieg, S., Peacock, W. J., Dennis, E. S., & Trevaskis, B. (2009). Regions associated with repression of the barley (*Hordeum vulgare*) VERNALIZATION1 gene are not required for cold induction. *Molecular Genetics and Genomics, 282*, 107–117.
Hemming, M. N., Peacock, W. J., Dennis, E. S., & Trevaskis, B. (2008). Low-temperature and daylength cues are integrated to regulate FLOWERING LOCUS T in barley. *Plant Physiology, 147*, 355–366.
Herrero, E., Kolmos, E., Bujdoso, N., Yuan, Y., Wang, M., Berns, M. C., et al. (2012). EARLY FLOWERING4 recruitment of EARLY FLOWERING3 in the nucleus sustains the *Arabidopsis* circadian clock. *Plant Cell, 24*, 428–443.
Higgins, J. A., Bailey, P. C., & Laurie, D. A. (2010). Comparative genomics of flowering time pathways using *Brachypodium distachyon* as a model for the temperate grasses. *PLoS ONE, 5*, e10065.
Huang, W., Pérez-García, P., Pokhilko, A., Millar, A. J., Antoshechkin, I., Riechmann, J. L., et al. (2012). Mapping the core of the *Arabidopsis* circadian clock defines the network structure of the oscillator. *Science, 336*, 75–79.
Ivandic, V., Hackett, C. A., Nevo, E., Keith, R., Thomas, W. T., & Forster, B. P. (2002). Analysis of simple sequence repeats (SSRs) in wild barley from the fertile crescent: associations with ecology, geography and flowering time. *Plant Molecular Biology, 48*(5–6), 511–527.
Iwaki, K., Haruna, S., Niwa, T., & Kato, K. (2001). Adaptation and ecological differentiation in wheat with special reference to geographical variation of growth habit and *Vrn* genotype. *Plant Breeding, 120*, 107–114.
Jang, S., Marchal, V., Pahigrahi, K. C. S., Wenkel, S., Soppe, W., Deng, X. W., et al. (2008). *Arabidopsis* COP1 shapes the temporal pattern of CO accumulation conferring a photoperiodic flowering response. *The EMBO Journal, 27*, 1277–1288.
Jia, Q., Zhang, J., Westcott, S., Zhang, X., Bellgard, M., Lance, R., et al. (2009). *GA-20 oxidase* as a candidate for the semidwarf gene sdw1/denso in barley. *Functional and Integrative Genomics, 9*, 255–262.

Jones, H., Leigh, F. J., Mackay, I., Bower, M. A., Smith, L. M. J., Charles, M. P., et al. (2008). Population-based resequencing reveals that the flowering time adaptation of cultivated barley originated east of the fertile crescent. *Molecular Biology and Evolution, 25*, 2211–2219.

Kamran, A., Iqbal, M., Navabi, A., Randhawa, H. S., Pozniak, C., & Spaner, D. (2013). Earliness per QTLs and their interaction with photoperiod insensitive allele *Ppd-D1a* in Cutler × AC Barrie spring wheat population. *Theoretical and Applied Genetics, 126*(8), 1965–1976.

Karsai, I., Szücs, P., Mészáros, K., Filichkina, T., Hayes, P. M., Skinner, J. S., et al. (2005). The Vrn-H2 locus is a major determinant of flowering time in a facultative winter growth habit barley (*Hordeum vulgare* L.) mapping population. *Theoretical and Applied Genetics, 110*, 1458–1466.

Khan, A. R., Enjalbert, J., Marsollier, A. C., Rousselet, A., Goldringer, I., & Vitte, C. (2013). Vernalization treatment induces site-specific DNA hypermethylation at the VERNALIZATION-A1 (VRN-A1) locus in hexaploid winter wheat. *BMC Plant Biology, 13*, 209.

Kikuchi, R., Kawahigashi, H., Ando, T., Tonooka, T., & Handa, H. (2009). Molecular and functional characterization of PEBP genes in barley reveal the diversification of their roles in flowering. *Plant Physiology, 149*, 1341–1353.

Kikuchi, R., Kawahigashi, H., Oshima, M., Ando, T., & Handa, H. (2011). The differential expression of HvCO9, a member of the CONSTANS-like gene family, contributes to the control of flowering under short-day conditions in barley. *Journal of Experimental Botany, 63*(2), 773–784.

Kinjo, H., Shitsukawa, N., Takumi, S., & Murai, K. (2012). Diversification of three APETALA1/FRUITFULL-like genes in wheat. *Molecular Genetics and Genomics, 287*(4), 283–294.

Kuchel, H., Hollamby, G., Langridge, P., Williams, K., & Jefferies, S. P. (2006). Identification of genetic loci associated with ear-emergence in bread wheat. *Theoretical and Applied Genetics, 113*, 1103–1112.

Laurie, D. A., Pratchett, N., Snape, J. W., & Bezant, J. H. (1995). RFLP mapping of five major genes and eight quantitative trait loci controlling flowering time in a winter × spring barley (*Hordeum vulgare* L.) cross. *Genome, 38*(3), 575–585.

Law, C. N., Sutka, J., & Worland, A. J. (1978). A genetic study of day length response in wheat. *Heredity, 41*, 575–585.

Lewis, S., Faricelli, M. E., Appendino, M. L., Valarik, M., & Dubcovsky, J. (2008). The chromosome region including the earliness per se locus *Eps-A^{m}1* affects the duration of early developmental phases and spikelet number in diploid wheat. *Journal of Experimental Botany, 59*, 3595–3607.

Li, C., & Dubcovski, J. (2008). Wheat FT protein regulates VRN1 transcription through interactions with FDL2. *The Plant Journal, 55*, 543–554.

Limin, A. E., & Fowler, D. B. (2006). Low-temperature tolerance and genetic potential in wheat (*Triticum aestivum* L.): response to photoperiod, vernalization, and plant development. *Planta, 224*, 360–366.

Loukoianov, A., Yan, L., Blechl, A., Sanchez, A., & Dubcovsky, J. (2005). Regulation of VRN-1 vernalization genes in normal and transgenic polyploid wheat. *Plant Physiology, 138*, 2364–2373.

Lundqvist, U. (2009). Eighty years of Scandinavian barley mutation genetics and breeding. In Q.Y. Shu (Ed.), *Induced mutations in the genomics era* (pp. 39–43). Rome: FAO.

Marquez-Cedillo, L. A., Hayes, P. M., Kleinhofs, A., Legge, W. G., Rossnagel, B. G., Sato, K., et al. (2001). QTL analysis of agronomic traits in barley based on the doubled haploid progeny of two elite North American varieties representing different germplasm groups. TNABGM Project *Theoretical and Applied Genetics, 103*, 625–637.

Miralles, D. J., & Richards, R. A. (2000). Responses of leaf and tiller emergence and primordium initiation in wheat and barley to interchanged photoperiod. *Annals of Botany, 85*, 655–663.

Mizuno, N., Nitta, M., Sato, K., & Nasuda, S. (2012). A wheat homologue of PHYTOCLOCK 1 is a candidate gene conferring the early heading phenotype to einkorn wheat. *Genes & Genetic Systems, 87*(6), 357–367.

Muñoz-Amatriaín, M., Moscou, M. J., Bhat, P. R., Svensson, J. T., Bartoš, J., Suchánková, P., et al. (2011). An improved consensus linkage map of barley based on flow-sorted chromosomes and single nucleotide polymorphism markers. *Plant Genome, 4*(3), 238–249.

Nemoto, Y., Kisaka, M., Fuse, T., Yano, M., & Ogihara, Y. (2003). Characterization and functional analysis of three wheat genes with homology to the CONSTANS flowering time gene in transgenic rice. *The Plant Journal, 36*, 82–93.

Nishida, H., Yoshida, T., Kawakami, K., Fujita, M., Long, B., Akashi, Y., et al. (2013). Structural variation in the 5' upstream region of the photoperiod-insensitive alleles Ppd-A1a and Ppd-B1a identified in hexaploid wheat (*Triticum aestivum* L.), and their effect on heading time. *Molecular Breeding, 31*, 27–37.

Nitcher, R., Distelfeld, A., Tan, C., Yan, L., & Dubcovsky, J. (2013). Increased copy number at the HvFT1 locus is associated with accelerated flowering time in barley. *Molecular Genetics and Genomics, 288*, 261–275.

Oliver, S. N., Deng, W., Casao, M. C., & Trevaskis, B. (2013). Low temperatures induce rapid changes in chromatin state and transcript levels of the cereal VERNALIZATION1 gene. *Journal of Experimental Botany, 13*, 2413–2422.

Oliver, S. N., Finnegan, E. J., Dennis, E. S., Peacock, W. J., & Trevaskis, B. (2009). Vernalization-induced flowering in cereals is associated with changes in histone methylation at the VERNALIZATION1 gene. *Proceedings of the National Academy of Sciences of the United States of America, 106*, 8386–8391.

Pillen, K., Zacharias, A., & Leon, J. (2003). Advanced backcross QTL analysis in barley (*Hordeum vulgare* L.). *Theoretical and Applied Genetics, 107*, 340–352.

Pillen, K., Zacharias, A., & Leon, J. (2004). Comparative AB-QTL analysis in barley using a single exotic donor of *Hordeum vulgare* ssp. *spontaneum*. *Theoretical and Applied Genetics, 108*(8), 1591–1601.

Pokhilko, A., Fernández, A. P., Edwards, K. D., Southern, M. M., Halliday, K. J., & Millar, A. J. (2012). The clock gene circuit in *Arabidopsis* includes a repressilator with additional feedback loops. *Molecular System Biology, 8*, 574.

Preston, J. C., & Kellogg, E. A. (2007). Conservation and divergence of APETALA1/FRUITFULL-like gene function in grasses: evidence from gene expression analyses. *The Plant Journal, 52*, 69–81.

Quail, P. H. (2002). Photosensory perception and signalling in plant cells: new paradigms? *Current Opinion in Plant Biology, 14*, 180–188.

Reinheimer, J. L., Barr, A. R., & Eglinton, J. K. (2004). QTL mapping of chromosomal regions conferring reproductive frost tolerance in barley (*Hordeum vulgare* L.). *Theoretical and Applied Genetics, 14*, 1267–1274.

Reynolds, M., Foulkes, M. J., Slafer, G. A., Berry, P., Parry, M. A. J., Snape, J. W., et al. (2009). Raising yield potential in wheat. *Journal of Experimental Botany, 60*(7), 1899–1918.

Rollins, J. A., Drosse, B., Mulki, M. A., Grando, S., Baum, M., Singh, M., et al. (2013). Variation at the vernalisation genes *Vrn-H1* and *Vrn-H2* determines growth and yield stability in barley (*Hordeum vulgare*) grown under dryland conditions in Syria. *Theoretical and Applied Genetics, 126*, 2803–2824.

Saisho, D., Ishii, M., Hori, K., & Sato, K. (2011). Natural variation of barley vernalization requirements: implication of quantitative variation of winter growth habit as an adaptive trait in East Asia. *Plant and Cell Physiology, 52*, 724–727.

Samach, A., Onouchi, H., Gold, S. E., Ditta, G. S., Schwarz-Sommer, Z., Yanogsky, M. F., et al. (2000). Distinct roles of CONSTANS target genes in reproductive development of *Arabidopsis*. *Science*, *288*, 1613–1616.

Sameri, M., Pourkheirandish, M., Chen, G., Takuji-Tonooka, T., & Komatsuda, T. (2011). Detection of photoperiod responsive and non-responsive flowering time QTL in barley. *Breeding Science*, *61*, 183–188.

Sasani, S., Hemming, M. N., Oliver, S. N., Greenup, A., Tavakkol-Afshari, R., Mahfoozi, S., et al. (2009). The influence of vernalization and daylength on expression of flowering-time genes in the shoot apex and leaves of barley (*Hordeum vulgare*). *Journal of Experimental Botany*, *60*, 2169–2178.

Scarth, R., & Law, C. N. (1983). The location of photoperiodic gene (*Ppd2*) and an additional genetic factor for ear emergence time on chromosome 2B of wheat. *Heredity*, *51*, 607–619.

Schmitz, J., Franzen, R., Ngyuen, T. H., Garcia-Maroto, F., Pozzi, C., Salamini, F., et al. (2000). Cloning, mapping and expression analysis of barley MADS-box genes. *Plant Molecular Biology*, *42*, 899–913.

Shah, M. M., Gill, K. S., Yen, Y., Kaeppler, S. M., & Ariyarathne, H. M. (1999). Molecular mapping of loci for agronomic traits on chromosome 3A of bread wheat. *Crop Science*, *39*, 1728–1732.

Shaw, L. M., Turner, A. S., Herry, L., Griffiths, S., & Laurie, D. A. (2013). Mutant alleles of *Photoperiod-1* in wheat (*Triticum aestivum* L.) that confer a late flowering phenotype in long days. *PLoS ONE*, *8*(11), e79459.

Shaw, L. M., Turner, A. S., & Laurie, D. A. (2012). The impact of photoperiod insensitive *Ppd-1a* mutations on the photoperiod pathway across the three genomes of hexaploid wheat (*Triticum aestivum*). *The Plant Journal*, *71*, 71–84.

Shimada, S., Ogawa, T., Kitagawa, S., Suzuki, T., Ikari, C., Shitsukawa, N., et al. (2009). A genetic network of flowering-time genes in wheat leaves, in which an *APETALA1/FRUITFULL*-like gene, VRN1, is upstream of FLOWERING LOCUS T. *The Plant Journal*, *58*, 668–681.

Shindo, C., Tsujimoto, H., & Sasakuma, T. (2003). Segregation analysis of heading traits in hexaploid wheat utilizing recombinant inbred lines. *Heredity*, *90*, 56–63.

Shin-Young, H., Sangmin, L., Pil, J. S., Moon-Sik, Y., & Chung-Mo, P. (2010). Identification and molecular characterization of a *Brachypodium distachyon GIGANTEA* gene: functional conservation in monocot and dicot plants. *Plant Molecular Biology*, *72*, 485–497.

Shitsukawa, N., Ikari, C., Shimada, S., Kitagawa, S., Sakamoto, K., Saito, H., et al. (2007). The einkorn wheat (*Triticum monococcum*) mutant, *maintained vegetative phase*, is caused by a deletion in the VRN1 gene. *Genes & Genetic Systems*, *82*, 167–170.

Slafer, G. A. (2003). Genetic basis of yield as viewed from a crop physiologists perspective. *Annals of Applied Biology*, *142*, 117–128.

Slafer, G. A., Abeledo, L. G., Miralles, D. J., Gonzalez, F. G., & Whitechurch, E. M. (2001). Photoperiod sensitivity during stem elongation as an avenue to raise potential yield in wheat. *Euphytica*, *119*, 191–197.

Stracke, S., & Börner, A. (1998). Molecular mapping of the photoperiod response gene *eam7* in barley. *Theoretical and Applied Genetics*, *97*, 797–800.

Szűcs, P., Karsai, I., von Zitzewitz, J., Mészáros, K., Cooper, L. L., Gu, Y. Q., et al. (2006). Positional relationships between photoperiod response QTL and photoreceptor and vernalization genes in barley. *Theoretical and Applied Genetics*, *112*, 1277–1285.

Szűcs, P., Skinner, J. S., Karsai, I., Cuesta-Marcos, A., Haggard, K. G., Corey, A. E., et al. (2007). Validation of the VRN-H2/VRN-H1 epistatic model in barley reveals that intron length variation in VRN-H1 may account for a continuum of vernalization sensitivity. *Molecular Genetics and Genomics*, *277*, 249–261.

Takata, N., Saito, S., Saito, C. T., & Uemura, M. (2010). Phylogenetic footprint of the plant clock system in angiosperms: evolutionary processes of pseudo-response regulators. *BMC Evolutionary Biology*, *10*, 126.

Tamaki, S., Matsuo, S., Wong, H. L., Yokoi, S., & Shimamoto, K. (2007). Hd3a Protein is a mobile flowering signal in Rice. *Science*, *316*, 1033–1036.

Teulat, B., Merah, O., Souyris, I., & This, D. (2001). QTLs for agronomic traits from a Mediterranean barley progeny grown in several environments. *Theoretical and Applied Genetics*, *103*, 774–787.

Thomashow, M. F. (1990). Molecular genetics of cold acclimation in higher plants. *Advances in Genetics*, *28*, 99–131.

Thomashow, M. F. (1999). Plant cold acclimation: freezing tolerance genes and regulatory mechanisms. *Annual Review of Plant Physiology*, *50*, 571–599.

Trevaskis, B., Bagnall, D. J., Ellis, M. H., Peacock, W. J., & Dennis, E. S. (2003). MADS box genes control vernalization-induced flowering in cereals. *Proceedings of the National Academy of Sciences of the United States of America*, *100*, 13099–13104.

Trevaskis, B., Hemming, M. N., Peacock, W. J., & Dennis, E. S. (2006). HvVRN2 responds to daylength, whereas HvVRN1 is regulated by vernalization and developmental status. *Plant Physiology*, *140*, 1397–1405.

Turner, A., Beales, J., Faure, S., Dunford, R. P., & Laurie, D. A. (2005). The Pseudo-response regulator *Ppd-H1* provides adaptation to photoperiod in barley. *Science*, *310*, 1031–1034.

Valverde, F., Mouradov, A., Soppe, W., Ravenscroft, D., Samach, A., & Coupland, G. (2004). Photoreceptor regulation of CONSTANS protein in photoperiodic flowering. *Science*, *303*(5660), 1003–1006.

von Korff, M., Grando, S., Del Greco, A., This, D., Baum, M., & Ceccarelli, S. (2008). Quantitative trait loci associated with adaptation to Mediterranean dryland conditions in barley. *Theoretical and Applied Genetics*, *117*(5), 653–669.

von Korff, M., Léon, J., & Pillen, K. (2010). Detection of epistatic interactions between exotic alleles introgressed from wild barley (*H. vulgare* ssp. *spontaneum*). *Theoretical and Applied Genetics*, *121*(8), 1455–1464.

von Korff, M., Wang, H., Leon, J., & Pillen, K. (2006). AB-QTL analysis in spring barley: II. Detection of favourable exotic alleles for agronomic traits introgressed from wild barley (*H. vulgare* ssp. *spontaneum*). *Theoretical and Applied Genetics*, *112*(7), 1221–1231.

von Zitzewitz, J., Szűcs, P., Dubcovsky, J., Yan, L. L., Francia, E., Pecchioni, N., et al. (2005). Molecular and structural characterization of barley vernalization genes. *Plant Molecular Biology*, *59*, 449–467.

Wang, G., Schmalenbach, I., von Korff, M., Léon, J., Kilian, B., Rode, J., et al. (2010). Association of barley photoperiod and vernalization genes with QTLs for flowering time and agronomic traits in a DH-population and a set of wild barley introgression lines. *Theoretical and Applied Genetics*, *120*(8), 1559–1574.

Welsh, J. J., Keim, D. L., Pirasteh, B., & Richards, R. D. (1973). Genomic control of photoperiod response in wheat. In E. S. Sears, & L. M. S. Sears (Eds.), *Proceedings of the international wheat genetic symposium* (pp. 879–884). Missouri, USA.

Weltzien, E. (1988). Evaluation of barley (*Hordeum vulgare* L) landrace populations originating from different growing regions in the Near East. *Plant Breeding*, *101*, 95–106.

Weltzien, E. (1989). Differentiation among barley landrace populations from the Near East. *Euphytica*, *43*, 29–39.

Wigge, P. A., Kim, M. C., Jaeger, K. E., Busch, W., Schmid, M., Lohmann, J. U., et al. (2005). Integration of spatial and temporal information during floral induction in *Arabidopsis*. *Science*, *309*, 1056–1059.

Wilhelm, E. P., Turner, A. S., & Laurie, D. A. (2009). Photoperiod insensitive *Ppd-A1a* mutations in tetraploid wheat (*Triticum durum* Desf.). *Theoretical and Applied Genetics*, *118*, 285–294.

Worland, A. J., Börner, A., Korzun, V., Li, W. M., Petrovíc, S., & Sayers, E. J. (1998). The influence of photoperiod genes on the adaptability of European winter wheats. *Euphytica, 100*, 385–394.
Xue, W., Xing, Y., Weng, X., Zhao, Y., Tang, W., Wang, L., et al. (2008). Natural variation in *Ghd7* is an important regulator of heading date and yield potential in rice. *Nature Genetics, 6*, 761–767.
Yan, L., Fu, D., Li, C., Blechl, A., Tranquilli, M., Bonafede, A., et al. (2006). The wheat and barley vernalization gene *VRN3* is an orthologue of FT. *Proceedings of the National Academy of Sciences of the United States of America, 103*, 19581–19586.
Yan, L., Loukoianov, A., Blechl, A., Tranquilli, G., Ramakrishna, W., SanMiguel, P., et al. (2004). The wheat *VRN2* gene is a flowering repressor down-regulated by vernalization. *Science, 303*, 1640–1644.
Yan, L., Loukoianov, A., Tranquilli, G., Helguera, M., Fahima, T., & Dubcovsky, J. (2003). Positional cloning of the wheat vernalization gene VRN1. *Proceedings of the National Academy of Sciences of the United States of America, 100*, 6263–6268.
Yano, M., Katayose, Y., Ashikari, M., Yamanouchi, U., Monna, L., Fuse, T., et al. (2000). Hd1, a major photoperiod sensitivity quantitative trait locus in rice, is closely related to the *Arabidopsis* flowering time gene CONSTANS. *Plant Cell, 12*, 2473–2483.
Zakhrabekova, S., Gough, S. P., Braumann, I., Müller, A. H., Lundqvist, J., Ahmann, K., et al. (2012). Induced mutations in circadian clock regulator Mat-a facilitated short-season adaptation and range extension in cultivated barley. *Proceedings of the National Academy of Sciences of the United States of America, 109*, 4326–4331.
Zhang, J., Wu, S., Yang, J., Liu, H., & Zhou, Y. (2012). A single nucleotide polymorphism at the Vrn-D1 promoter region in common wheat is associated with vernalization response. *Theoretical and Applied Genetics, 125*(8), 1697–1704.
Zhao, X. Y., Liu, M. S., Li, J. R., Guan, C. M., & Zhang, X. S. (2005). The wheat TaGI1, involved in photoperiodic flowering, encodes an *Arabidopsis* GI ortholog. *Plant Molecular Biology, 58*, 53–64.

CHAPTER SIX

Genetic Control of *Arabidopsis* Flower Development

Diarmuid S. O'Maoileidigh*, Emmanuelle Graciet*,† and Frank Wellmer*,1

*Smurfit Institute of Genetics, Trinity College Dublin, Dublin, Ireland
†Department of Biology, National University of Ireland, Maynooth, Ireland
[1]Corresponding author: e-mail address: wellmerf@tcd.ie

Contents

Abstract

The formation of flowers in the model plant *Arabidopsis thaliana* is one of the main systems to study the genetic mechanisms underlying the growth and development of angiosperms. Research conducted over the past 30 years led to the discovery of key regulatory genes for many of the processes that take place during flower morphogenesis. Although it is currently not always known how the products of these genes function at the molecular level, recent work has led, in many cases, to substantial progress in our understanding of their activities. In fact, knowledge obtained with *Arabidopsis* is now so plentiful that it can be used as a basis for a comparative analysis of flower development to unravel the genetic mechanisms underlying flower evolution and to explain the astounding variations in flower architecture that can be found among the ~250,000 extant angiosperm species. In this chapter, we summarise the main findings and concepts relating to *Arabidopsis* flower development. We describe how technological progress has led to big steps forward in our understanding of how flowers are formed from a small number of undifferentiated stem cells, and we outline current gaps in our knowledge and discuss possible strategies that could be used to fill them in the future.

Advances in Botanical Research, Volume 72
ISSN 0065-2296
http://dx.doi.org/10.1016/B978-0-12-417162-6.00006-7

6.1 INTRODUCTION

The formation of flowers has been an area of interest for botanists and naturalists long before twentieth century plant biology provided the experimental approaches to begin to decipher the genetic and molecular basis of this important developmental process. This initial interest was probably sparked not only by the fact that flowers contain the reproductive organs of angiosperms and thus are of exceptional importance for the life of this largest group of land plants, but also by the aesthetic beauty of many flowers and the astounding variations in flower size, colour, scent and architecture. Further interest in flower development stems from the fact that flowers are of great agricultural and economical importance because they give rise to much of the food humans and their livestock consume, and elucidating how they form can thus have a major impact on the development of strategies for increased agricultural production and a more sustainable world food supply.

Over the past three decades, the formation of flowers has been one of the favourite model systems of biologists to unravel the mechanisms underlying plant growth and development. Flower morphogenesis has been studied extensively using both genetic and molecular approaches, leading to impressive progress in our understanding of how flowers form. This work has been largely done with a small number of angiosperms such as the mouse-ear cress *Arabidopsis thaliana*, snapdragon (*Antirrhinum majus*) and petunia (*Petunia hybrida*). Of these species, *Arabidopsis* was of particular importance because it is amenable to a wide range of experimental approaches and was almost universally adopted as the main model for plant research throughout most of the 1990s and the first decade of the third millennium, leading to an unparalleled number of publically available resources and reagents. A search of the *PubMed.gov* literature database for '*Arabidopsis* flower development' results in almost 3500 hits, demonstrating the wealth of information that is now available about this key developmental process in the model plant. In fact, our knowledge about *Arabidopsis* flowers is now so plentiful that evolutionary biologists have begun in recent years to use it for studying the mechanisms underlying floral evolution through comparative analyses with other angiosperm species. A key aim of this work is to explain what Charles Darwin called an 'abominable mystery', i.e. the origin of flowers and the apparent rapid diversification of angiosperms in the Cretaceous as derived from the fossil record (Friedman, 2009).

In this chapter, we will provide an overview of what is known about flower development in *Arabidopsis*. Given the wealth of available information, such a summary must inevitably be widely incomplete. We will therefore focus mainly on key findings and concepts and emphasise recent progress in the field. We will also point out knowledge gaps that remain to be filled and, whenever pertinent, outline strategies that might be suitable to address these open questions. We will focus especially on the early steps of flower development because they have been particularly well studied. Specifically, we will discuss the specification and termination of floral meristems (FMs), FM patterning and floral organ specification and development.

6.2 STAGES OF FLOWER DEVELOPMENT AND FLOWER MORPHOLOGY

Flowering in *Arabidopsis* is induced in response to both endogenous signals and environmental cues to allow a precise timing of reproduction under favourable conditions and to ensure reproductive success. Several genetic pathways have been identified that mediate the responses of the plant to changes in day length (reviewed in Chapters 1 and 4) and ambient temperature; to prolonged cold treatment (reviewed in Chapter 2); and to its metabolic and developmental state (reviewed in Chapter 3). These pathways, which cannot be strictly separated because they show extensive cross talk (Srikanth & Schmid, 2011), converge on a small number of genes, the so-called floral pathway integrators (Simpson & Dean, 2002), whose expression state is crucial for the control of flowering time.

Morphologically, the switch from the vegetative to the reproductive phase of plant development is characterised by the shoot apical meristem transitioning from making leaves with associated secondary axillary meristems to an inflorescence meristem, which produces floral primordia. During this transition, the apical meristem forms a small number of cauline leaves, which are borne by the stem, and subtend associated axillary meristems that give rise to branches. Once the switch to the reproductive phase of development is complete, flowers are continuously produced on the flanks of the inflorescence meristem and, in *Arabidopsis*, are initiated in a spiral pattern and sequentially (under standard growth conditions, approximately two flowers are initiated each day; (Smyth, Bowman, & Meyerowitz, 1990)), so that all flowers in an inflorescence are of a different developmental age (Figure 6.1(A)).

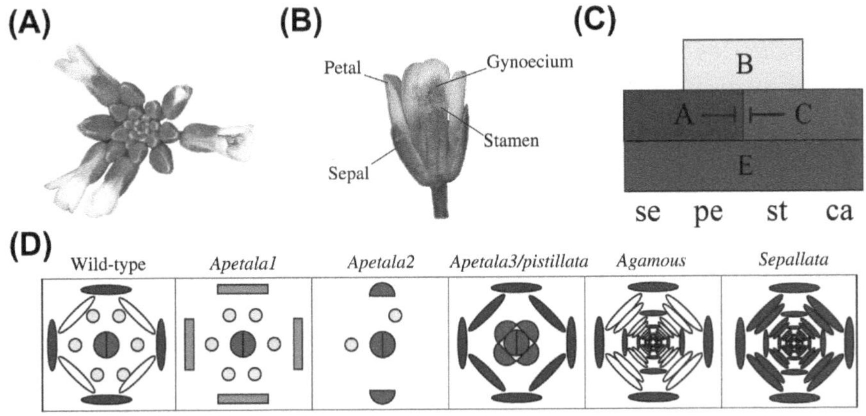

Figure 6.1 Overview of flower development and the ABCE model. (A) Top view on an *Arabidopsis* wild-type inflorescence. Because floral primordia are sequentially initiated from the inflorescence meristem in a spiral pattern, no two flowers in an inflorescence are at the exact same stage of development (Smyth et al., 1990). (B) Four distinct organ types are present on *Arabidopsis* flowers. These organ types are arranged in concentric circles termed whorls. Typically, four sepals are present in the outermost whorl (the first whorl), four petals are present in the second whorl, six stamens are present in the third whorl and two fused carpels, which form the gynoecium that houses the ovules, are present in the fourth whorl (Smyth et al., 1990). (C) The ABCE model states that the overlapping activities of four classes of homeotic genes specify the four types of floral organs. A and E class genes are required for sepal identity; A, B and E class genes are required for petal identity; B, C and E class genes specify stamens; and C and E class genes specify carpels (Coen & Meyerowitz, 1991; Krizek & Fletcher, 2005). (D) Schematic of the phenotypes of the floral homeotic mutants. In strong *ap1* alleles, sepals are transformed into bract-like organs while petals are mostly absent (Irish & Sussex, 1990). In strong *ap2* alleles, sepals are transformed to carpels, while petals are absent and stamen numbers are reduced (Bowman et al., 1991). Strong *ap3* and *pi* alleles have sepals in place of petals and carpels in place of stamens (Bowman et al., 1991; Jack et al., 1992). Strong mutant alleles of the C function gene *AG* have petals in place of stamens and sepals in place of carpels while the floral meristem fails to terminate resulting in the indefinite reiteration of sepals and petals (Bowman et al., 1989). Quadruple *sep1 sep2 sep3 sep4* flowers reiterate leaf-like organs indefinitely (Ditta et al., 2004; Pelaz et al., 2000). (See the colour plate.)

Flowers of *Arabidopsis* typically contain four sepals, four petals, six stamens and two fused carpels that form the gynoecium (Figure 6.1(B)). These four different types of organs are positioned in a largely invariant pattern in four concentric domains called floral whorls. Organ numbers are often variable among the first few flowers of a plant, perhaps because the developmental programme that is required for flower formation is not fully fixed immediately after the transition to flowering. Organ numbers (especially that of stamens) can also vary somewhat between different *Arabidopsis* accessions (Smyth et al., 1990), indicating a certain degree of natural variation in this trait. In contrast to many other angiosperms, flowers of *Arabidopsis* (and other members of the Brassicaceae) are not subtended by specialised leaves, called bracts. However, it has been shown through molecular markers that bract *Anlagen* (i.e. incipient primordia) are actually present but that they fail to develop further (Long & Barton, 2000). These so-called 'cryptic bracts' grow out in certain mutant backgrounds, for example, in plants in which the *JAGGED* (*JAG*) gene, which codes for a C2H2-type zinc-finger transcription factor, is constitutively expressed (Dinneny, Yadegari, Fischer, Yanofsky, & Weigel, 2004; Ohno, Reddy, Heisler, & Meyerowitz, 2004).

Based on morphological landmark events, flower development has been divided into 20 distinct stages (Smyth et al., 1990). The formation of flowers begins with a bulge of cells that grows out from the inflorescence meristem. These emerging floral primordia or FMs are composed of cells that are thought to be entirely undifferentiated. At stage 3, organ formation commences with the formation of sepal primordia on the flanks of the FM. This is followed by the emergence of petal and stamen primordia in whorls 2 and 3 and finally the initiation of carpels in whorl 4 in the centre of the FM around stage 6. After approximately 14 days from the time of initiation, flowers are mature and anthesis occurs at stage 13. Stages 14–20 summarise the phase of flower development after fertilisation during which fruit development takes place (leading to the formation of a ripe silique that contains ~50 seeds) and all other floral organs wither and ultimately fall off. Notably, certain processes during flower development have been assigned their own staging system to allow a more precise description of morphological events. These include the formation of anthers (Sanders et al., 1999) and ovules (Schneitz, Hulskamp, & Pruitt, 1995).

6.3 EARLY EVENTS DURING FLOWER FORMATION

As described above, the commencement of flower development is marked by the appearance of a FM on the flanks of the inflorescence meristem. Outgrowth of the primordia is mediated largely by the activity of the phytohormone auxin, which is thought to be concentrated by the action of the auxin efflux carrier PIN-FORMED1 (Galweiler et al., 1998) in regions of the meristem, where primordia initiation takes place (Heisler et al., 2005; Reinhardt et al., 2003). Recently, it was shown that components of the auxin signal transduction pathway are also involved in upregulating the expression of the gene *LEAFY* (*LFY*) (Yamaguchi et al., 2013), which plays an important role in the specification of cells in the emerging FM, resulting in the start of the developmental programme that leads to the formation of flowers and floral organs (Weigel, Alvarez, Smyth, Yanofsky, & Meyerowitz, 1992). This so-called FM identity specification is mediated by a complex interplay of many regulatory genes, but *LFY* and another gene, *APETALA1* (*AP1*) (Irish & Sussex, 1990; Mandel, Gustafson-Brown, Savidge, & Yanofsky, 1992), are arguably most important and have been particularly well studied. While plants mutant for *LFY* exhibit (incomplete) flower-to-shoot conversions (Schultz & Haughn, 1991; Weigel et al., 1992), mutations in the *AP1* gene lead to the formation of flowers that have leaf-like organs in place of sepals and that largely lack petals (Bowman, Alvarez, Weigel, Meyerowitz, & Smyth, 1993; Irish & Sussex, 1990; Mandel & Yanofsky, 1995) (Figure 6.1(D)). Furthermore, additional mutant flowers arise in the axils of the first whorl organs of *ap1* flowers, a phenotype that is thought to be a direct consequence of an impairment in FM identity specification. The *ap1 lfy* double-mutant plants show much more severe phenotypes than either of the single mutants alone (Weigel et al., 1992), suggesting that the specification of FM identity is controlled by *AP1* and *LFY* in partially redundant manner. In fact, ectopic expression of either of these genes results in the transformation of shoot apical meristems into FMs, and thus to a conversion from shoot to flower characteristics (Mandel & Yanofsky, 1995; Weigel & Nilsson, 1995).

LFY encodes a plant-specific transcription factor (Weigel et al., 1992) whose DNA binding domain has a structure that resembles a helix-turn-helix motif (Hames et al., 2008). *LFY* is expressed in incipient floral primordia and throughout FMs during early stages (Weigel et al., 1992) (Figure 6.2). *AP1* encodes a member of the large family of MADS domain transcription factors (see Section 6.5 below for a more detailed description of this family) and begins to be expressed in stage 1 flowers (Mandel et al., 1992). Activation of *AP1* expression is mediated by *LFY* through direct control (Wagner, Sablowski, & Meyerowitz, 1999) as well as by a number of other regulators (reviewed in (O'Maoileidigh, Graciet, & Wellmer, 2014; Wellmer & Riechmann, 2010)), including components of the 'florigen' (Abe et al., 2005; Wigge et al., 2005), which controls flowering in response to changes in the photoperiod. *LFY* also activates expression of the *AP1* paralogue, *CAULIFLOWER* (William et al., 2004), which controls the onset of flower development together with *AP1* in a redundant manner (Bowman et al., 1993; Ferrandiz, Gu, Martienssen, & Yanofsky, 2000).

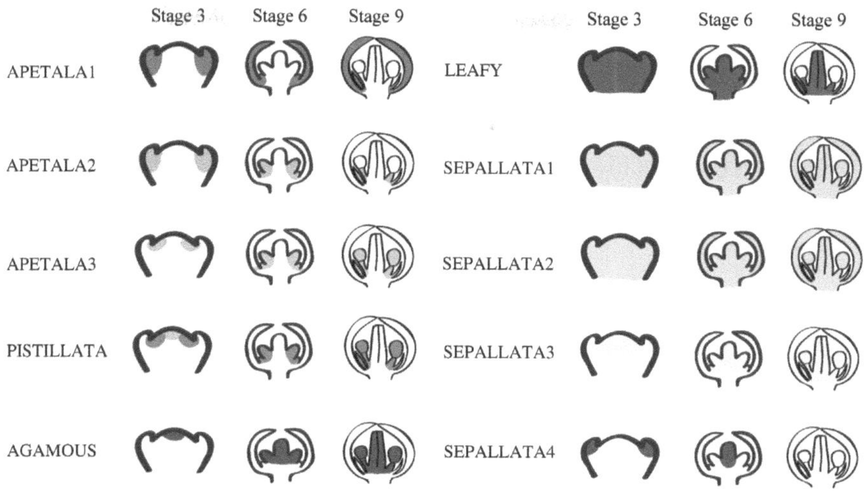

Figure 6.2 Expression patterns of key floral regulators at three stages of flower development. *AP1* is expressed in the outer two floral whorls at stages 3, 6 and 9 (Mandel et al., 1992). *AP2* is expressed in the outer two whorls at stage 3. By stage 6, *AP2* is expressed throughout the second and third whorls before being restricted from the anthers at stage 9 (Wollmann et al., 2010). *AP3* and *PI* are expressed throughout the second and third whorls from stage 3 to stage 9 (Goto & Meyerowitz, 1994; Jack et al., 1992). *AP3* is expressed at the base of sepals at stage 6 (Jack et al., 1992), while *PI* is expressed at low levels in the centre of stage 3 floral meristems (Goto & Meyerowitz, 1994). *AG* is expressed throughout the third and fourth whorls from stages 3–9 (Gustafson-Brown et al., 1994; Yanofsky et al., 1990). *SEP1* and *SEP2* are expressed throughout the flower from stage 3–9 (Flanagan & Ma, 1994; Savidge et al., 1995). *SEP3* is expressed in the inner three whorls from stage 3 to 9 (Mandel & Yanofsky, 1998). *SEP4* is expressed in the first whorl at stage 3, the fourth whorl at stage 6, and expression is not detected at stage 9 (Ditta et al., 2004). The floral meristem identity gene *LFY* is expressed in all whorls at stage 3 before it becomes gradually restricted to petals and stamen filaments at stage 9 (Weigel et al., 1992). (See the colour plate.)

Through the combination of genome-wide localisation studies and transcriptomics approaches, the genes controlled by *AP1* and *LFY* have been identified on a global scale (Kaufmann et al., 2010; Moyroud et al., 2011; Wagner et al., 2004; William et al., 2004; Winter et al., 2011). This work showed that these master regulators act to a large extent by controlling the expression of an entire suite of genes with regulatory functions, of which many have known roles in flower development. *LFY* and *AP1* appear to have dual activities in that they promote the expression of genes (such as the floral homeotic genes discussed in Section 6.4 below) that are required for later stages of flower development and simultaneously repress genes that function as so-called floral repressors or shoot identity genes and that normally prevent flowering when conditions for reproduction are not favourable. *LFY*, whose expression is promoted by auxin, also appears to modulate the activity of the auxin response pathway in emerging floral primordia (Li et al., 2013; Yamaguchi et al., 2013). Surprisingly, *LFY* has also been shown to be involved in the repression of responses to certain biotic stresses (Winter et al., 2011), but the biological implications of this finding are currently not fully understood.

Once floral primordia have been specified, they are functionally subdivided into specific zones from which the different floral whorls arise. Within the floral whorls, floral organ primordia are initiated in largely invariant positions. How this crucial patterning event is controlled and how organ numbers are fixed is currently not well understood. It is known, though, that mutants for components of the response pathways for auxin frequently exhibit organ number defects, suggesting that auxin plays an important role in this process (summarised in: (Cheng & Zhao, 2007)). Further evidence for a role of auxin in mediating the phyllotactic pattern observed in flowers comes from the analysis of the trihelix transcription factor PETAL LOSS (PTL) (Brewer et al., 2004). *Ptl* mutants are characterised (as the name would suggest) by a decreased number of petals in the second whorl, and molecular analyses showed that PTL, at least in part, exerts its function by controlling auxin availability in the perianth whorls (Lampugnani, Kilinc, & Smyth, 2013). Recently, it has been demonstrated that mutants for the cytokinin signalling inhibitor ARABIDOPSIS HISTIDINE PHOSPHOTRANSFER PROTEIN 6 show organ number defects in the whorls of the perianth (Besnard et al., 2014), suggesting that cytokinin is another hormone involved in the control of the phyllotactic pattern of flowers.

Floral organ primordia are derived from small sets of so-called organ founder cells in the uppermost (the L1) layer of the developing FM (Bossinger & Smyth, 1996). How these founder cells are specified is not clear as of yet, but at least one regulator, the *APETALA2* (*AP2*)-family transcription factor DORNRÖSCHEN-LIKE (DRNL), has been identified (Chandler, Jacobs, Cole, Comelli, & Werr, 2011; Nag, Yang, & Jack, 2007). Whether DRNL acts strictly upstream of auxin or whether auxin is a bona fide inductive signal for founder cell specification is still a matter of active debate.

Another crucial step during early flower development is the establishment of boundaries between whorls and organ primordia, ensuring that they can develop separately from each other. Consequently, defects in boundary specification often result in the fusion of floral organs. Key players in this process are the *CUP-SHAPED COTYLEDON* (*CUC*) genes, *CUC1, CUC2* and *CUC3*, which encode NAC domain transcription factors (Aida, Ishida, Fukaki, Fujisawa, & Tasaka, 1997; Vroemen, Mordhorst, Albrecht, Kwaaitaal, & de Vries, 2003). They are expressed throughout boundary regions between organ primordia and floral whorls. Interestingly, *CUC1* and *CUC2* (but not *CUC3*) are regulated by the three members of the MIR164 family of microRNAs (miRNAs). Mutations in one of these genes, *MIR164c*, leads to the formation of extra petals in early arising flowers (Baker, Sieber, Wellmer, & Meyerowitz, 2005). When all three *MIR164* genes are knocked down, flowers have more sepals and petals but fewer stamens than those of the wild type. In addition, floral organ size is variable and carpels are often unfused (Sieber, Wellmer, Gheyselinck, Riechmann, & Meyerowitz, 2007). The overexpression of miRNA-resistant versions of *CUC1* and *CUC2* led to floral organs that are strongly reduced in size due to smaller cell numbers, suggesting the CUC transcription factors may act by limiting growth in the boundary domains in which they are normally expressed (Sieber et al., 2007). It has been shown recently that the C2H2-type zinc-finger transcription factor RABBIT EARS (RBE) (Takeda, Matsumoto, & Okada, 2004) regulates all three *MIR164* genes, and acts primarily via the repression of *MIR164c* (Huang, Lopez-Giraldez, Townsend, & Irish, 2012). Flowers of *rbe* mutants show a loss or the formation of aberrant petals, as well as fused sepals and resemble to some extent those of *cuc1 cuc2* double mutants. Thus, it appears likely that most of the *rbe* mutant phenotypes are a consequence of reduced *CUC1/CUC2* expression levels caused by an upregulation of *MIR164c*. Other genes, such as the aforementioned *PTL*, also play a role in boundary formation in specific whorls. For example, *ptl* mutant flowers are not only characterised by a loss of petals (as described above) but also by a fusion of sepals. In agreement with this mutant phenotype, *PTL* is expressed in-between sepal boundaries (Brewer et al., 2004), suggesting that it contributes to boundary formation within the 1st floral whorl.

6.4 THE SPECIFICATION OF FLORAL ORGANS

A major breakthrough in the understanding of flower development came in the late 1980s from the characterisation of mutants that lack certain types of floral organs (Bowman, Smyth, & Meyerowitz, 1989). The missing organs in these mutants are typically replaced by other floral organs types or by leaflike structures (Figure 6.1(D)). These mutants were termed floral homeotic mutants because their organ transformation phenotypes resemble those of the classic homeotic mutants of the fruitfly *Drosophila melanogaster*. Based on the regions of the flower that show the primary defects in the different mutants, the gene activities affected were assigned to three groups, termed A, B and C. In *Arabidopsis*, A function is mediated by *AP1* (which is also involved in the specification of FMs; see Section 6.3 above) and *AP2*. Mutants in these genes show primarily defects in the organs of the perianth in the outer two floral whorls. As the name of the mutants suggest, they typically lack petals although *ap1* mutants do occasionally form one or two petals per flower, suggesting that *AP1* might not be absolutely essential for petal formation. Sepals in the first whorl are also affected. In *ap2* mutants, they are converted into carpel-like structures or carpelloid leaves bearing ovules, while in *ap1* mutant flowers, they are replaced by organs that are leaf- or bract-like. Mutants for the B function genes *APETALA3* (*AP3*) and *PISTILLATA* (*PI*) show defects in the second and third whorl where petals and stamens are replaced by sepals and carpels (or filamentous structures), respectively. Strong mutant alleles of *AGAMOUS* (*AG*), the only C function gene that was identified, lack all reproductive organs (and thus form no gametes, giving rise to its name). While stamens are replaced by petals, in the centre of the flower a new flower arises that again looks like that of an *ag* mutant so that many iterations of whorls with a sepal-petal-petal pattern are observed. This overgrowth phenotype suggested that the termination of meristematic activity, which normally occurs once all floral organs have been initiated (see Section XII below) is considerably delayed or entirely absent in this mutant background.

The morphological analysis of the floral homeotic mutants as well as of different double- and triple-mutant combinations (Bowman et al., 1993; Bowman, Smyth, & Meyerowitz, 1989, 1991) led to the proposition of the now widely known ABC model of floral organ identity specification (Coen & Meyerowitz, 1991) (Figure 6.1(C)). This model is based on the idea that the genes affected in the mutants act in a combinatorial manner to control the developmental programmes required for floral organ development. According to the model, A function activity alone is sufficient to specify sepals, while the formation of petals requires A function as well as B function activities. The model further proposed that the development of stamens is based on the combined activities of B and C function and that C function alone is sufficient for carpel formation. Another important assumption of the ABC model was that A and C function activities are mutually antagonistic and prevent each other's activity in the inner or outer two whorls, respectively. This antagonism is mostly mediated by *AP2* and *AG*, as in *ap2* mutants, carpel-like structures appear in the first whorl instead of sepals (suggesting a spread of C function into the perianth), while in *ag* mutants, all organs produced by the much enlarged flower are sepals or petals (a phenotype that is in agreement with a spread of A function throughout the FM).

The ABC model has been invaluable in guiding work on flower development for almost 25 years, and thanks to its relative simplicity has been widely introduced into undergraduate teaching in genetics and plant biology (Bowman, Smyth, & Meyerowitz, 2012). While the model has been largely confirmed at the molecular level since its inception, it has been modified, revised and expanded over the years based on a number of new discoveries.

The molecular characterisation of the floral homeotic mutants during the early to mid-1990s led to the identification of the affected genes, which were found to all code for transcription factors (Goto & Meyerowitz, 1994; Jack, Brockman, & Meyerowitz, 1992; Jofuku, den Boer, Van Montagu, & Okamuro, 1994; Mandel et al., 1992; Yanofsky et al., 1990). In agreement with the tenets of the ABC model, these genes are expressed in general in the organs in which they are active and thus in partially overlapping domains (Figure 6.2). The phenotypes of the floral homeotic mutants clearly indicate that the affected genes are necessary for the specification of floral organ identity, but are they also sufficient? To test this, the floral homeotic genes were ectopically expressed in plants (Jack, Fox, & Meyerowitz, 1994; Krizek & Meyerowitz, 1996; Mizukami & Ma, 1992). While the corresponding transgenic plants showed floral phenotypes largely as predicted by the ABC model, this overexpression did not induce organ transformation in any other part of the plant. Thus, it was assumed that flower-specific co-factors are required for the activity of the floral homeotic genes. These co-factors were identified through reverse genetics and named SEPALLATA1 (SEP1) through SEP4. While single *sep* mutants and different double-mutant combinations have no or only minor effects on flower development, *sep1 sep2 sep3* triple mutants form sepals in all floral whorls (Pelaz, Ditta, Baumann, Wisman, & Yanofsky, 2000). When the fourth *SEP* gene, *SEP4*, is also inactivated, quadruple-mutant *sep1–4* plants generate leaves instead of floral organs (Ditta, Pinyopich, Robles, Pelaz, & Yanofsky, 2004). This phenotype is similar to what has been described for mutants in which A, B and C function genes have been disrupted simultaneously (Bowman et al., 1991) and indicate that the *SEP* genes are necessary for floral organ specification in all four floral whorls. Indeed, their expression domains are in general much broader than that of the other floral homeotic genes, although they vary somewhat between the different members of this small group of genes (Figure 6.2). Because of their pivotal role in floral organ specification, the *SEP* genes were incorporated into the ABC model as so-called 'E function' genes, leading to an ABCE model (Figure 6.1(C)). This revised model was put to the test by expressing certain combinations of all four gene activities ectopically in plants (Goto, Kyozuka, & Bowman, 2001; Honma & Goto, 2001; Pelaz, Tapia-Lopez, Alvarez-Buylla, & Yanofsky, 2001). These experiments showed that leaves could be converted into floral organs, indicating that the ABCE genes together are both necessary and sufficient for the specification of floral organs.

A question that the identification of the floral homeotic genes did not immediately answer was when during development the specification of organ identity takes place. The analysis of their temporal expression patterns did not help much in answering this question, because it showed that the expression of these genes commences at early stages but then typically continues in the different floral organs for much of their development (Figure 6.2). However, stage-specific perturbation and activation experiments of the B and C function genes (Bowman et al., 1989; Ito, Ng, Lim, Yu, & Meyerowitz, 2007; O Maoileidigh et al., 2013; Wuest et al., 2012) revealed that stamens and carpels are specified immediately after the initiation of their primordia. In contrast, petals are specified considerably later, at more intermediate stages. These experiments further provided evidence that the function of the floral homeotic genes extends well beyond the early phase of flower development and is also required for the control of processes during organ maturation.

6.5 MOLECULAR BASIS AND MODIFICATIONS OF THE ABC MODEL

With the exception of AP2, which is the founding member of the AP2/ERF family of transcription factors (Okamuro, Caster, Villarroel, Van Montagu, & Jofuku, 1997), all floral organ identity factors, including the SEPs, belong to the MADS domain family, which is greatly enlarged in flowering plants compared to other eukaryotes (Gramzow & Theissen, 2010; Smaczniak, Immink, Angenent, & Kaufmann, 2012). The name 'MADS' stems from the four founding members of this family, i.e. MINICHROMOSOME MAINTENANCE 1 from *Saccharomyces cerevisiae*, AG from *Arabidopsis*, DEFICIENS from *A. majus*, and SERUM RESPONSE

FACTOR from *Homo sapiens*, while 'MADS domain' refers to the shared DNA binding motif (which is encoded by the MADS box sequence) of these transcription factors. The MADS box family, which contains over 100 members in *Arabidopsis*, can be divided into type I and type II genes based on characteristic differences in the domain structure of the corresponding proteins (Gramzow & Theissen, 2010; Smaczniak et al., 2012). The MADS box genes involved in floral organ identity specification all belong to type II, and more specifically, to a subgroup of type II genes, termed MIKC^c. Besides an N-terminal MADS domain, these transcriptional regulators contain additional domains referred to as I (for intervening), K (for keratin-like) and C (for C-terminal). The I domain is thought to determine the specificities for MADS domain protein dimer formation. The K and C domains are also involved in mediating protein–protein interactions, between different MADS domain proteins in case of the K domain, and for the formation of multimeric complexes in case of the C domain. The C domain also mediates, at least in some cases, transactivation activity, while other MADS domain proteins appear to lack such activity and may have to interact with other transcription factors to promote or inhibit gene expression.

MADS domain transcription factors bind to DNA either as homo- or as heterodimers in certain combinations (de Folter et al., 2005; Riechmann, Krizek, & Meyerowitz, 1996; Riechmann, Wang, & Meyerowitz, 1996). The DNA binding specificities of MADS domain transcription factors have been well characterised, leading to the discovery of the so-called CArG box (consensus: CCW_6GG) as the canonical binding motif. Although it would appear from this shared motif that the large number of MADS domain proteins in *Arabidopsis* have the potential to bind to the same sites in the genome, it has been recently demonstrated that different MADS domain proteins involved in flower development do show differences in their binding patterns, likely due to differences in structural characteristics between different CArG boxes (Muino, Smaczniak, Angenent, Kaufmann, & van Dijk, 2014).

In the early 2000s, it was proposed that the floral organ identity factors act in tetrameric complexes, leading to the postulation of the so-called 'quartet model' (Theissen, 2001; Theissen & Saedler, 2001). This model is based on the idea that four regulatory complexes, each composed of combinations of four MADS domain proteins ('floral quartets'), control the developmental programmes required for the formation of the four different types of floral organs (Figure 6.3). Through yeast two- and three-hybrid assays, it was shown that the proposed higher-order complexes can indeed form, at

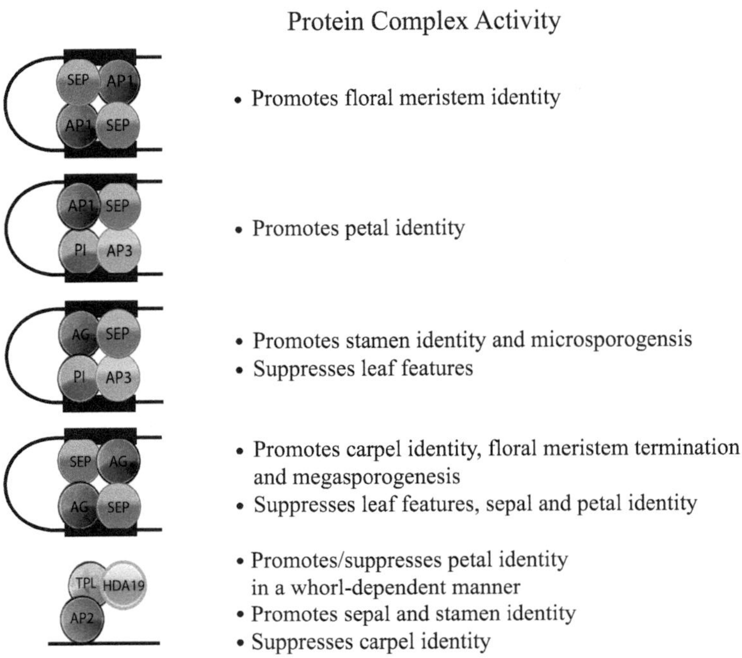

Figure 6.3 Functions of the complexes formed by the floral organ identity factors. The MADS domain-containing floral organ identity factors can form higher-order complexes, termed floral quartets, that are thought to bind two CArG boxes simultaneously via DNA looping to regulate their downstream targets (Honma & Goto, 2001; Riechmann, Krizek et al., 1996). Functions can be ascribed to these putative complexes based on molecular and genetic analysis (as reviewed in (Krizek & Fletcher, 2005; O'Maoileidigh et al., 2014)). The non-MADS domain-containing complex AP2–TPL–HDA19 transcriptionally regulates *SEP3*, *AP3* and *AG* to control floral patterning (Krogan et al., 2012). (See the colour plate).

least in a heterologous system (Honma & Goto, 2001). More evidence for the existence of larger MADS domain transcription factor complexes came only recently from proteomics analyses in which proteins interacting with the floral organ identity factors were identified in vivo (Smaczniak, Immink et al., 2012). It was shown that the different transcription factors under study interact with those MADS domain proteins that were predicted by the quartet model. This work also revealed that a number of other, non-MADS domain proteins are associated with the floral organ identity factors. These include other transcription factors as well as proteins involved in the epigenetic control of gene expression. These findings may be especially significant because genome-wide localisation studies have demonstrated that the binding sites of the different floral organ identity factors show a considerable degree of overlap even in cases where they are not thought to be part of the same regulatory complex (Kaufmann et al., 2009; Kaufmann et al., 2010; O Maoileidigh et al., 2013; Wuest et al., 2012). At the same time, it was found that only a relatively small portion of genes that are bound by these factors respond to a perturbation of their activities. Furthermore, despite the similar DNA binding activities of these factors, their sets of target genes are only partially overlapping. All this leads to the question of how the floral organ identity factors achieve specificity in the context of gene regulation. The results of the proteomics analysis of the MADS domain protein complexes could provide an answer to this question, for example, if further studies were to show that the additional transcription factors identified serve as another input for the transcriptional control of target genes of the floral organ identity factors. The discovery of proteins involved in the control of chromatin states, on the other hand, could provide further insights into how MADS domain transcription factor activity is linked to changes in chromatin structure, which then ultimately lead to the suppression or activation of gene expression.

Based on the quartet model, it was predicted that the two MADS domain protein dimers that form one regulatory complex would bind to two CArG boxes in the promoters of target genes, and that this dual binding event may induce looping of the DNA that is found in-between the CArG boxes as a consequence of the physical interaction between the two dimers. Recent evidence from both in vitro and in vivo studies provided strong support for this idea (Melzer, Verelst, & Theissen, 2009; Mendes et al., 2013). However, whether DNA looping itself contributes to the transcriptional control of target genes of the floral organ identity factors, or whether it is merely a side effect of the binding of the floral quartets, is still unknown.

6.6 FLORAL PATTERNING

As mentioned above, the molecular mechanisms underlying floral patterning, i.e. the division of the FM into distinct functional regions from which whorls and ultimately floral organs arise in a stereotypic pattern, are currently not well understood. However, it is known that one important step during floral patterning is the establishment of the correct spatiotemporal expression domains of the floral homeotic genes, which is mediated by a highly complex gene regulatory network. During the earliest floral stages, B, C and E function genes are actively suppressed to prevent their precocious expression, which would result in a premature onset of the differentiation programme for floral organ development (Liu, Thong, & Yu, 2009).

This repression is mediated by the combined activities of several MADS domain proteins, including AGAMOUS-LIKE24 (AGL24), SHORT VEGETATIVE PHASE (SVP), SUPPRESSOR OF OVEREXPRESSION OF CONSTANS 1 (SOC1) and *AP1* (Gregis, Sessa, Colombo, & Kater, 2006; Gregis, Sessa, Dorca-Fornell, & Kater, 2009; Liu, Xi, Shen, Tan, & Yu, 2009; Sridhar, Surendrarao, & Liu, 2006). The results of several studies indicate that these factors function by recruiting chromatin modifiers to target gene promoters (Liu, Thong, et al., 2009). For example, it has been shown that a transcriptional corepressor complex, composed of LEUNIG (LEU) (or the closely related LEUNIG_HOMOLOG) and SEUSS (SEU), is required for the repression of B and C function genes and interacts directly with selected MADS domain proteins, including *AP1* (Conner & Liu, 2000; Franks, Wang, Levin, & Liu, 2002; Gregis et al., 2006; Gregis et al., 2009; Sridhar, Surendrarao, Gonzalez, Conlan, & Liu, 2004; Sridhar et al., 2006). This corepressor complex likely acts by recruiting histone deacytelases to target promoters as well as by interactions with the plant's Mediator complex (Gonzalez, Bowen, Carroll, & Conlan, 2007). There is also ample evidence that components of Polycomb Group (PcG) complexes, which often act, through chromatin modifications, as repressors of entire gene expression programmes in plants (Kohler & Hennig, 2010), are involved in the suppression of the floral homeotic genes (Calonje, Sanchez, Chen, & Sung, 2008; Goodrich et al., 1997; Liu, Xi, et al., 2009). This mechanism appears to be aimed primarily at silencing these floral regulators during vegetative development, as mutants in PcG components typically show ectopic expression of these genes. The multifaceted suppression of the floral homeotic genes begins to be lifted around stage 2, in part through the activities of the trithorax group proteins ULTRAPETALA1 (ULT1) and ULT2 (Carles, Choffnes-Inada, Reville, Lertpiriyapong, & Fletcher, 2005; Carles & Fletcher, 2009; Monfared, Carles, Rossignol, Pires, & Fletcher, 2013) as well as ARABIDOPSIS HOMOLOG OF TRITHORAX1 (Alvarez-Venegas et al., 2003) that counteract the PcG protein activities described above.

During the early phase of flower development, the regulation of *SEP3* expression appears to be a crucial control mechanism. *SEP3* is repressed during the earliest stages of flower development but is activated in stage 2 FMs (Mandel & Yanofsky, 1998). The initial repression of *SEP3* is mediated by the activities of AGL24, SOC1 and SVP (Liu, Xi, et al., 2009). In contrast, *AP1* promotes *SEP3* expression through two mechanisms: first, it directly downregulates *AGL24*, *SOC1* and *SVP* expression (Liu et al., 2007) and thus eliminates these repressors over time, and, second, it activates *SEP3* by directly binding to its promoter (Kaufmann et al., 2010). This activation of *SEP3* is then thought to lead to the formation of AP1/SEP3 heterodimers that are involved in the activation of floral homeotic genes, most notably *AP3* (Kaufmann et al., 2010; Ng & Yanofsky, 2001). SEP3 might also function as a co-factor of LFY (Liu, Xi, et al., 2009), which is pivotal for the activation of the floral homeotic genes (Parcy, Nilsson, Busch, Lee, & Weigel, 1998). Together with AP1 and the F-box protein UNUSUAL FLORAL ORGANS (Lee, Wolfe, Nilsson, & Weigel, 1997), which is a component of an SKP1-cullin-F-box (SCF)-type ubiquitin ligase, LFY activates the B function genes *AP3* and *PI* (Ng & Yanofsky, 2001). The expression of *UFO* in a ringlike domain in the FM appears hereby to provide the spatial information for the activation of the B function genes in whorls 2 and 3. How UFO acts in this process has long been enigmatic, but it has been shown more recently that it physically interacts with LFY and may be directly required for its activity, possibly by removing 'spent' LFY proteins from target gene promoters and by allowing new copies of the transcription factor to bind (Chae, Tan, Hill, & Irish, 2008). However, it also seems possible that UFO is involved in the ubiquitin-dependent degradation of one or more repressors that block target genes from an activation by LFY (Zhao, Yu, Chen, & Ma, 2001).

An additional mechanism for B function gene regulation is mediated by the C2H2-type zinc-finger transcription factor SUPERMAN (SUP) (Bowman et al., 1992; Sakai, Medrano, & Meyerowitz, 1995). Mutations in the *SUP* gene, which is expressed in a ringlike domain around the fourth whorl (Ito, Sakai, & Meyerowitz, 2003; Sakai et al., 1995), lead to supernumerary stamens and carpel defects, as well as a slight overgrowth of the FM, indicating a role in the control of floral determinacy. In *sup* mutant flowers, B function gene expression expands further into the centre of the flower than in the wild type (Bowman et al., 1992), and consequently, SUP is thought to set the boundary between the third and fourth floral whorls and to prevent B function from entering into the centre of the FM. However, its exact mode of action remains unknown.

The expression of *AP1*, which is initially found throughout the emerging FM, is (around stage 3) confined to whorls 1 and 2 (Mandel et al., 1992), where AP1 contributes to the formation of perianth organs. As predicted by the ABC model, this suppression is mediated to a large extent by the C function regulator AG (Gustafson-Brown, Savidge, & Yanofsky, 1994; O Maoileidigh et al., 2013), although the B function regulators AP3/PI also appear to play a role in this process (Sundstrom, Nakayama, Glimelius, & Irish, 2006).

The establishment of the *AG* expression domain has been particularly well studied and shown to be highly complex. The onset of *AG* expression in whorls 3 and 4 around stage 3 (Drews, Bowman, & Meyerowitz, 1991) appears to be largely dependent on the combined activities of *LFY* and the homeodomain transcription factor WUSCHEL (WUS) (Lenhard, Bohnert, Jurgens, & Laux, 2001; Lohmann et al., 2001), which bind directly to the large second intron of the *AG* gene that contains important regulatory elements (Busch, Bomblies, & Weigel, 1999; Deyholos & Sieburth, 2000; Sieburth & Meyerowitz, 1997). However, because *LFY* is expressed throughout early-stage flowers (Weigel et al., 1992) and the expression of *WUS* is limited to a small domain underneath the stem cells (Mayer et al., 1998) (to the so-called 'organising centre'; see below), this finding did not immediately explain how the expression of *AG* is established in, and confined to, whorls 3 and 4. The recent discovery that WUS protein can move within the meristem (Yadav et al., 2011) may help to solve this conundrum and explain how *AG* expression is established in a domain that is considerably larger than that one of its direct activators.

In addition to LFY and WUS, several other factors play roles in establishing the *AG* expression domain. These include the putative transcription factor STERILE APETALA, which represses *AG* expression in the whorls of the perianth (Byzova et al., 1999), the aforementioned C2H2 zinc-finger transcription factor RBE, which also acts as a repressor of *AG* in the perianth whorls (Krizek, Lewis, & Fletcher, 2006), and the basic leucine zipper transcription factor PERIANTHIA (Chuang, Running, Williams, & Meyerowitz, 1999) that promotes *AG* expression through direct regulation (Das et al., 2009; Maier et al., 2009).

The most important negative regulator of *AG* is arguably AP2, which directly suppresses its expression in the outer floral whorls (Dinh et al., 2012; Drews et al., 1991; Yant et al., 2010), thus fulfilling one of the central tenets of the ABC model, namely that A function suppresses C function. *AP2* was initially described to be almost uniformly expressed throughout young FMs, leading to the question of how it could specifically act in the organs of the perianth. A closer examination showed that *AP2* is in fact expressed predominantly in whorls 1 and 2 (Wollmann, Mica, Todesco, Long, & Weigel, 2010), with weaker expression being detected in the third floral whorl. Discovery of the miR172 miRNA showed that it can target and suppress *AP2* as well as several related genes through both transcriptional and translational control (Aukerman & Sakai, 2003; Chen, 2004; Schwab et al., 2005). Because miR172 is most strongly expressed in the fourth whorl

(Wollmann et al., 2010), it reduces AP2 activity in this region. AP2, in turn, binds the *MIR172* promoter (Yant et al., 2010) and repress its expression in whorls 1 and 2 through interactions with the corepressors SEU and LEU (Grigorova et al., 2011), ensuring that the expression of the miRNA is mainly found in the centre of the flower.

According to the ABC model, A function genes are required for the specification of the organs of the perianth and for the repression of C function in the outer two floral whorls. However, bona fide A function genes could not be identified in other angiosperms (Litt & Kramer, 2010). This led to a re-evaluation of the concept of A function and to the proposition of a modified model for floral organ identity specification (Causier, Schwarz-Sommer, & Davies, 2010). This so-called (A)BCE model proposes that A function (together with E function) genes are mainly required for FM identity specification and that sepals (which are specified by a combination of A and E function activities) are the ground state of a floral organ. In agreement with this revised view of the initial ABC model, it has been demonstrated that both AP1 and AP2 are important regulators of FM identity (Ferrandiz et al., 2000; Okamuro, Szeto, Lotys-Prass, & Jofuku, 1997). Furthermore, it has been suggested recently that AP2 acts predominantly in the control of floral homeotic gene expression, and not organ specification (Krogan, Hogan, & Long, 2012). It was shown that AP2 not only represses *AG* but also *SEP3* and *AP3* in whorl 1. This repression is mediated by interactions between AP2 and the transcriptional corepressor TOPLESS, which recruits the histone deacetylase HDA19 (Krogan et al., 2012). Taken together, this work and the required re-evaluation of the ABCE model have shown that the mechanisms underlying A function are much more complex than what had been initially thought, and that further efforts may be needed to devise a truly universal model for floral organ identity specification in angiosperms.

As described in Section 6.4, once the expression of the floral homeotic genes has been established, it is maintained throughout most of floral organ development. Although not much is known about the regulation of these genes at later stages of flower development, it is clear that autoregulation plays an important part in ensuring their continued expression (Gomez-Mena, de Folter, Costa, Angenent, & Sablowski, 2005; Honma & Goto, 2000). Furthermore, it has been shown for *AP3* that its expression is promoted by either AP1 or AG activity (Gomez-Mena et al., 2005). In addition to these mechanisms, the phytohormone gibberellin appears to be an important stimulant of floral homeotic gene expression at more intermediate to later stages of flower development (Yu et al., 2004).

6.7 TERMINATION OF FM ACTIVITY

In contrast to apical meristems, which are indeterminate, FMs are determinate structures. They cease their activity around stage 6 once carpel primordia have been initiated in the centre of the flower. Stem cell maintenance in FMs appears to depend largely on the activity of the CLAVATA signalling pathway (Aichinger, Kornet, Friedrich, & Laux, 2012), which is also active in shoot apical meristems. This pathway allows the communication of the stem cells in the very tip of the meristem with an underlying domain, the so-called organising centre. Cells of the organising centre are characterised by the expression of the aforementioned homeodomain transcription factor WUS (Mayer et al., 1998). *WUS* is pivotal for the maintenance of stem cells and appears to be the primary target during FM termination (Lenhard et al., 2001; Lohmann et al., 2001). Indeed, *WUS* expression has been found to be prolonged in mutants in which FM determinacy is delayed, leading to an overgrowth of the flower. The repression of *WUS* during FM termination is directly connected to the C function regulator AG, whose expression in the centre of the flower leads to a loss of stem cells (Lenhard et al., 2001; Lohmann et al., 2001). In agreement with the central importance of AG in FM termination, many genes whose loss of function causes FM determinacy defects act upstream of *AG* and their mutant phenotypes are likely a consequence of altered AG activity.

Given the crucial role of AG for FM termination, how can one reconcile the differences in the developmental stage at which *AG* expression commences (stage 3) and at which the termination of floral stem cells occurs (stage 6)? Part of the answer to this question was provided by the identification of *KNUCKLES* (*KNU*), which encodes a C2H2-type zinc-finger transcription factor (Payne, Johnson, & Koltunow, 2004) that acts downstream of AG. *Knu* mutant flowers have determinacy defects and show a prolonged expression of *WUS* (Sun, Xu, Ng, & Ito, 2009). *KNU* is expressed from stage 6 onward in the centre of the flower (Payne et al., 2004; Sun et al., 2009) and thus, the onset of its expression coincides with the termination of FMs. AG has been shown to promote *KNU* expression by directly binding to its promoter (Sun et al., 2009). Surprisingly, this binding is already observed in stage 3 flowers, but the activation of *KNU* expression was shown to be prevented initially by repressive histone marks that cover the *KNU* locus. These marks are removed over time, leading to delayed activation of the *KNU* genes. Once KNU is expressed, it represses *WUS* either directly or indirectly (Sun et al., 2009).

It has been suggested that AG may also be able to bind directly to the *WUS* promoter (Liu et al., 2011). In this scenario, binding of AG to *WUS* at stage 3 leads to the recruitment of PcG proteins, which then terminate *WUS* expression over time. Both proposed mechanisms for the repression of *WUS* by AG may act in parallel and constitute an efficient timing mechanism to ensure that FM activity is terminated immediately after the initiation of floral organ primordia has been completed.

6.8 FLORAL ORGAN DEVELOPMENT

As proposed by Goethe in the late eighteenth century (von Goethe, 1790), floral organs are directly derived from leaves. This was elegantly shown in experiments in which either A, B and C functions were disrupted simultaneously (Bowman et al., 1991) or in which all four *SEP* genes are inactivated (Ditta et al., 2004), leading in both cases to the transformation of floral organs into leaves. Furthermore, as described above, the ectopic expression of certain combinations of the floral homeotic genes results in the transformation of leaves into floral organs (Goto et al., 2001). Although these experiments unequivocally showed that the floral organ identity factors are necessary and sufficient for floral organ development, they did not immediately explain how a leaflike organ is converted into a floral organ with specialised functions. Transcriptomics approaches aimed at characterising the gene expression programmes underlying floral organ development revealed many genes with predominant or even floral organ-specific expression (Alves-Ferreira et al., 2007; Jiao & Meyerowitz, 2010; Peiffer et al., 2008; Schmid et al., 2005; Wellmer, Alves-Ferreira, Dubois, Riechmann, & Meyerowitz, 2006; Wellmer, Riechmann, Alves-Ferreira, & Meyerowitz, 2004) especially in the case of the reproductive floral organs that are morphologically much more complex than the organs of the perianth. While the functions of many of these genes are still unknown, they are suspected to mediate key aspects of floral organ growth and maturation. The global identification of genes regulated by SEP3, AP1, AP2, AP3/PI and AG, through a combination of genome-wide localisation studies and global gene expression profiling, showed for many of these 'flower genes' that their expression is controlled by one or more of the floral organ identity factors through direct or indirect control mechanisms (Gomez-Mena et al., 2005; Kaufmann et al., 2009; Kaufmann et al., 2010; O Maoileidigh et al., 2013; Wuest et al., 2012; Yant et al., 2010). Thus, the floral organ identity factors activate an entire suite of genes that is required for floral organ formation (Wellmer, Graciet, & Riechmann, 2014).

Recent evidence suggests that the B and C function regulators are also involved in the direct repression of the leaf development programme (O Maoileidigh et al., 2013; Wuest et al., 2012). Through gene perturbation experiments, it was shown that AG and AP3/PI suppress the appearance of characteristic leaf features on floral organs. These include the formation of branched trichomes, which are normally absent from floral organs of *Arabidopsis*, but appear on carpel valves when AG activity is specifically disrupted at intermediate developmental stages (O Maoileidigh et al., 2013). This repression of trichome formation is dependent on the transcriptional control of several repressors and activators of trichome initiation.

The B and C function regulators were also found to have partially antagonistic activities, which appear to be important for the proper development of the reproductive floral organs (O Maoileidigh et al., 2013; Wuest et al., 2012). This is especially notable in the case of several genes that encode key regulators of carpel development. While such genes are in general activated by AG in the fourth floral whorl, they are suppressed by AP3/PI in the third whorl, thus preventing a spread of their activities outside of the domain where carpel primordia are initiated.

Because many of the identified direct target genes of the floral organ identity factors were found to encode other transcription factors, it appears possible that individual regulators could mediate certain aspects of the functions that were initially attributed to the floral homeotic genes based on their mutant phenotypes. A striking example in support of this idea involves the putative transcription factor NOZZLE/SPOROCYTELESS (NZZ/SPL), a master regulator of sporogenesis (Schiefthaler et al., 1999; Yang, Ye, Xu, & Sundaresan, 1999), whose expression is directly activated by AG (Ito et al., 2004) as well as by the B function regulators AP3/PI (O Maoileidigh et al., 2013; Wuest et al., 2012). It was shown that the ectopic expression of *NZZ/SPL* in *ag* mutant flowers can induce microsporogenesis in petals (Ito et al., 2004). Thus, it appears that the activation of a single regulatory gene by floral organ identity factors can trigger a key process during reproductive development.

AG also directly activates the MADS box genes *SHATTERPROOF1* (*SHP1*), *SHP2* and *SEEDSTICK* (*STK*) (Gomez-Mena et al., 2005; O Maoileidigh et al., 2013) that form a monophyletic clade with *AG*. The combined activities of these genes (which are also together referred to as 'D function' genes) were shown to be required for the specification of ovule identity (Favaro et al., 2003; Pinyopich et al., 2003), and *SHP1/2* and *STK* were further found to play important roles in the control of different aspects of gynoecium and fruit development (Liljegren et al., 2000; Pinyopich et al., 2003).

In addition to identifying the regulatory genes that mediate cellular differentiation during flower development, it is important to obtain insights into how organ size is controlled. In flowers, this process is currently not well understood but a number of key regulators have been identified mainly through forward genetics. These include, among others, the AP2-type transcription factor AINTEGUMENTA (Elliott et al., 1996), the aforementioned C2H2-type zinc-finger transcription factor JAG (which acts in a partially redundant manner with the closely related NUBBIN) (Dinneny, Weigel, & Yanofsky, 2006; Dinneny et al., 2004; Ohno et al., 2004), and the E3 ubiquitin ligase BIG BROTHER (BB) (Disch et al., 2006).

6.9 CONCLUSIONS AND PERSPECTIVES

Almost 30 years of work on *Arabidopsis* flower development has yielded a truly rich harvest. Many important regulators of this key developmental process have been identified and first detailed insights into the molecular mechanisms underlying their activities have been obtained. Furthermore, spatial and temporal patterns of gene expression during flower formation have been determined on a global scale, providing a detailed map of gene activities. Because there is now such a large body of knowledge available for flower development, researchers have begun to use computational methods and mathematical modelling approaches to simulate the molecular and morphological events during the formation of flowers and to delineate the gene regulatory networks that guide processes, such as floral organ specification (Espinosa-Soto, Padilla-Longoria, & Alvarez-Buylla, 2004) and development (La Rota et al., 2011). While such approaches have already been shown to be extremely useful, they have certainly not yet reached their full potential, and further development of these methods should lead to more predictive models that can be used to drive experimentation.

Despite the considerable progress in the area of flower development, many open questions remain (Wellmer,Bowman et al., 2014). For example, we currently know very little about how the different cell types of which flowers are composed are established. Although this problem is now actively addressed by researchers in the field (Roeder et al., 2010), we are still a long way from understanding the mechanisms underlying cellular differentiation during flower development. This work is made particularly difficult by the fact that, unlike in roots where cell division patterns and cell fates are highly predictable (owing to the well defined roles of the different stem cells in the root apical meristems), tracking of cells and cell fates in FMs is not straightforward. However, with the further development of advanced imaging technologies and methods that allow the isolation of specific cells from developing flowers, this work should become much more feasible in the not too distant future.

The development of next-generation DNA sequencing methods, in particular, has allowed scientists in recent years to establish species from different groups of the plant kingdom as new models for botanical research. These efforts have already led to a much more detailed and broader view of plant development than what could have been obtained with *Arabidopsis* alone. As this work continues, it will provide the unprecedented opportunity to compare what is known about *Arabidopsis* flower development to the regulation of flower formation in other species and to identify the genetic basis of the dramatic variations in flower and inflorescence architecture that can be found among the angiosperms. Because the specification of the reproductive organs of all seed plants has a common basis and appears to depend on the activities of B and C function genes (Theissen & Melzer, 2007), this comparative analysis must be extended to gymnosperms to learn more about the origin of flowers and to shed light on Darwin's 'abominable mystery'.

ACKNOWLEDGEMENTS

Work in our laboratories is funded by grants from Science Foundation Ireland. We apologise to colleagues whose work we could not cite in this chapter because of space constraints.

REFERENCES

Abe, M., Kobayashi, Y., Yamamoto, S., Daimon, Y., Yamaguchi, A., Ikeda, Y., et al. (2005). FD, a bZIP protein mediating signals from the floral pathway integrator FT at the shoot apex. *Science*, *309*(5737), 1052–1056.

Aichinger, E., Kornet, N., Friedrich, T., & Laux, T. (2012). Plant stem cell niches. *Annual Review of Plant Biology*, *63*, 615–636.

Aida, M., Ishida, T., Fukaki, H., Fujisawa, H., & Tasaka, M. (1997). Genes involved in organ separation in *Arabidopsis*: an analysis of the cup-shaped cotyledon mutant. *Plant Cell*, *9*(6), 841–857.

Alvarez-Venegas, R., Pien, S., Sadder, M., Witmer, X., Grossniklaus, U., & Avramova, Z. (2003). ATX-1, an *Arabidopsis* homolog of trithorax, activates flower homeotic genes. *Current Biology*, *13*(8), 627–637.

Alves-Ferreira, M., Wellmer, F., Banhara, A., Kumar, V., Riechmann, J. L., & Meyerowitz, E. M. (2007). Global expression profiling applied to the analysis of *Arabidopsis* stamen development. *Plant Physiology*, *145*(3), 747–762.

Aukerman, M. J., & Sakai, H. (2003). Regulation of flowering time and floral organ identity by a MicroRNA and its APETALA2-like target genes. *Plant Cell, 15*(11), 2730–2741.
Baker, C. C., Sieber, P., Wellmer, F., & Meyerowitz, E. M. (2005). The early extra petals1 mutant uncovers a role for microRNA miR164c in regulating petal number in *Arabidopsis*. *Current Biology, 15*(4), 303–315.
Besnard, F., Refahi, Y., Morin, V., Marteaux, B., Brunoud, G., Chambrier, P., et al. (2014). Cytokinin signalling inhibitory fields provide robustness to phyllotaxis. *Nature, 505*(7483), 417–421.
Bossinger, G., & Smyth, D. R. (1996). Initiation patterns of flower and floral organ development in *Arabidopsis thaliana*. *Development, 122*(4), 1093–1102.
Bowman, J. L., Alvarez, J., Weigel, D., Meyerowitz, E. M., & Smyth, D. R. (1993). Control of flower development in *Arabidopsis thaliana* by APETALA1 and interacting genes. *Development, 119*, 721–743.
Bowman, J. L., Sakai, H., Jack, T., Weigel, D., Mayer, U., & Meyerowitz, E. M. (1992). SUPERMAN, a regulator of floral homeotic genes in *Arabidopsis*. *Development, 114*(3), 599–615.
Bowman, J. L., Smyth, D. R., & Meyerowitz, E. M. (1989). Genes directing flower development in *Arabidopsis*. *Plant Cell, 1*(1), 37–52.
Bowman, J. L., Smyth, D. R., & Meyerowitz, E. M. (1991). Genetic interactions among floral homeotic genes of *Arabidopsis*. *Development, 112*(1), 1–20.
Bowman, J. L., Smyth, D. R., & Meyerowitz, E. M. (2012). The ABC model of flower development: then and now. *Development, 139*(22), 4095–4098.
Brewer, P. B., Howles, P. A., Dorian, K., Griffith, M. E., Ishida, T., Kaplan-Levy, R. N., et al. (2004). PETAL LOSS, a trihelix transcription factor gene, regulates perianth architecture in the *Arabidopsis* flower. *Development, 131*(16), 4035–4045.
Busch, M. A., Bomblies, K., & Weigel, D. (1999). Activation of a floral homeotic gene in *Arabidopsis*. *Science, 285*(5427), 585–587.
Byzova, M. V., Franken, J., Aarts, M. G., de Almeida-Engler, J., Engler, G., Mariani, C., et al. (1999). *Arabidopsis* STERILE APETALA, a multifunctional gene regulating inflorescence, flower, and ovule development. *Genes & Development, 13*(8), 1002–1014.
Calonje, M., Sanchez, R., Chen, L., & Sung, Z. R. (2008). EMBRYONIC FLOWER1 participates in polycomb group-mediated AG gene silencing in *Arabidopsis*. *Plant Cell, 20*(2), 277–291.
Carles, C. C., Choffnes-Inada, D., Reville, K., Lertpiriyapong, K., & Fletcher, J. C. (2005). ULTRAPETALA1 encodes a SAND domain putative transcriptional regulator that controls shoot and floral meristem activity in *Arabidopsis*. *Development, 132*(5), 897–911.
Carles, C. C., & Fletcher, J. C. (2009). The SAND domain protein ULTRAPETALA1 acts as a trithorax group factor to regulate cell fate in plants. *Genes & Development, 23*(23), 2723–2728.
Causier, B., Schwarz-Sommer, Z., & Davies, B. (2010). Floral organ identity: 20 years of ABCs. *Seminars in Cell and Developmental Biology, 21*(1), 73–79.
Chae, E., Tan, Q. K., Hill, T. A., & Irish, V. F. (2008). An *Arabidopsis* F-box protein acts as a transcriptional co-factor to regulate floral development. *Development, 135*(7), 1235–1245.
Chandler, J. W., Jacobs, B., Cole, M., Comelli, P., & Werr, W. (2011). DORNROSCHEN-LIKE expression marks *Arabidopsis* floral organ founder cells and precedes auxin response maxima. *Plant Molecular Biology, 76*(1–2), 171–185.
Chen, X. (2004). A microRNA as a translational repressor of APETALA2 in *Arabidopsis* flower development. *Science, 303*(5666), 2022–2025.
Cheng, Y. F., & Zhao, Y. D. (2007). A role for auxin in flower development. *Journal of Integrative Plant Biology, 49*(1), 99–104.
Chuang, C. F., Running, M. P., Williams, R. W., & Meyerowitz, E. M. (1999). The PERIANTHIA gene encodes a bZIP protein involved in the determination of floral organ number in *Arabidopsis thaliana*. *Genes Development, 13*(3), 334–344.

Coen, E. S., & Meyerowitz, E. M. (1991). The war of the whorls: genetic interactions controlling flower development. *Nature*, *353*(6339), 31–37.
Conner, J., & Liu, Z. (2000). LEUNIG, a putative transcriptional corepressor that regulates AGAMOUS expression during flower development. *Proceedings of the National Academy of Sciences of the United States of America*, *97*(23), 12902–12907.
Das, P., Ito, T., Wellmer, F., Vernoux, T., Dedieu, A., Traas, J., et al. (2009). Floral stem cell termination involves the direct regulation of AGAMOUS by PERIANTHIA. *Development*, *136*(10), 1605–1611.
de Folter, S., Immink, R. G., Kieffer, M., Parenicova, L., Henz, S. R., et al. (2005). Comprehensive Interaction Map of the *Arabidopsis* MADS Box Transcription Factors. *Plant Cell*, *17*(5), 1424–1433.
Deyholos, M. K., & Sieburth, L. E. (2000). Separable whorl-specific expression and negative regulation by enhancer elements within the AGAMOUS second intron. *Plant Cell*, *12*(10), 1799–1810.
Dinh, T. T., Girke, T., Liu, X., Yant, L., Schmid, M., & Chen, X. (2012). The floral homeotic protein APETALA2 recognizes and acts through an AT-rich sequence element. *Development*, *139*, 1978–1986.
Dinneny, J. R., Weigel, D., & Yanofsky, M. F. (2006). NUBBIN and JAGGED define stamen and carpel shape in *Arabidopsis*. *Development*, *133*, 1645–1655.
Dinneny, J. R., Yadegari, R., Fischer, R. L., Yanofsky, M. F., & Weigel, D. (2004). The role of JAGGED in shaping lateral organs. *Development*, *131*(5), 1101–1110.
Disch, S., Anastasiou, E., Sharma, V. K., Laux, T., Fletcher, J. C., & Lenhard, M. (2006). The E3 ubiquitin ligase BIG BROTHER controls arabidopsis organ size in a dosage-dependent manner. *Current Biology*, *16*(3), 272–279.
Ditta, G., Pinyopich, A., Robles, P., Pelaz, S., & Yanofsky, M. F. (2004). The SEP4 gene of *Arabidopsis thaliana* functions in floral organ and meristem identity. *Current Biology*, *14*(21), 1935–1940.
Drews, G. N., Bowman, J. L., & Meyerowitz, E. M. (1991). Negative regulation of the *Arabidopsis* homeotic gene AGAMOUS by the APETALA2 product. *Cell*, *65*(6), 991–1002.
Elliott, R. C., Betzner, A. S., Huttner, E., Oakes, M. P., Tucker, W. Q., Gerentes, D., et al. (1996). AINTEGUMENTA, an APETALA2-like gene of *Arabidopsis* with pleiotropic roles in ovule development and floral organ growth. *Plant Cell*, *8*(2), 155–168.
Espinosa-Soto, C., Padilla-Longoria, P., & Alvarez-Buylla, E. R. (2004). A gene regulatory network model for cell-fate determination during *Arabidopsis thaliana* flower development that is robust and recovers experimental gene expression profiles. *Plant Cell*, *16*(11), 2923–2939.
Favaro, R., Pinyopich, A., Battaglia, R., Kooiker, M., Borghi, L., Ditta, G., et al. (2003). MADS-box protein complexes control carpel and ovule development in *Arabidopsis*. *Plant Cell*, *15*(11), 2603–2611.
Ferrandiz, C., Gu, Q., Martienssen, R., & Yanofsky, M. F. (2000). Redundant regulation of meristem identity and plant architecture by FRUITFULL, APETALA1 and CAULIFLOWER. *Development*, *127*(4), 725–734.
Flanagan, C. A., & Ma, H. (1994). Spatially and temporally regulated expression of the MADS-box gene AGL2 in wild-type and mutant arabidopsis flowers. *Plant Molecular Biology*, *26*(2), 581–595.
Franks, R. G., Wang, C., Levin, J. Z., & Liu, Z. (2002). SEUSS, a member of a novel family of plant regulatory proteins, represses floral homeotic gene expression with LEUNIG. *Development*, *129*(1), 253–263.
Friedman, W. E. (2009). The meaning of Darwin's 'abominable mystery'. *American Journal of Botany*, *96*(1), 5–21.
Galweiler, L., Guan, C., Muller, A., Wisman, E., Mendgen, K., Yephremov, A., et al. (1998). Regulation of polar auxin transport by AtPIN1 in *Arabidopsis* vascular tissue. *Science*, *282*(5397), 2226–2230.

Gomez-Mena, C., de Folter, S., Costa, M. M., Angenent, G. C., & Sablowski, R. (2005). Transcriptional program controlled by the floral homeotic gene AGAMOUS during early organogenesis. *Development, 132*(3), 429–438.

Gonzalez, D., Bowen, A. J., Carroll, T. S., & Conlan, R. S. (2007). The transcription corepressor LEUNIG interacts with the histone deacetylase HDA19 and mediator components MED14 (SWP) and CDK8 (HEN3) to repress transcription. *Molecular Cell Biology, 27*(15), 5306–5315.

Goodrich, J., Puangsomlee, P., Martin, M., Long, D., Meyerowitz, E. M., & Coupland, G. (1997). A Polycomb-group gene regulates homeotic gene expression in *Arabidopsis*. *Nature, 386*(6620), 44–51.

Goto, K., Kyozuka, J., & Bowman, J. L. (2001). Turning floral organs into leaves, leaves into floral organs. *Current Opinion in Genetics and Development, 11*(4), 449–456.

Goto, K., & Meyerowitz, E. M. (1994). Function and regulation of the *Arabidopsis* floral homeotic gene PISTILLATA. *Genes & Development, 8*(13), 1548–1560.

Gramzow, L., & Theissen, G. (2010). A hitchhiker's guide to the MADS world of plants. *Genome Biology, 11*(6), 214.

Gregis, V., Sessa, A., Colombo, L., & Kater, M. M. (2006). AGL24, SHORT VEGETATIVE PHASE, and APETALA1 redundantly control AGAMOUS during early stages of flower development in *Arabidopsis*. *Plant Cell, 18*(6), 1373–1382.

Gregis, V., Sessa, A., Dorca-Fornell, C., & Kater, M. M. (2009). The *Arabidopsis* floral meristem identity genes AP1, AGL24 and SVP directly repress class B and C floral homeotic genes. *Plant Journal, 60*(4), 626–637.

Grigorova, B., Mara, C., Hollender, C., Sijacic, P., Chen, X., & Liu, Z. (2011). LEUNIG and SEUSS co-repressors regulate miR172 expression in *Arabidopsis* flowers. *Development, 138*(12), 2451–2456.

Gustafson-Brown, C., Savidge, B., & Yanofsky, M. F. (1994). Regulation of the arabidopsis floral homeotic gene APETALA1. *Cell, 76*(1), 131–143.

Hames, C., Ptchelkine, D., Grimm, C., Thevenon, E., Moyroud, E., Gerard, F., et al. (2008). Structural basis for LEAFY floral switch function and similarity with helix-turn-helix proteins. *Embo Journal, 27*(19), 2628–2637.

Heisler, M. G., Ohno, C., Das, P., Sieber, P., Reddy, G. V., Long, J. A., et al. (2005). Patterns of auxin transport and gene expression during primordium development revealed by live imaging of the *Arabidopsis* inflorescence meristem. *Current Biology, 15*(21), 1899–1911.

Honma, T., & Goto, K. (2000). The *Arabidopsis* floral homeotic gene PISTILLATA is regulated by discrete cis-elements responsive to induction and maintenance signals. *Development, 127*(10), 2021–2030.

Honma, T., & Goto, K. (2001). Complexes of MADS-box proteins are sufficient to convert leaves into floral organs. *Nature, 409*(6819), 525–529.

Huang, T., Lopez-Giraldez, F., Townsend, J. P., & Irish, V. F. (2012). RBE controls microRNA164 expression to effect floral organogenesis. *Development, 139*, 2161–2169.

Irish, V. F., & Sussex, I. M. (1990). Function of the apetala-1 gene during *Arabidopsis* floral development. *Plant Cell, 2*(8), 741–753.

Ito, T., Ng, K. H., Lim, T. S., Yu, H., & Meyerowitz, E. M. (2007). The homeotic protein AGAMOUS controls late stamen development by regulating a jasmonate biosynthetic gene in *Arabidopsis*. *Plant Cell, 19*(11), 3516–3529.

Ito, T., Sakai, H., & Meyerowitz, E. M. (2003). Whorl-specific expression of the SUPERMAN gene of *Arabidopsis* is mediated by cis elements in the transcribed region. *Current Biology, 13*(17), 1524–1530.

Ito, T., Wellmer, F., Yu, H., Das, P., Ito, N., Alves-Ferreira, M., et al. (2004). The homeotic protein AGAMOUS controls microsporogenesis by regulation of SPOROCYTELESS. *Nature, 430*(6997), 356–360.

Jack, T., Brockman, L. L., & Meyerowitz, E. M. (1992). The homeotic gene APETALA3 of *Arabidopsis thaliana* encodes a MADS box and is expressed in petals and stamens. *Cell, 68*(4), 683–697.

Jack, T., Fox, G. L., & Meyerowitz, E. M. (1994). *Arabidopsis* homeotic gene APETALA3 ectopic expression: transcriptional and posttranscriptional regulation determine floral organ identity. *Cell, 76*(4), 703–716.

Jiao, Y., & Meyerowitz, E. M. (2010). Cell-type specific analysis of translating RNAs in developing flowers reveals new levels of control. *Molecular Systems Biology, 6*, 419.

Jofuku, K. D., den Boer, B. G., Van Montagu, M., & Okamuro, J. K. (1994). Control of *Arabidopsis* flower and seed development by the homeotic gene APETALA2. *Plant Cell, 6*(9), 1211–1225.

Kaufmann, K., Muino, J. M., Jauregui, R., Airoldi, C. A., Smaczniak, C., Krajewski, P., et al. (2009). Target genes of the MADS transcription factor SEPALLATA3: integration of developmental and hormonal pathways in the *Arabidopsis* flower. *PLoS Biology,* 7(4), e1000090.

Kaufmann, K., Wellmer, F., Muino, J. M., Ferrier, T., Wuest, S. E., Kumar, V., et al. (2010). Orchestration of floral initiation by APETALA1. *Science, 328*(5974), 85–89.

Kohler, C., & Hennig, L. (2010). Regulation of cell identity by plant Polycomb and trithorax group proteins. *Current Opinion in Genetics and Development, 20*(5), 541–547.

Krizek, B. A., & Fletcher, J. C. (2005). Molecular mechanisms of flower development: an armchair guide. *Natural Reviews Genetics, 6*(9), 688–698.

Krizek, B. A., Lewis, M. W., & Fletcher, J. C. (2006). RABBIT EARS is a second-whorl repressor of AGAMOUS that maintains spatial boundaries in *Arabidopsis* flowers. *Plant Journal, 45*(3), 369–383.

Krizek, B. A., & Meyerowitz, E. M. (1996). The *Arabidopsis* homeotic genes APETALA3 and PISTILLATA are sufficient to provide the B class organ identity function. *Development, 122*(1), 11–22.

Krogan, N. T., Hogan, K., & Long, J. A. (2012). APETALA2 negatively regulates multiple floral organ identity genes in *Arabidopsis* by recruiting the co-repressor TOPLESS and the histone deacetylase HDA19. *Development, 139*(22), 4180–4190.

La Rota, C., Chopard, J., Das, P., Paindavoine, S., Rozier, F., Farcot, E., et al. (2011). A data-driven integrative model of sepal primordium polarity in *Arabidopsis*. *Plant Cell, 23*(12), 4318–4333.

Lampugnani, E. R., Kilinc, A., & Smyth, D. R. (2013). Auxin controls petal initiation in *Arabidopsis*. *Development, 140*(1), 185–194.

Lee, I., Wolfe, D. S., Nilsson, O., & Weigel, D. (1997). A LEAFY co-regulator encoded by UNUSUAL FLORAL ORGANS. *Current Biology,* 7(2), 95–104.

Lenhard, M., Bohnert, A., Jurgens, G., & Laux, T. (2001). Termination of stem cell maintenance in *Arabidopsis* floral meristems by interactions between WUSCHEL and AGAMOUS. *Cell, 105*(6), 805–814.

Li, W., Zhou, Y., Liu, X., Yu, P., Cohen, J. D., & Meyerowitz, E. M. (2013). LEAFY controls auxin response pathways in floral primordium formation. *Science Signaling, 6*(277). ra23cr.

Liljegren, S. J., Ditta, G. S., Eshed, Y., Savidge, B., Bowman, J. L., & Yanofsky, M. F. (2000). SHATTERPROOF MADS-box genes control seed dispersal in *Arabidopsis*. *Nature, 404*(6779), 766–770.

Litt, A., & Kramer, E. M. (2010). The ABC model and the diversification of floral organ identity. *Seminars in Cell and Development Biology, 21*(1), 129–137.

Liu, C., Thong, Z., & Yu, H. (2009). Coming into bloom: the specification of floral meristems. *Development, 136*(20), 3379–3391.

Liu, C., Xi, W., Shen, L., Tan, C., & Yu, H. (2009). Regulation of floral patterning by flowering time genes. *Developmental Cell, 16*(5), 711–722.

Liu, C., Zhou, J., Bracha-Drori, K., Yalovsky, S., Ito, T., & Yu, H. (2007). Specification of *Arabidopsis* floral meristem identity by repression of flowering time genes. *Development, 134*(10), 1901–1910.

Liu, X., Kim, Y. J., Muller, R., Yumul, R. E., Liu, C., Pan, Y., et al. (2011). AGAMOUS terminates floral stem cell maintenance in *Arabidopsis* by directly repressing WUSCHEL through recruitment of Polycomb Group proteins. *Plant Cell, 23*(10), 3654–3670.

Lohmann, J. U., Hong, R. L., Hobe, M., Busch, M. A., Parcy, F., Simon, R., et al. (2001). A molecular link between stem cell regulation and floral patterning in *Arabidopsis*. *Cell, 105*(6), 793–803.

Long, J., & Barton, M. K. (2000). Initiation of axillary and floral meristems in *Arabidopsis*. *Developmental Biology, 218*(2), 341–353.

Maier, A. T., Stehling-Sun, S., Wollmann, H., Demar, M., Hong, R. L., Haubeiss, S., et al. (2009). Dual roles of the bZIP transcription factor PERIANTHIA in the control of floral architecture and homeotic gene expression. *Development, 136*(10), 1613–1620.

Mandel, M. A., Gustafson-Brown, C., Savidge, B., & Yanofsky, M. F. (1992). Molecular characterization of the *Arabidopsis* floral homeotic gene APETALA1. *Nature, 360*(6401), 273–277.

Mandel, M. A., & Yanofsky, M. (1998). The *Arabidopsis AGL9* MADS-box gene is expressed in young flower promordia. *Sexual Plant Reproduction, 11*, 22–28.

Mandel, M. A., & Yanofsky, M. F. (1995). A gene triggering flower formation in *Arabidopsis*. *Nature, 377*(6549), 522–524.

Mayer, K. F., Schoof, H., Haecker, A., Lenhard, M., Jurgens, G., & Laux, T. (1998). Role of WUSCHEL in regulating stem cell fate in the *Arabidopsis* shoot meristem. *Cell, 95*(6), 805–815.

Melzer, R., Verelst, W., & Theissen, G. (2009). The class E floral homeotic protein SEPALLATA3 is sufficient to loop DNA in 'floral quartet'-like complexes in vitro. *Nucleic Acids Research, 37*(1), 144–157.

Mendes, M. A., Guerra, R. F., Berns, M. C., Manzo, C., Masiero, S., Finzi, L., et al. (2013). MADS domain transcription factors mediate short-range DNA looping that is essential for target gene expression in *Arabidopsis*. *Plant Cell, 25*(7), 2560–2572.

Mizukami, Y., & Ma, H. (1992). Ectopic expression of the floral homeotic gene AGAMOUS in transgenic *Arabidopsis* plants alters floral organ identity. *Cell, 71*(1), 119–131.

Monfared, M. M., Carles, C. C., Rossignol, P., Pires, H. R., & Fletcher, J. C. (2013). The ULT1 and ULT2 trxG genes play overlapping roles in *Arabidopsis* development and gene regulation. *Molecular Plant, 6*(5), 1564–1579.

Moyroud, E., Minguet, E. G., Ott, F., Yant, L., Pose, D., Monniaux, M., et al. (2011). Prediction of regulatory interactions from genome sequences using a biophysical model for the *Arabidopsis* LEAFY transcription factor. *Plant Cell, 23*(4), 1293–1306.

Muino, J. M., Smaczniak, C., Angenent, G. C., Kaufmann, K., & van Dijk, A. D. (2014). Structural determinants of DNA recognition by plant MADS-domain transcription factors. *Nucleic Acids Research, 42*(4), 2138–2146.

Nag, A., Yang, Y., & Jack, T. (2007). DORNROSCHEN-LIKE, an AP2 gene, is necessary for stamen emergence in *Arabidopsis*. *Plant Molecular Biology, 65*(3), 219–232.

Ng, M., & Yanofsky, M. F. (2001). Activation of the *Arabidopsis* B class homeotic genes by APETALA1. *Plant Cell, 13*(4), 739–753.

O Maoileidigh, D. S., Wuest, S. E., Rae, L., Raganelli, A., Ryan, P. T., Kwasniewska, K., et al. (2013). Control of reproductive floral organ identity specification in *Arabidopsis* by the C function regulator AGAMOUS. *Plant Cell, 25*(7), 2482–2503.

O'Maoileidigh, D. S., Graciet, E., & Wellmer, F. (2014). Gene networks controlling *Arabidopsis thaliana* flower development. *New Phytologist, 201*(1), 16–30.

Ohno, C. K., Reddy, G. V., Heisler, M. G., & Meyerowitz, E. M. (2004). The *Arabidopsis* JAGGED gene encodes a zinc finger protein that promotes leaf tissue development. *Development, 131*(5), 1111–1122.

Okamuro, J. K., Caster, B., Villarroel, R., Van Montagu, M., & Jofuku, K. D. (1997). The AP2 domain of APETALA2 defines a large new family of DNA binding proteins in *Arabidopsis*. *Proceedings of the National Academy of Sciences of the United States of America, 94*(13), 7076–7081.

Okamuro, J. K., Szeto, W., Lotys-Prass, C., & Jofuku, K. D. (1997). Photo and hormonal control of meristem identity in the *Arabidopsis* flower mutants apetala2 and apetala1. *Plant Cell, 9*(1), 37–47.

Parcy, F., Nilsson, O., Busch, M. A., Lee, I., & Weigel, D. (1998). A genetic framework for floral patterning. *Nature, 395*(6702), 561–566.

Payne, T., Johnson, S. D., & Koltunow, A. M. (2004). KNUCKLES (KNU) encodes a C2H2 zinc-finger protein that regulates development of basal pattern elements of the *Arabidopsis* gynoecium. *Development, 131*(15), 3737–3749.

Peiffer, J. A., Kaushik, S., Sakai, H., Arteaga-Vazquez, M., Sanchez-Leon, N., Ghazal, H., et al. (2008). A spatial dissection of the *Arabidopsis* floral transcriptome by MPSS. *BMC Plant Biology, 8*, 43.

Pelaz, S., Ditta, G. S., Baumann, E., Wisman, E., & Yanofsky, M. F. (2000). B and C floral organ identity functions require SEPALLATA MADS-box genes. *Nature, 405*(6783), 200–203.

Pelaz, S., Tapia-Lopez, R., Alvarez-Buylla, E. R., & Yanofsky, M. F. (2001). Conversion of leaves into petals in *Arabidopsis*. *Current Biology, 11*(3), 182–184.

Pinyopich, A., Ditta, G. S., Savidge, B., Liljegren, S. J., Baumann, E., Wisman, E., & Yanofsky, M. F. (2003). Assessing the redundancy of MADS-box genes during carpel and ovule development. *Nature, 424*(6944), 85–88.

Reinhardt, D., Pesce, E. R., Stieger, P., Mandel, T., Baltensperger, K., Bennett, M., et al. (2003). Regulation of phyllotaxis by polar auxin transport. *Nature, 426*(6964), 255–260.

Riechmann, J. L., Krizek, B. A., & Meyerowitz, E. M. (1996). Dimerization specificity of *Arabidopsis* MADS domain homeotic proteins APETALA1, APETALA3, PISTILLATA, and AGAMOUS. *Proceedings of the National Academy of Sciences of the United States of America, 93*(10), 4793–4798.

Riechmann, J. L., Wang, M., & Meyerowitz, E. M. (1996). DNA-binding properties of *Arabidopsis* MADS domain homeotic proteins APETALA1, APETALA3, PISTILLATA and AGAMOUS. *Nucleic Acids Research, 24*(16), 3134–3141.

Roeder, A. H., Chickarmane, V., Cunha, A., Obara, B., Manjunath, B. S., & Meyerowitz, E. M. (2010). Variability in the control of cell division underlies sepal epidermal patterning in *Arabidopsis thaliana*. *PLoS Biology, 8*(5), e1000367.

Sakai, H., Medrano, L. J., & Meyerowitz, E. M. (1995). Role of SUPERMAN in maintaining *Arabidopsis* floral whorl boundaries. *Nature, 378*(6553), 199–203.

Sanders, P. M., Bui, A. Q., Weterings, K., McIntire, K. N., Hsu, Y. C., Lee, P. Y., et al. (1999). Anther developmental defects in *Arabidopsis thaliana* male-sterile mutants. *Sexual Plant Reproduction, 11*, 297–322.

Savidge, B., Rounsley, S. D., & Yanofsky, M. F. (1995). Temporal relationship between the transcription of two *Arabidopsis* MADS box genes and the floral organ identity genes. *Plant Cell, 7*(6), 721–733.

Schiefthaler, U., Balasubramanian, S., Sieber, P., Chevalier, D., Wisman, E., & Schneitz, K. (1999). Molecular analysis of NOZZLE, a gene involved in pattern formation and early sporogenesis during sex organ development in *Arabidopsis thaliana*. *Proceedings of the National Academy of Sciences of the United States of America, 96*(20), 11664–11669.

Schmid, M., Davison, T. S., Henz, S. R., Pape, U. J., Demar, M., Vingron, M., et al. (2005). A gene expression map of *Arabidopsis thaliana* development. *Nature Genetics, 37*(5), 501–506.

Schneitz, K., Hulskamp, M., & Pruitt, R. E. (1995). Wild-type ovule development in *Arabidopsis thaliana*–a light microscope study of cleared whole-mount tissue. *Plant Journal, 7*, 731–749.

Schultz, E. A., & Haughn, G. W. (1991). LEAFY, a homeotic gene that regulates inflorescence development in *Arabidopsis*. *Plant Cell, 3*(8), 771–781.

Schwab, R., Palatnik, J. F., Riester, M., Schommer, C., Schmid, M., & Weigel, D. (2005). Specific effects of microRNAs on the plant transcriptome. *Developmental Cell*, *8*(4), 517–527.
Sieber, P., Wellmer, F., Gheyselinck, J., Riechmann, J. L., & Meyerowitz, E. M. (2007). Redundancy and specialization among plant microRNAs: role of the MIR164 family in developmental robustness. *Development*, *134*(6), 1051–1060.
Sieburth, L. E., & Meyerowitz, E. M. (1997). Molecular dissection of the AGAMOUS control region shows that cis elements for spatial regulation are located intragenically. *Plant Cell*, *9*(3), 355–365.
Simpson, G. G., & Dean, C. (2002). *Arabidopsis*, the Rosetta stone of flowering time? *Science*, *296*(5566), 285–289.
Smaczniak, C., Immink, R. G., Angenent, G. C., & Kaufmann, K. (2012). Developmental and evolutionary diversity of plant MADS-domain factors: insights from recent studies. *Development*, *139*(17), 3081–3098.
Smaczniak, C., Immink, R. G., Muino, J. M., Blanvillain, R., Busscher, M., Busscher-Lange, J., et al. (2012). Characterization of MADS-domain transcription factor complexes in *Arabidopsis* flower development. *Proceedings of the National Academy of Sciences of the United States of America*, *109*(5), 1560–1565.
Smyth, D. R., Bowman, J. L., & Meyerowitz, E. M. (1990). Early flower development in *Arabidopsis*. *Plant Cell*, *2*(8), 755–767.
Sridhar, V. V., Surendrarao, A., Gonzalez, D., Conlan, R. S., & Liu, Z. (2004). Transcriptional repression of target genes by LEUNIG and SEUSS, two interacting regulatory proteins for *Arabidopsis* flower development. *Proceedings of the National Academy of Sciences of the United States of America*, *101*(31), 11494–11499.
Sridhar, V. V., Surendrarao, A., & Liu, Z. (2006). APETALA1 and SEPALLATA3 interact with SEUSS to mediate transcription repression during flower development. *Development*, *133*(16), 3159–3166.
Srikanth, A., & Schmid, M. (2011). Regulation of flowering time: all roads lead to Rome. *Cellular and Molecular Life Sciences*, *68*(12), 2013–2037.
Sun, B., Xu, Y. F., Ng, K. H., & Ito, T. (2009). A timing mechanism for stem cell maintenance and differentiation in the *Arabidopsis* floral meristem. *Genes & Development*, *23*(15), 1791–1804.
Sundstrom, J. F., Nakayama, N., Glimelius, K., & Irish, V. F. (2006). Direct regulation of the floral homeotic APETALA1 gene by APETALA3 and PISTILLATA in *Arabidopsis*. *Plant Journal*, *46*(4), 593–600.
Takeda, S., Matsumoto, N., & Okada, K. (2004). RABBIT EARS, encoding a SUPERMAN-like zinc finger protein, regulates petal development in *Arabidopsis thaliana*. *Development*, *131*(2), 425–434.
Theissen, G. (2001). Development of floral organ identity: stories from the MADS house. *Current Opinion in Plant Biology*, *4*(1), 75–85.
Theissen, G., & Melzer, R. (2007). Molecular mechanisms underlying origin and diversification of the angiosperm flower. *Annals of Botany*, *100*(3), 603–619.
Theissen, G., & Saedler, H. (2001). Plant biology. Floral quartets. *Nature*, *409*(6819), 469–471.
von Goethe, J. W. (1790). *Versuch die Metamorphose der Pflanzen zu erklären*. Gotha, Germany: Ettinger.
Vroemen, C. W., Mordhorst, A. P., Albrecht, C., Kwaaitaal, M. A., & de Vries, S. C. (2003). The CUP-SHAPED COTYLEDON3 gene is required for boundary and shoot meristem formation in *Arabidopsis*. *Plant Cell*, *15*(7), 1563–1577.
Wagner, D., Sablowski, R. W., & Meyerowitz, E. M. (1999). Transcriptional activation of APETALA1 by LEAFY. *Science*, *285*(5427), 582–584.
Wagner, D., Wellmer, F., Dilks, K., William, D., Smith, M. R., Kumar, P. P., et al. (2004). Floral induction in tissue culture: a system for the analysis of LEAFY-dependent gene regulation. *Plant Journal*, *39*(2), 273–282.

Weigel, D., Alvarez, J., Smyth, D. R., Yanofsky, M. F., & Meyerowitz, E. M. (1992). LEAFY controls floral meristem identity in *Arabidopsis*. *Cell*, *69*(5), 843–859.

Weigel, D., & Nilsson, O. (1995). A developmental switch sufficient for flower initiation in diverse plants. *Nature*, *377*(6549), 495–500.

Wellmer, F., Alves-Ferreira, M., Dubois, A., Riechmann, J. L., & Meyerowitz, E. M. (2006). Genome-wide analysis of gene expression during early *Arabidopsis* flower development. *PLoS Genetics*, *2*(7), e117.

Wellmer, F., Bowman, J. L., Davies, B., Ferrandiz, C., Fletcher, J. C., Franks, R. G., et al. (2014). Flower development: open questions and future directions. *Methods in Molecular Biology*, *1110*, 103–124.

Wellmer, F., Graciet, E., & Riechmann, J. L. (2014). Specification of floral organs in *Arabidopsis*. *Journal of Exprimental Botany*, *65*(1), 1–9.

Wellmer, F., & Riechmann, J. L. (2010). Gene networks controlling the initiation of flower development. *Trends in Genetics*, *26*(12), 519–527.

Wellmer, F., Riechmann, J. L., Alves-Ferreira, M., & Meyerowitz, E. M. (2004). Genome-wide analysis of spatial gene expression in *Arabidopsis* flowers. *Plant Cell*, *16*(5), 1314–1326.

Wigge, P. A., Kim, M. C., Jaeger, K. E., Busch, W., Schmid, M., Lohmann, J. U., et al. (2005). Integration of spatial and temporal information during floral induction in *Arabidopsis*. *Science*, *309*(5737), 1056–1059.

William, D. A., Su, Y., Smith, M. R., Lu, M., Baldwin, D. A., & Wagner, D. (2004). Genomic identification of direct target genes of LEAFY. *Proceedings of the National Academy of Sciences of the United States of America*, *101*(6), 1775–1780.

Winter, C. M., Austin, R. S., Blanvillain-Baufume, S., Reback, M. A., Monniaux, M., Wu, M. F., et al. (2011). LEAFY target genes reveal floral regulatory logic, cis motifs, and a link to biotic stimulus response. *Developmental Cell*, *20*(4), 430–443.

Wollmann, H., Mica, E., Todesco, M., Long, J. A., & Weigel, D. (2010). On reconciling the interactions between APETALA2, miR172 and AGAMOUS with the ABC model of flower development. *Development*, *137*(21), 3633–3642.

Wuest, S. E., O'Maoileidigh, D. S., Rae, L., Kwasniewska, K., Raganelli, A., Hanczaryk, K., et al. (2012). Molecular basis for the specification of floral organs by APETALA3 and PISTILLATA. *Proceedings of the National Academy of Science of the United States of America*, *109*(33), 13452–13457.

Yadav, R. K., Perales, M., Gruel, J., Girke, T., Jonsson, H., & Reddy, G.V. (2011). WUSCHEL protein movement mediates stem cell homeostasis in the *Arabidopsis* shoot apex. *Genes & Development*, *25*(19), 2025–2030.

Yamaguchi, N., Wu, M. F., Winter, C. M., Berns, M. C., Nole-Wilson, S., Yamaguchi, A., et al. (2013). A molecular framework for auxin-mediated initiation of flower primordia. *Developmental Cell*, *24*(3), 271–282.

Yang, W. C., Ye, D., Xu, J., & Sundaresan, V. (1999). The SPOROCYTELESS gene of *Arabidopsis* is required for initiation of sporogenesis and encodes a novel nuclear protein. *Genes & Development*, *13*(16), 2108–2117.

Yanofsky, M. F., Ma, H., Bowman, J. L., Drews, G. N., Feldmann, K. A., & Meyerowitz, E. M. (1990). The protein encoded by the Arabidopsis homeotic gene agamous resembles transcription factors. *Nature*, *346*(6279), 35–39.

Yant, L., Mathieu, J., Dinh, T.T., Ott, F., Lanz, C., Wollmann, H., et al. (2010). Orchestration of the floral transition and floral development in Arabidopsis by the bifunctional transcription factor APETALA2. *Plant Cell*, *22*(7), 2156–2170.

Yu, H., Ito, T., Zhao, Y., Peng, J., Kumar, P., & Meyerowitz, E. M. (2004). Floral homeotic genes are targets of gibberellin signaling in flower development. *Proceedings of the National Academy of Sciences of the United States of America*, *101*(20), 7827–7832.

Zhao, D., Yu, Q., Chen, M., & Ma, H. (2001). The ASK1 gene regulates B function gene expression in cooperation with UFO and LEAFY in Arabidopsis. *Development*, *128*(14), 2735–2746.

CHAPTER SEVEN

Grass Inflorescence: Basic Structure and Diversity

Junko Kyozuka
Graduate School of Agriculture and Life Sciences, University of Tokyo, Yayoi, Bunkyo, Tokyo, Japan
E-mail address: akyozuka@mail.ecc.u-tokyo.ac.jp

Contents

Abstract

Inflorescence structure directly affects the yield of grain crops. Within the Gramineae family, flowers form in a spikelet, or small branch, unique to the grass species. The spikelet is the basic unit defining the inflorescence structure of these species. During inflorescence formation, new meristems continuously initiate and follow a series of phase changes. Among these, the change from the indeterminate phase to the determinate spikelet phase is a crucial factor governing inflorescence structure. Temporal

Advances in Botanical Research, Volume 72
ISSN 0065-2296
http://dx.doi.org/10.1016/B978-0-12-417162-6.00007-9

regulation of the phase change in each meristem determines the spatial arrangement of spikelets in the inflorescence. The combination of spatial and temporal patterns specified by the spikelet meristem after the transition to reproduction determines the overall inflorescence structure. Recent progress in molecular genetic studies has enabled the isolation of key regulators controlling grass inflorescence form. Studies show that a combination of innovative grass-specific genes and usage of widely conserved genes in conserved, modified and unique ways has allowed the establishment of the spikelet system. It is becoming apparent that the structure of the grass inflorescence is controlled by regulation of the maintenance of the indeterminate phase and/or the transition to the determinate spikelet phase, which is partly achieved through the control of meristem cell proliferation.

7.1 INFLORESCENCE

7.1.1 Inflorescence Development

Inflorescence structure is one of the most important traits of plants because it determines the number of seeds produced. An inflorescence is a group of flowers or clusters of flowers arranged on a stem. The most important feature used to describe the inflorescence structure is the pattern by which flowers are arranged on the main stem and branches (Hake, 2008; Weberling, 1989). The inflorescence form is further modified by variations in many factors such as the length of internodes, phyllotaxis, proportions, angle, compressions and swellings, etc. In this chapter, the inflorescence form is considered to be the same as the inflorescence architecture, which is determined by the patterns of the arrangement of flowers.

The inflorescence structure results from the activities of shoot apical meristems (SAMs). The SAM is a population of stem cells maintained at the tip of the shoot axis (Sussex & Kerk, 2002). The growth and development of the aerial part of the plant depends on the continuous activity of the SAM. In most plants, in addition to the SAM produced during embryogenesis, SAMs are generated post-embryonically. Each SAM adopts a specific identity depending on the timing and position of its occurrence and the developmental stage of the plant. The meristem identity, which is defined by the type of organs produced by the meristem, is not permanent but changes as the plant development progresses. A critical change in meristem identity takes place at the floral transition, the transition from the vegetative to the reproductive phase. After floral transition, the vegetative SAM becomes the reproductive SAM, and is called an inflorescence meristem (IM). The IM produces flowers or flowering shoots, and starts to produce an inflorescence. In contrast to SAM and IM whose growth is indeterminate, the floral meristem is a determinate meristem whose activity is terminated after producing a certain number of floral organs. The transition to the floral meristem is characterized by the change from indeterminate to determinate growth, and thus is crucial to the inflorescence form.

7.1.2 Inflorescences in Grass Species

The grass family contains more than 10,000 taxa, including a number of important crop species, such as rice, maize, sorghum, wheat and barley (Gibson, 2009; Kellogg & Buell, 2009). Inflorescence form is one of the most important traits in the characterization of grass species. In these species, an inflorescence is formed at the top of each vegetative shoot branch or tiller. In other words, the tiller terminates with an inflorescence. A remarkable characteristic of the grass inflorescence is that flowers are formed in a small, highly modified flowering branch called a spikelet (Bell & Bryan, 1991; Clark & Pohl, 1979; Doust, Devos, Gadberry, Gale, & Kellogg, 2005; Kellogg et al., 2013) (Figure 7.1). The number of flowers in a spikelet varies amongst grass species from one to more than 50. For example, only a single flower is formed in a rice spikelet while in a wheat spikelet, the spikelet meristem displays indeterminate growth and

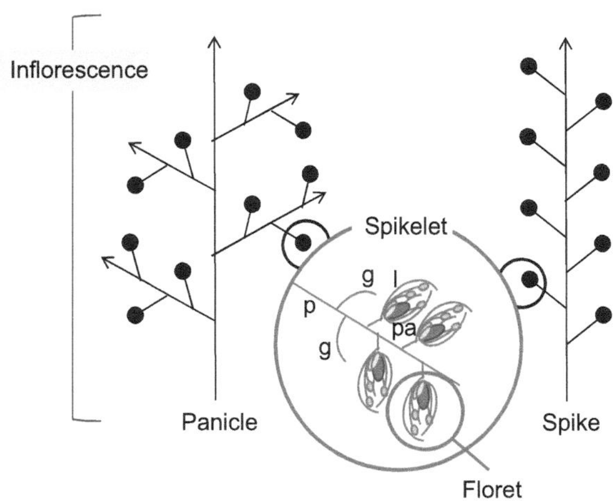

Figure 7.1 Structure of the grass inflorescence. A schematic representation of a grass inflorescence is shown. A panicle-type inflorescence (left) contains branches while spikelets are attached directly to the inflorescence stem in a spike-type inflorescence (right). A spikelet is subtended by a pair of suppressed leaves called glumes. A flower (floret) contains a lemma, palea, lodicules, stamens and a pistil. Closed circle, spikelet; p, pedicel; g, glume; l, lemma; pa, palea. (See the colour plate.)

can potentially produce an indeterminate number of flowers. Although the spikelet meristem maintains its ability to produce floral meristems, the developmental events within a single spikelet do not affect the structure of the whole inflorescence. Therefore, the spikelet is considered a terminal structure and the inflorescence form is described based on the arrangement of spikelets, regardless of how flowers are arranged within a spikelet. Evolution of spikelet structure may have been an innovation allowing greater complexity in inflorescence development, enabling the production of more seeds.

Grass inflorescences are classified as either a spike or a panicle (Figure 7.1). In spike-type inflorescences, spikelets are attached directly to the inflorescence stem. Variations in the number of spikelets produced at a node and the length of the inflorescence stem, defined by the number of nodes and the elongation of the internodes, determines the basic pattern of the spike-type inflorescence. The panicle-type inflorescences contain primary, secondary and sometimes higher-order branches, with spikelets forming on these branches. This inflorescence structure is primarily determined by factors defining the branching pattern, such as the extent of reiteration of branch formation, the length of branches (defined by the number of organs produced on each branch) and the gradient of the length of the lateral organs on each branch.

7.1.3 Inflorescence Development in Grass Species

In principle, the inflorescence form is defined by the arrangement of terminal structures (Prusinkiewicz et al., 2007). Since the spikelets are considered to be the terminal structure in grass species, the grass inflorescence form is explained by the spatial arrangement of spikelets. This is determined by the temporal control of the meristem phase change from an indeterminate branch meristem or IM to the determinate spikelet meristem. Upon floral transition, the vegetative SAM of grass species is transformed into the IM, which becomes the source of the whole inflorescence. The IM starts to produce suppressed leaves called bracts and new meristems are generated in the axil of these bracts. A newly formed meristem acquires either indeterminate branch meristem identity, which reiterates new meristem formation, or determinate spikelet meristem identity, depending upon the timing and position of its occurrence (Figure 7.2). Delays in spikelet meristem specification lead to iterations of branching, resulting in larger inflorescences that could potentially produce more spikelets. Conversely, the acceleration of

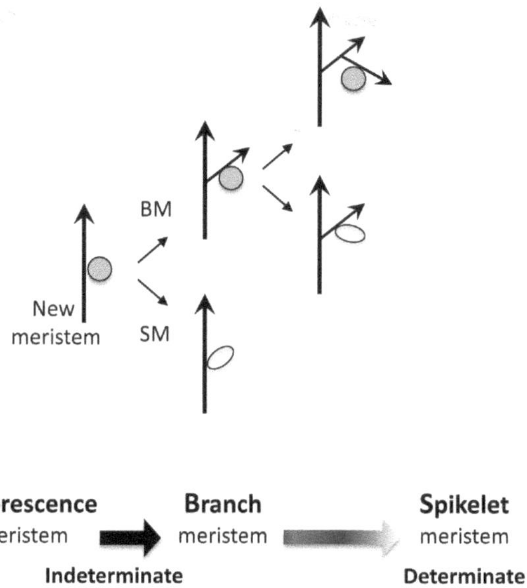

Figure 7.2 Concept of meristem identity specification. A new meristem (green circle) acquires the identity of a indeterminate branch meristem (BM, green arrow) or a determinate spikelet meristem (SM, purple oval). The BM further produces new meristems that acquire either SM or BM identities. (See the colour plate.)

spikelet meristem specification results in smaller inflorescences with fewer spikelets.

A schematic of a developing grass inflorescence is shown in Figure 7.3. The number of primary branches produced in the inflorescence is determined by the timing of the transition of the IM to the spikelet meristem. Similarly, the number of organs produced on each branch is determined by the time when the branch meristem is transformed into the spikelet meristem, which will grow as a terminal spikelet. On a branch, the proximal lateral organs grow as next order branches, while distal organs that initiate in the later stage grow as lateral spikelets. The ratio of next order branches to spikelets in lateral organs on a branch reflects the timing of the programme change from branch meristem formation to spikelet formation along the branch. Any type of grass inflorescence can be produced by the combinatorial exploitation of these factors. The difference between spike- and panicle-type inflorescences is explained by the timing of the meristem phase change (Figure 7.3). When the ratio of next order branches to the whole number of lateral organs on the stem is zero, the inflorescence becomes a spike.

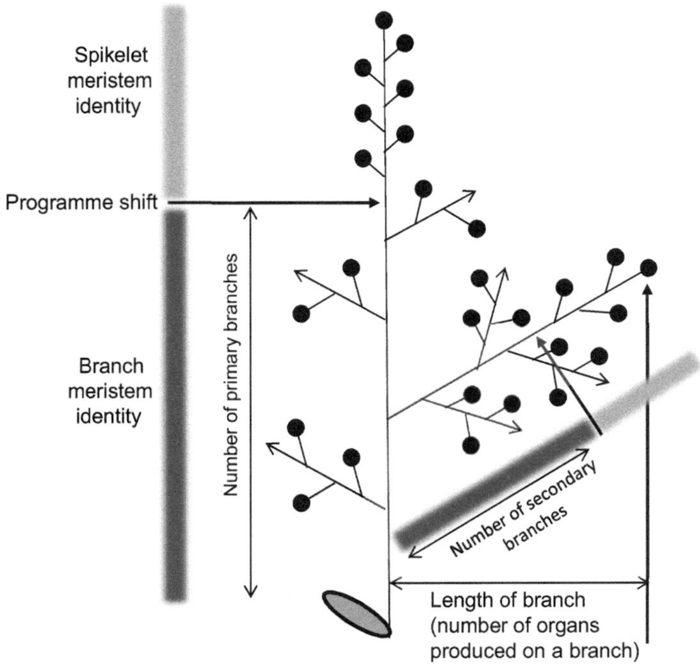

Figure 7.3 Basic concept of grass inflorescence architecture. The pattern of inflorescence branching is determined by the timing of the meristem phase shift from the BM to the SM. A new meristem acquires the identity of a branch meristem or a spikelet meristem. Among meristems produced by a branch meristem, early lateral meristems acquire a BM identity and grow as next order branches at proximal positions (green region), while later lateral meristems are specified as spikelet meristems (purple region). The timing of the programme shift determines the number of branches and the shape of the branch. Timing of the transition to spikelet identity at the top of each branch determines the total number of organs produced on the branch. (See the colour plate.)

7.2 THE GENETIC BASIS OF GRASS INFLORESCENCE DEVELOPMENT

7.2.1 Initiation of New Meristems in the Inflorescence

Initiation of new meristems is a prerequisite for grass inflorescence development. In most plant species, post-embryonic SAMs initiate as an axillary meristem in the axil of leaves. In grasses, leaf growth is severely suppressed after floral transition. Despite the suppression of growth, axillary meristems are produced in the axil of each rudimentary leaf, called bracts, and grow as axillary buds. There is a notable difference in the behaviour of the axillary buds produced in the vegetative phase and the buds produced in reproductive phases. The axillary buds produced during the vegetative phase become dormant and grow only when the environmental and internal conditions are suitable for their growth, whereas the buds produced after floral transition are not subjected to this regulation and start to grow immediately after their initiation (McSteen & Leyser, 2005; Wang & Li, 2011). *LAX PANICLE1* (*LAX1*) of rice and *BARREN STALK1* (*BA1*) of maize are essential for axillary meristem formation in both vegetative and reproductive phases (Gallavotti et al., 2004; Komatsu, Maekawa, Shimamoto, & Kyozuka, 2001; Komatsu et al., 2003; Oikawa & Kyozuka 2009; Ritter, Padilla, & Schmidt, 2002). *LAX1* and *BA1* are orthologous basic helix–loop–helix (bHLH) transcription factors. Formation of all types of shoot branches, including tillers, inflorescence branches and spikelets, is severely suppressed when *LAX1*/*BA1* function is lost in their loss-of-function mutants. *LAX1* and *BA1* mRNA are expressed in the dorsal margin of the newly initiating axillary meristems (Gallavotti et al., 2004; Komatsu et al., 2003). Interestingly, mRNA expression is observed in all types of axillary meristems irrespective of their future identity, whereas no expression was observed in the SAM generated at embryogenesis. Accordingly, the embryo-derived SAM is not affected in *lax1* and *ba1* mutants. In weak alleles of *lax1*, normal spikelets are formed at the top of each inflorescence branch whereas lateral spikelets are almost completely suppressed. Thus, *LAX1* function is required for new meristem initiation in the axil of leaves but not for meristem function itself. From the analysis of *BA1* function, it is proposed that *BA1* works to integrate a signal from the leaves with the polar auxin transport pathway to create the auxin levels required for new axillary meristem formation (Gallavotti et al., 2010; Skirpan, Wu, & McSteen, 2008; Wu & McSteen, 2007). Phylogenetic analysis reveals that *LAX1* and *BA1* genes are conserved not only among monocots but also in dicot species, with the origin for the lineage at least near the origin of flowering plants (Kaur et al., 2008; Woods, Hope, & Malcomber, 2011; Yang, Wang, Schmitz, Müller, & Theres, 2012). In maize, the *Barren stalk fastigiate1* (*Baf1*) gene was identified as a regulator of axillary meristem initiation (Gallavotti et al., 2011). *Baf1* encodes a putative transcription factor containing an AT-hook DNA binding motif and a conserved domain of unknown function, called the Plant and Prokaryote Conserved (PPC) domain. *Baf1* mutants show phenotypes resembling those of *ba1* and they are expressed in an overlapping narrow domain demarcating a boundary region between the SAM and an initiating axillary meristem. The *ba1 baf1* double mutant plants show similar phenotypes as *ba1* single mutant and *Baf1* expression was abolished in *ba1* mutants. Based on genetic analysis, it was proposed that *Baf1* functions upstream of *Ba1* in the initiation of axillary meristems. *Baf1* orthologues are likely present in all cereals. The rice *Baf1* orthologue was identified as a gene showing an expression pattern similar to that of *LAX1* by a microarray analysis and in situ hybridization analysis (Furutani, Sukegawa, & Kyozuka, 2006). However, it was suggested that the rice *Baf1* orthologue works independently from *LAX1*. In rice, the *LAX2* gene, which works to initiate axillary meristems in the same genetic pathway as *LAX1*, was also identified (Tabuchi et al., 2011). Despite the significant contribution of *LAX1* and *BA1* function to axillary meristem formation, some axillary meristems are still formed in their complete loss-of-function alleles, suggesting that other genes work redundantly with the *LAX1/BA1* pathway. One candidate is *MONOCULM1 (MOC1)*, an orthologue of *LATERAL SUPPRESSOR* (*LAS*) of *Arabidopsis* and *ls* of tomato, which encode putative transcription factors of the GRAS family (Greb et al., 2003; Li et al., 2003). Comparative sequence analysis of *MONOCULM1*-orthologous regions in 14 *Oryza* genomes revealed highly conserved gene colinearity and structure in the *MOC1* region (Lu et al., 2009). *LAX1* was reported as a candidate for qSPB1, a QTL locus responsive for secondary inflorescence branch number (Gao et al., 2013).

7.2.2 Regulators of Spikelet Meristem Identity

To date, several genes have been identified as the regulators of spikelet meristem identity, mainly from rice and maize. Analyses of these genes have provided valuable knowledge in understanding the molecular and genetic basis of grass inflorescence development.

7.2.2.1 APO1 *and* APO2 *are Key Regulators of Meristem Phase Change in Rice Inflorescence Development*

ABERRANT PANICLE ORGANIZATION 1 (*APO1*) and *APO2* of rice encode an F-box protein orthologue of *UNUSUAL FLORAL ORGAN* (*UFO*), and an orthologue of *Arabidopsis LEAFY* (*LFY*), respectively (Ikeda, Ito, Nagasawa, Kyozuka, & Nagato, 2007; Ikeda, Nagasawa, & Nagato, 2005; Ikeda-Kawakatsu, Maekawa, Izawa, Itoh, & Nagato, 2011; Ikeda-Kawakatsu et al., 2009). Phenotypes of loss-of-function mutations of *APO1* and *APO2* genes are very similar to each other. *apo1* and *apo2* mutants produce small inflorescences with reduced branching. In addition, in *apo1* and *apo2* loss-of-function mutants, the IM is reduced in size, the duration of leaf initiation (plastochrone) becomes shorter and development of floral organs is disturbed. The inflorescence branching phenotype is interpreted as the precocious specification of spikelet meristem identity, thus, *APO1* and *APO2* function to maintain branch meristem identity and to suppress the transition to spikelet meristem identity. *APO1* was first isolated as a causal gene of a recessive *apo1* mutant (Ikeda et al., 2005, 2007). Subsequently, it was shown that *Ur1*, a dominant mutant producing increased inflorescence branching, is a dominant gain-of-function mutant of the *APO1* gene (Ikeda-Kawakatsu et al., 2009; Imai, Murai, Hao, Chiba, Chiba, & Ishikawa, 2009; Murai & Izawa 1994). Increases in inflorescence branching were observed in other dominant alleles of *apo1* mutants (Ikeda-Kawakatsu et al., 2009). *APO1* was also identified as *Primary Branch Number/PRIMARY RACHIS BRANCHES (PRBs)*, a QTL locus for grain number and stem size in independent studies (Ookawa et al., 2010; Terao, Nagata, Morino, & Hirose, 2010). Analyses of *apo1* loss-of-function and gain-of-function mutants showed that the branching phenotypes are dependent on the expression level of *APO1*. When *APO1* was over-expressed using the CaMV35S promoter, transgenic rice plants exhibited an extreme phenotype, in which the inflorescence consists of enormous numbers of branches (Ikeda-Kawakatsu et al., 2009). APO1 and APO2 proteins interact directly and *APO2* function is essential for *APO1* function, while the level of *APO1*, but not *APO2*, activity acts as a determinant for the timing of meristem phase change (Ikeda-Kawakatsu et al., 2011). The elevation of *APO2* expression alone is not sufficient to cause any changes in inflorescence form. A similar dependency was reported for *LFY* and *UFO* homologues in petunia (Souer, Rebocho, Bliek, Kusters, de Bruin, & Koes, 2008). Although *APO1* and its orthologue play crucial roles in determining the inflorescence form, their molecular function still remains unknown.

The rice orthologue of *LFY* was also reported as *RFL* (*Rice FLORICAULA LEAFY*), and its role was analysed in RNAi knockdown plants prior to the isolation of *apo2* (Kyozuka, Konishi, Nemoto, Izawa, & Shimamoto, 1998; Rao, Prasad, Kumar, & Vijayraghavan, 2008). These studies and the analysis of *apo2* mutants demonstrated that *APO2/RFL* of rice suppresses spikelet meristem identity and promotes indeterminate BM identity. The action of *APO2/RFL* is seemingly opposite to that known for *LFY* orthologues in dicots, that is, the promotion of the determinate floral fate (reviewed in Irish, 2010; Moyroud, Kusters, Monniaux, Koes, & Parcy,

2010). *LFY* orthologues were identified from other grass species, including maize, wheat, and *Lolium temulentum* (Bomblies et al., 2003; Gocal et al., 2001; Shitsukawa, Takagishi, Ikari, Takumi, & Murai, 2006), but its functions were reported only for maize *LFY* orthologues, namely, *ZFL1* (Zea maize *FLO/LFY1*) and *ZFL2*. *zfl1 zfl2* double mutants show defects in floret determinacy and organ identity in male and female inflorescences (Bomblies et al., 2003). Expression of *ZFL1* and *ZFL2* in early floral meristems and developing floral organ primordia also resemble the expression patterns of *LFY* homologues in dicots. Mutant phenotypes and expression patterns of *ZFL1* and *ZFL2* suggest that *LFY/FLO* function in the control of meristem determinacy seems to be conserved between maize and dicot species. Currently, rice is the only species whose LFY homologue functions to promote meristem indeterminacy. *APO2/RFL* is also different from *ZFL1* and *ZFL2* in its expression pattern. *ZFL1* and *ZFL2* are expressed in the floral meristems and floral organs, whereas *APO2/RFL* expression begins in the whole region of the IM, then proceeds to become excluded from the meristem regions in the inflorescence. Interestingly, *WFL*, a wheat *LFY/FLO* orthologue, shows expression patterns very similar to that of *APO2/RFL* (Shitsukawa et al., 2006).

LFY exists in all land plants studied to date and functions as a major regulator of floral meristem identity and flower development in eudicot species (reviewed in Moyroud et al., 2010). Studies in moss (*Physcomitrella patens*) showed that *LFY* orthologues might play a role in cell division and meristem development and that this function is more ancestral than its function in floral meristem and floral organ development (Tanahashi, Sumikawa, Kato, & Hasebe, 2005). In addition to the suppression of spikelet meristem identity, *APO1* and *APO2* are involved in various other developmental processes (Ikeda-Kawakatsu et al., 2011). During the vegetative stage, they positively regulate leaf growth and negatively regulate the intervals of leaf initiation. They are also involved in the proliferation of the meristem. After the transition to the reproductive phase, *APO1* and *APO2* expression begins in the meristem where they stimulate cell division. They also regulate floral organ development (Ikeda et al., 2005, 2007; Ikeda-Kawakatsu et al., 2009, 2012). Regulation of floral organ development in the second and third whorl organs is well conserved between maize, rice and dicots. *APO1* and *APO2* probably use both the ancestral and newer roles of *UFO* and *LFY* orthologues. Elucidation of the functions of *APO1* and *APO2* at the molecular level will be required for a further understanding of meristem phase change.

7.2.2.2 SPL Transcription Factor Enhances the Indeterminate Branch Phase in Maize and Rice

SQUAMOSA PROMOTER BINDING PROTEIN (*SPL*) encodes a plant-specific transcription factor which binds to the promoter sequence of the SQUAMOSA-type MADS-box genes. The *SPL* genes were first reported in *Antirrhinum* (Klein, Saedler, & Huijser, 1996). Subsequently, functional studies, largely performed in *Arabidopsis*, revealed that *SPL* genes regulate the juvenile-to-adult phase transition, the duration of leaf initiation (plastochron), flowering time, shoot branching and inflorescence form (Schwarz, Grande, Bujdoso, Saedler, & Huijser, 2008; Wang, Czech, & Weigel, 2009; Wang, Schwab, Czech, Mica, & Weigel, 2008; Wu & Poethig, 2006). *SPL* genes are composed of a small gene family (Preston & Hileman 2013; Yang, Wang, Gu, Hu, Xu, & Xu, 2007). Some of the members of the SPL family contain a target sequence for miR156 and are negatively regulated by miR156 at the level of translation as well as by the cleavage of mRNA (reviewed in Poethig, 2013; Spanudakis & Jackson 2014). The functions of miR156 and its target *SPL* genes are well conserved among diverse plant species (Chuck, Cigan, Saeteurn, & Hake, 2007; Schwab et al., 2005; Wu & Poethig 2006). *SPL* genes are involved in the control of the inflorescence structure through the control of spikelet meristem identity in grass species. In rice, *OsSPL14* was identified as the QTL loci *WEALTHY FARMER'S PANICLE* (*WFP*) and *Ideal Plant Architecture1* (*IPA1*), which control grain number, by two independent studies (Jiao et al., 2010; Miura et al., 2010). These studies demonstrated that an increase in the expression level of *OsSPL14* causes an increase in inflorescence branching, thus leading to the increase in spikelet number. In the ST-12 line, which produces an inflorescence with more branches, expression of *OsSPL14* is increased by a heritable epigenetic change (Miura et al., 2010). The *OsSPL14* gene contains a target sequence for miR156 and indeed is down-regulated by miR156. Shaoniejing (SNJ), a japonica cultivar with high grain number, contains an *OsSPL14/ipa1* allele, in which a mutation in the target sequence of miR156 caused an up-regulation of *SPL14* expression. This rise in *OsSPL14* expression led to an increase in inflorescence branching, thus, also leading to an increase in the number of spikelet meristems. It was also demonstrated that a japonica cultivar, Aikawa1 contains a natural mutation in the miR156 target sequence. All these findings indicate that higher *OsSPL14/WFP/IPA1* activity promotes indeterminate branch meristem phase and delays the transition to spikelet meristem identity.

TASSEL SHEATH 4 (*TSH4*) of maize encoding zmSBP6, which is the closest orthologue of *OsSPL14*, controls meristem determinacy in the maize inflorescence (Chuck, Whipple, Jackson, & Hake, 2010; Preston & Hileman, 2013). A wild-type tassel, the male inflorescence of maize, produces more than 10 long tassel branches, whereas, less than 5 tassel branches are produced in the *tsh4* mutant. In maize, an additional meristem phase exists before spikelet meristem identity: a pair of spikelets is formed by the spikelet pair meristem. The spikelet pair meristems are frequently replaced by spikelets in *tsh4* mutants. The defects indicate that the transition is accelerated in *tsh4*, indicating that wild-type *TSH4* positively regulates meristem indeterminacy or suppresses the transition to the determinate phase. Thus, the functions of the *SPL* genes are well conserved between rice and maize. In dominant mutants of the *Corngrass1* (*Cg1*) gene of maize, higher levels of miR156 accumulate due to an insertion of transposons in the regulatory region of the miR156 precursor gene (Chuck et al., 2007).

OsSPL14 is expressed in the bracts of inflorescence branches but not in the meristems that undergo phase change (Luo, Li, Miura, Ashikari, & Kyozuka, 2012). This suggests that *OsSPL14* works in a non-cell autonomous manner. In *Arabidopsis*, a non-cell autonomous function of SPL9, and probably SPL10 and SPL15, in inhibiting initiation of new leaves was proposed (Wang et al., 2008). Wang et al. (2008) proposed that a mobile factor, which acts in the meristem to inhibit leaf initiation is synthesized in the young leaf primordia, although the factor has not yet been identified. A plausible possibility is that the SPL proteins themselves move from leaf primordia to the meristem. A similar non-cell autonomous action of the SPL gene, *TSH4*, was also observed in maize but interpreted differently (Chuck et al., 2010). TSH4 protein, the orthologue of OsSPL14, localizes at the base of lateral meristems, but not within the meristem itself in the inflorescence. Precise analysis of TSH4 localization revealed that TSH4 forms a boundary between the bract and lateral meristems. On the other hand, *miR156* is expressed in the meristem. The complementary expression patterns of *TSH4* and *miR156* support the negative regulation of *THS4* expression by *miR156*. Chuck et al. (2010) proposed that the repression of *TSH4* expression in the meristem allows the establishment of lateral meristems, and the balance between the proliferation of the lateral meristem and suppression of leaf growth determines the timing of meristem phase transition. They also found that the *tsh4* gene works through the regulation of the *ramosa2* (*ra2*) meristem determinacy gene encoding a protein encoding LATERAL ORGAN BOUNDARY (LOB) domain (Bortiri et al., 2006).

OsSPL14 enhances the formation of inflorescence branches as well as suppressing the growth of tillers, i.e. vegetative branches (Jiao et al., 2010; Miura et al., 2010). At a glance, it seems odd that one gene can affect branching in seemingly opposite ways; however, these two steps are the consequences of different developmental contexts. In the case of inflorescence branching, the number of branches is determined by the timing of the meristem phase change to spikelet meristem identity, whereas the number of tillers is determined by the regulation of the outgrowth of the axillary buds. The latter is regulated by an intricate network of hormones including strigolactone, auxin and cytokinin. Currently, it is unclear how *OsSPL14* controls these two developmental processes. Elucidation of the molecular mechanism by which OsSPL14 exerts its effect on these different developmental steps will be of great interest. Interestingly, *OsSPL14*, *APO1* and *APO2* functions largely overlap. All three genes promote indeterminate branch identity, increase plastochrone and suppress the outgrowth of tiller buds. Moreover, there is a similarity in the expression patterns of *OsSPL14* and *APO2* (Ikeda-Kawakatsu et al., 2011; Kyozuka et al., 1998; Luo et al., 2012). Furthermore, plastochrone and inflorescence branching are affected in all *apo1*, *apo2* mutants and *OsSPL14* knockdown plants (Ikeda et al., 2005; Ikeda-Kawakatsu et al., 2011; Luo et al., 2012). Genetic analysis showed that *OsSPL14* works downstream of the *APO1-APO2* pathway, at least in the control of inflorescence branching (Luo et al., 2012). In *Arabidopsis*, *SPL3* directly binds the promoter of *LFY* (Yamaguchi et al., 2009). These observations imply that the genetic modules SPL-LFY and OsSPL14-APO1/APO2 are conserved between rice and *Arabidopsis*, but that these modules work in opposite directions. It would be very interesting to investigate the interactions between *OsSPL14*, *APO1* and *APO2* at the molecular level and identify downstream target genes involved in the control of inflorescence branching.

Isolation of *OsSPL14* was a significant achievement in terms of rice breeding. Recent rice breeding approaches to improved yield are described as the new plant type concept in which the ideal plant architecture for rice is defined as having a relatively small number of highly productive tillers (Khush, 2001). The phenotype conferred by the altered expression level and pattern of *OsSPL14* and *APO1*, of fewer tillers with a stronger culm and a panicle bearing more grains, indicates the potential of these genes in rice breeding.

7.2.2.3 ALOG Family Genes Mediate Meristem Phase Changes

TAWAWA1 (*TAW1*) was identified as a negative regulator of spikelet meristem identity in rice (Yoshida et al., 2013). In *taw1-D1* and *taw1-D2*, two independent dominant mutants of *TAW1*, *TAW1* mRNA accumulation is enhanced due to the insertion of a transposon in the 3′ regulatory regions. This leads to a delay in spikelet meristem identity specification and results in the formation of inflorescences with increased numbers of branches. In both dominant mutants, the spatial pattern of *TAW1* mRNA expression is largely unaffected, however, the level of *TAW1* mRNA expression is increased and the duration of *TAW1* expression is prolonged, suggesting that the sequence around the insertion region is important for the fine-tuning of *TAW1* expression pattern. In plants carrying the homozygous *taw1-D1*, a more severe allele due to the strong suppression of the progression to spikelet meristem identity, branch meristem formation is reiterated and the inflorescence appears as an aggregation of undifferentiated meristems. In contrast, a reduced number of inflorescence branches are produced in a weak loss-of-function mutant allele of *TAW1*. These observations indicate that *TAW1* is a negative regulator of the transition to determinate spikelet meristem identity and functions to maintain indeterminate branch meristem identity.

TAW1 is a member of a small gene family named ALOG after *Arabidopsis LSH* and *Oryza G1* (Yoshida, Suzaki, Tanaka, & Hirano, 2009; Zhao et al., 2004). Based on their conserved amino acid sequences and their localization in the nucleus, it is proposed that ALOG proteins play a role as DNA binding proteins (Iyer & Aravind 2012; MacAlister et al., 2012; Takeda et al., 2011; Yoshida et al., 2009, 2013). However, TAW1 and G1 show only low levels of activity in stimulating transcription, indicating a possibility that the LSH/ALOG proteins may function as transcriptional cofactors. An elucidation of the molecular functions of the ALOG family of proteins will be a crucial step towards understanding the control of meristem phase changes.

The mechanisms by which TAW1 regulates meristem phase change also remain to be elucidated. The founder members of the ALOG family, *Arabidopsis LSH* genes, were identified as causal genes of gain-of-function mutants that show a hypersensitivity to continuous light (Zhao et al., 2004). Constitutive expression of *TAW1* in rice also causes development of green pigmented roots (Yoshida et al., 2013). These findings may indicate a possible link between light signalling and meristem phase change. Consistent with this view, constitutive expression of *LSH4* resulted in ectopic meristem formation in *Arabidopsis* (Takeda et al., 2011), implying that *TAW1* and *LSH1* function to maintain meristematic activity. In spite of the functional conservation, no clear defects are observed in loss-of-function mutants of *LSH* genes in *Arabidopsis* (Takeda et al., 2011). On the other hand, mutations in the *TERMINATING FLOWER* (*TMF*) gene of tomato show clear defects (MacAlister et al., 2012). In loss-of-function *tmf* mutants, the primary SAMs terminate to a single flower instead of making sympodial inflorescences. The different responses to the loss or reduction of *TAW1* and its homologues among rice, tomato and *Arabidopsis* may be related to their inflorescence structures. In the panicle and sympodial inflorescences of rice and tomato, respectively, the indeterminate meristems are maintained for a defined developmental time and the flowers are produced in a cluster. On the other hand, in the *Arabidopsis* inflorescence, a flower is directly attached to the main stem, implying a lack of the indeterminate phase in the meristem.

Two members, *LSH3* and *LSH4*, of the *Arabidopsis* ALOG family were shown to be direct downstream targets of *CUP SHAPED COTYLEDON1* (*CUC1*) (Takeda et al., 2011). *CUC1* defines the boundary of the meristem redundantly with both *CUC2* and *CUC3* (Aida, Ishida, Fukaki, Fujisawa, & Tasaka, 1997; Hibara et al., 2006; Hibara, Takada, & Tasaka, 2003). Mutations in *CUC1* and *CUC2* compromise maintenance of meristem cells during embryogenesis in *Arabidopsis* (Aida et al., 1997). These results suggest that *LSH4* may play roles in the control of meristem function through the maintenance of undifferentiated cells. A close homologue of *TAW1* was identified in a transcriptome study of embryogenesis in maize (Takacs et al., 2012). The maize *TAW1* homologue demarcates the boundary between lateral organs and the SAM during embryogenesis. In combination, these results indicate the possibility that the *TAW1*-related ALOG family genes perform a function related to meristem cells. Indeed, *TAW1* expression occurs in meristems with the greatest expression in the branch meristems and disappears when meristems acquire spikelet meristem identity. If the primary function of ALOG family proteins is the maintenance of stem cells in the SAM, the next interesting question to answer is how does the meristem activity link to the control of meristem phase transition.

Genetic analysis in tomato demonstrated that the precocious expression of *ANANTHA* (*AN*) and *FALSIFLORA* (*FA*), which are the tomato orthologues of *UFO* and *LFY*, respectively, was prevented by TMF (Lippman et al., 2008; Molinero-Rosales et al., 1999). In tomato inflorescence development, *AN* and *FA* promote floral meristem identity. These results suggest that the role of TMF is to fine-tune the timing of the meristem phase change to a determinate fate by suppressing the transition to the floral fate. As discussed in the previous section, the *UFO-LFY* module in rice works to suppress floral fate. Therefore, it is unlikely that the relationship between *TMF-AN/FA* observed in tomato is also conserved in rice.

Transcriptome analysis in *taw1* mutants identified *SHORT VEGETATIVE PHASE* (*SVP*)-type MADS-box genes as candidates downstream of TAW1 (Arora et al., 2007; Lee et al., 2007; Yoshida et al., 2013). These genes function to suppress meristem phase transition to spikelet meristem identity. The SVP gene delays floral transition in *Arabidopsis* (Hartmann et al., 2000). The rice genome has three SVP-like genes, *OsMADS22*, *OsMADS47* and *OsMADS55* (Arora et al., 2007). Over-expression of these rice SVP genes promotes vegetative development in both rice and *Arabidopsis* (Fornara, Gregis, Pelucchi, Colombo, & Kater, 2008; Sentoku, Kato, Kitano, & Imai, 2005). The function of SVP-like MADS-box genes to suppress determinate spikelet identity is also conserved in barley (Trevaskis et al., 2007). *BM1* and *BM2*, SVP-like genes in barley, are expressed in vegetative tissues. They inhibited spikelet development and caused partial reversion to vegetative growth when ectopically expressed (Trevaskis et al., 2007). These results indicate a possibility that they play roles to prevent precocious spikelet formation in normal development. Therefore, SVP genes function to prevent the precocious transition to a determinate fate in grasses, ensuring the appropriate degree of inflorescence branch formation. *TAW1* may exert its function partially through the fine-tuning of *SVP* function. Interestingly, all three rice SVP genes work as negative regulators of brassinosteroid (BR) responses (Lee, Choi, & An, 2008). It is expected that future studies on the role of BR in the control of meristem phase change will reveal new aspects of the control of grass inflorescence development.

The morphological mutant of maize, Pod corn has a dominant mutation in the *Tunicate1* (*Tu1*) gene (Han, Jackson, & Martienssen, 2012; Wingen et al., 2012). In this mutant, vegetative traits are enhanced and glumes are extremely elongated. In addition, branch determinacy is affected. The peculiar phenotype of the *Tu1* mutant is a result of the ectopic expression of *ZMM19*, a member of the *SVP*-like MADS-box genes, in the inflorescences. The inflorescence branching phenotype of *Tu1* resembles that of *ramosa* mutants. Based on genetic experiments, the idea that *Ramosa1* and *Tu1* activity is mutually repressive has been proposed (Han et al., 2012). Further analysis of the involvement of SVP genes, conferring vegetative traits and other genetic pathways at the genetic and molecular level, may be important in understanding how the progression of meristem phase transition is controlled.

7.2.2.4 Positive Regulators of Spikelet Meristem Identity

RAMOSA1 (*ra1*), *ra2* and *ra3* of maize are well-studied regulators of inflorescence branching (Bortiri et al., 2006; Satoh-Nagasawa, Nagasawa, Malcomber, Sakai, & Jackson, 2006; Vollbrecht, Springer, Goh, Buckler, & Martienssen, 2005). Loss-of-function mutants show increases in inflorescence branching, indicating they function in the wild type to promote the transition to the determinate meristem phase. *RA1*, *RA2* and *RA3* encode a C2H2 zinc finger transcription factor, a LOB domain transcription factor and a trehalose-6-phosphate phosphatase, respectively. Interestingly, *RA1* exists only in Panicoideae. *RA2* and *RA3* are found in other grasses, but their functions remain unclear.

Most regulators of inflorescence branching isolated from rice are negative regulators, such as *APO1*, *APO2*, *OsSPL14* and *TAW1*. On the contrary, *PANICLE PHYTOMER 2* (*PAP2*) of rice, which encodes OsMADS34, a member of the SEPALLATA (SEP) subfamily of MADS-box proteins, works as a positive regulator of spikelet meristem identity (Gao et al., 2010; Kobayashi, Maekawa, Miyao, Hirochika, & Kyozuka, 2010; Kobayashi et al., 2012). *PAP2/OsMADS34* belongs to the *LOF-SEP* subgroup of MADS-box genes that show grass-specific diversification caused by gene duplication events. In the *pap2-1* mutant, the inflorescence produces an increased number of branches (Gao et al., 2010; Kobayashi et al., 2010). Although their functions remain to be analysed, the maize and wheat orthologues of *PAP2*, *MADS4* and *AGLG*, respectively, exhibit expression patterns similar to that of *PAP2*, suggesting that their function may be conserved in grass species (Danilevskaya et al., 2008; Yan et al., 2003). In addition to the suppression of transition to spikelet meristem identity, rudimentary glumes and sterile lemmas are elongated in *pap2-1* (Gao et al., 2010; Kobayashi et al., 2010). Usually, leaf growth is extensively suppressed after the transition to the reproductive phase in grass species. Assuming that suppression of glumes is likely to be essential to secure spikelet meristem identity, this defect may be an indication of reduced determinacy, that is the spikelet fate. Recently, it was demonstrated that *PAP2* promotes spikelet meristem identity through suppression of *RCN*s, rice orthologues of *TERMINAL FLOWER1* (*TFL1*) (Liu et al., 2013). Ectopic expression of *RCN* delays floral transition and spikelet meristem identity, resulting in formation of highly branched inflorescences (Nakagawa, Shimamoto, & Kyozuka, 2002). These data provide a clear link between the progression of developmental phases and the inflorescence form.

7.2.3 Contribution of Flowering Time Genes to Inflorescence Development

7.2.3.1 Florigen May Be the Ultimate Target of the Phase Change to Determinate Meristems

As discussed, the basic architecture of the inflorescence form is determined as a result of the temporal regulation of meristem phase changes from an indeterminate branch fate to the determinate spikelet fate. On the other hand, floral transition is the change from vegetative to reproductive development. Flowering is controlled by genetic pathways that integrate environmental and endogenous signals. The day-length signals are integrated by the levels of the flowering hormone florigen (reviewed in Pin & Nilsson, 2012; Srikanth & Schmid, 2011). The expression of *FT*, the florigen of *Arabidopsis*, is induced in leaves when plants are exposed to inductive day-length conditions. Then, the FT protein is transported from the leaf to the shoot apex, where it is perceived by the 14-3-3 proteins and triggers the change from a vegetative to a reproductive phase (Taoka et al., 2011).

In grass species, the first visible change accompanying floral transition is an increase in meristem size, both vertically and horizontally. This is triggered by the FT–FD complex as a consequence of the induction of *FT* expression in leaves by flowering time genes. The flowering pathway is extensively studied in rice because of the importance of controlling flowering time in its cultivation and breeding (reviewed in Brambilla & Fornara, 2013). A number of studies confirmed that the basic mechanisms controlling flowering are well conserved in higher plants. In rice, two *FT* genes, *Hd3a* and *RFT*, work as major integrators of the day-length signal (Komiya, Ikegami, Tamaki, Yokoi, & Shimamoto, 2008; Komiya, Yokoi, & Shimamoto, 2009). In addition, rice-specific genes, such as *Ghd7* and *Ehd1*, work in complex regulatory networks to strictly regulate flowering time (reviewed in Chapter 4).

Most studies of flowering time focus on the transition of the meristem phase from the vegetative to the reproductive stage. Other types of meristem phase changes tend to attract little attention. However, floral transition may not be the only change affected by the florigen genes and other types of meristem change are also affected in mutants of flowering time genes. Over-expression of *Hd3a* led to the production of a single terminal spikelet (Izawa et al., 2002). Similarly extreme phenotypes are observed in various species that ectopically express florigen genes in other species too (Abe et al., 2005). This implies that *FT* accelerates not only the transition from the vegetative SAM to the IM, but also allows spikelet meristems to by-pass the branch meristem phase. An FT-GFP protein fusion was observed in the young developing rice inflorescence (Tamaki, Matsuo, Wong, Yokoi, & Shimamoto, 2007). This indicates the possibility that FT promotes meristem phase changes during inflorescence development. Inflorescence development is affected by some flowering time genes, supporting this idea. *Ghd7* is a rice-specific repressor of flowering which works upstream of *Ehd1* and *Hd3a*. Enhanced expression of *Ghd7* causes an increase in grain number because the inflorescence produces more branches (Brambilla & Fornara, 2013; Xue et al., 2008). Another QTL locus, *Ghd8* also regulates floral transition and inflorescence architecture (Yan et al., 2011).

7.2.4 Meristem Activity and Meristem Phase Transition

7.2.4.1 Genes that Promote IM Transition

In rice, *OsMADS14* and *OsMADS15*, members of the *AP1/FUL* subgroup of the MADS-domain genes, are implicated as direct targets of the florigen complex and are induced in the SAM upon reproductive transition (Komiya et al., 2009; Taoka et al., 2011). Genes induced in the SAM upon the phase change from the vegetative SAM to the IM were identified by transcriptome profiling analyses, along with laser micro-dissection (Kobayashi et al., 2012). Five MADS-box genes were dramatically induced in the meristem at phase transition. They included *PAP2/OsMADS34* and four MADS-box genes, *OsMAD14*, *OsMADS15*, *OsMADS18* and *OsMADS20* that belong to the *AP1/FUL* subfamily (Kobayashi et al., 2012). The simultaneous knockdown of *OsMADs14*, *OsMADS15* and *OsMADS18* did not show any defects, whereas the addition of the *pap2-1* mutation severely hampered the phase change to the IM. This indicates that *PAP2* and the three *AP1/FUL*-like genes act redundantly to promote IM identity (Kobayashi et al., 2012).

7.2.4.2 Control of the Activity of the Apical Meristem

In grass species, the SAM is transformed into the IM, which then begins to produce an inflorescence. After production of the branches, the IM itself is transformed into a spikelet or aborted. In rice, the IM activity degenerates and it remains as a scar called a degenerate point. Therefore, in rice, the timing of degeneration determines the number of primary branches produced in an inflorescence, which is a crucial factor affecting yield. The number of primary branches is basically under genetic regulation. This indicates that the timing of the abortion of the IM is genetically controlled. In rice *apo1* and *apo2* mutants, the IM is converted into a spikelet. This is because the IM is programmed to become a spikelet, but, in a wild-type background, the degeneration occurs precociously, thus, the terminal spikelet is not formed. In an *apo1* and *apo2* background, due to the accelerated spikelet meristem specification, the transition to spikelet identity precedes meristem degeneration (Ikeda et al., 2005). On the other hand, in *apo1-D* mutants, in which *APO1* expression is enhanced, the number of primary branches is greater than in the wild-type and the IM degenerates. It seems that the genetic programmes for the abortion of the IM and for the specification of terminal spikelet formation are independent. However, *APO1* function is required to maintain IM activity until the time of abortion. The delay of abortion in the case when APO1 activity is enhanced indicates a possible interaction between APO1 activity and the duration of meristem activity.

There is a positive correlation between APO1 activity and meristem proliferation. In gain-of-function mutants of *APO1*, the meristem proliferates more vigorously than the wild type, leading to production of more primary branches. The enhanced proliferation is first observed during the reproductive transition concomitant with the onset of *APO1* and *APO2* expression in the meristem (Ikeda-Kawakatsu et al., 2009). In contrast, in *apo1* and *apo2* loss-of-function mutants, the meristem size is much reduced, indicating that meristem cell proliferation is suppressed. The positive correlation between APO1 and APO2 activity, meristem cell proliferation and the timing of IM degeneration implies that meristem degeneration is determined through unknown mechanisms that sense meristem activity.

A positive interaction between meristem cell proliferation and maintenance of meristem indeterminacy was suggested in other studies. *GRAIN NUMBER1a* (*GN1a*) of rice was identified as a QTL gene that confers increased inflorescence branching, thus grain number. *GN1a* encodes a cytokinin oxidase/dehydrogenase, which degrades cytokinin (Ashikari et al., 2005). There is a clear positive correlation between *GN1a* mRNA levels and the IM size, which leads to the increased branches due to elevated cytokinin levels. The IM in these cultivars is also larger than normal. The *DROUGHT AND SALT TOLERANCE* (*DST*) gene was shown to be a direct upstream activator of *GN1a* expression (Huang, Chao, et al., 2009; Li et al., 2013). The correlation between the expression level of *DST* and the meristem size was also observed. Furthermore, a similar function for *DST* was also observed in barley and wheat. The correlation between IM size and increased branching was observed in plants constitutively over-expressing *RCN* genes, the rice orthologues of *Arabidopsis TERMINAL FLOWER1* and in *DEP1* mutant plants (Huang, Qian, et al., 2009; Nakagawa et al.,

2002). It will be interesting to reveal the basis for the correlation observed between the prolonged indeterminate fate (delays in specifying SM identity) and the enhanced meristem activity.

7.2.5 Control of Inflorescence Architecture in the Triticeae

Barley (*Hordeum vulgare* ssp. *vulgare* L.) belongs to the tribe *Triticeae*, along with bread and durum wheat (*Triticum aestivum* L. and *Triticum durum* Desf.), rye (*Secale cereal* L.) and about 350 other species (reviewed in Sakuma, Salomon, & Komatsuda, 2011). The *Triticeae* species produce spike-type inflorescences. Most species in the *Triticeae* produce one spikelet per node whereas barley initiates three spikelets at each alternating node in a distichous manner on the central main stem of the spike, the inflorescence. Among the three spikelets at each node, the central spikelet fully develops while the two lateral spikelets either develop or become sterile. The barley inflorescence is classified into two types based on the size and fertility of the lateral spikelets. All three spikelets are fertile and produce grains in six-rowed barley and only the central spikelets grow in two-rowed barley. The six-rowed type inflorescence is predicted to have arisen from domesticated two-rowed barley 8600 to 8000'years ago by means of spontaneous mutations that allow restoration of the fertility in the lateral spikelets (Palmer, Moore, Clapham, Rose, & Allaby, 2009). It is likely that the six-rowed trait became dominant because of the higher yields achieved. The six-rowed trait is controlled by at least five independent loci including *Six-rowed spike1* (*vrs1*), *vrs2*, *vrs3*, *vrs4* and *Intermedium-C* (*Int-c*) (Sakuma et al., 2011). The *Vrs1* gene was identified as a major regulator of the six-rowed spikelet trait (Komatsuda et al., 2007; Sakuma et al., 2013; Sakuma, Pourkheirandish, Matsumoto, Koba, & Komatsuda, 2010). *Vrs1* negatively regulates lateral spikelet fertility, and loss of function of *Vrs1* results in the restoration of fertility in the lateral spikelets, leading to the development of the two-rowed inflorescence. *Vrs1* encodes a homeodomain-leucine zipper class I transcription factor. The expression pattern of *Vrs1*, which is limited to the lateral spikelets, and the dominant nature of *Vrs1* responsible for the two-rowed inflorescence trait, indicate that Vrs1 works as a repressor protein that binds to the DNA of target genes that positively regulate the development of lateral spikelets. Strict temporal and spatial regulation of *Vrs1* expression led to the production of the six-rowed inflorescence in barley. It is believed that the inflorescence architecture of the grasses continuously evolved towards a simpler form, the spike-type inflorescence, from the more ancient panicle-type inflorescence in species such as rice, maize and sorghum, etc. In the spike, a single spikelet per node is produced in wheat and rye. In barley, the restoration of spikelet number was achieved through the suppression of the function of *Vrs1* during domestication. Future analysis of *Vrs1* is expected to improve our understanding of grass inflorescence evolution.

Genetic analyses indicate that the effect of the *Vrs1* gene can be modified by up to 10 *INTERMEDIUM* (*INT*) genes. Among them, *INT-C*, an orthologue of maize *TEOSINTE BRANCHED1* (*TB1*), shows the largest effect (Doebley, Stec, & Hubbard, 1997; Ramsay et al., 2011). *TB1* was isolated as one of the major genes contributing to the domestication of maize (Doebley et al., 1997). Thus, two homologous genes were used during the domestication of barley and maize, leading Ramsay et al. (2011) to propose that only limited numbers of genetic points are amenable for human selection.

Vrs4 was also recently isolated (Koppolu et al., 2013). Unlike *vrs1* and *int-c*, in which the fertility of the lateral spikelets is affected, meristem determinacy is affected. In addition, in *vrs4* mutants, fertility has been restored in the lateral spikelets. More than three spikelets are frequently formed at one node. If the three spikelets produced in a node are interpreted as a short branch containing one terminal spikelet and two lateral spikelets, the phenotype in *vrs4* can be explained as having increased branch meristem determinacy. *Vrs4* encodes a lateral organ boundary (LOB) domain protein and is orthologous to maize *RAMOSA2* (*RA2*) (Bortiri et al., 2006; Koppolu et al., 2013). In maize, the *RAMOSA* pathway plays a major role in the control of inflorescence branching patterns (Bortiri et al., 2006; Gallavotti et al., 2010; Satoh-Nagasawa et al., 2006; Vollbrecht et al., 2005). It is expected that future analysis of *Vrs4* will provide a clue to understand the link between meristem determinacy and organ development.

7.3 PERSPECTIVES AND CONCLUSIONS

Recent progress in molecular genetic studies has enabled the isolation of key regulators controlling grass inflorescence form, mostly from rice, maize, wheat and barley. Studies of these genes have shown that both grass-specific genes and widely conserved genes are used to control development of spikelets, the structures unique to grass species. These regulators control inflorescence form by fine-tuning the timing of the transition to the spikelet meristem phase. It has also been shown that these regulators are important in terms of grain yield. Future research will provide opportunities to better understand the basic mechanisms of meristem phase change, a fundamental aspect of plant biology, as well as to develop strategies to feed our ever-growing population.

ACKNOWLEDGEMENTS

JK is supported in part by Grants-in-Aid from the Ministry of Education, Culture, Sports, Science, and Technology, Japan (22119008, 22247004 and 21027012).

REFERENCES

Abe, M., Kobayashi, Y., Yamamoto, S., Daimon, Y., Yamaguchi, A., Ikeda, Y., et al. (2005). FD, a bZIP protein mediating signals from the floral pathway integrator FT at the shoot apex. *Science, 309*, 1052–1056.

Aida, M., Ishida, T., Fukaki, H., Fujisawa, H., & Tasaka, M. (1997). Genes involved in organ separation in *Arabidopsis*: an analysis of the cup-shaped cotyledon mutant. *Plant Cell, 9*, 841–857.

Arora, R., Agarwal, P., Ray, S., Singh, A. K., Singh, V. P., Tyagi, A. K., et al. (2007). MADS-box gene family in rice: genome-wide identification, organization and expression profiling during reproductive development and stress. *BMC Genomics, 8*, 242.

Ashikari, M., Sakakibara, H., Lin, S., Yamamoto, T., Takashi, T., Nishimura, A., et al. (2005). Cytokinin oxidase regulates rice grain production. *Science, 309*, 741–745.

Bell, A. D., & Bryan, A. (1991). *Plant form, an illustrated guide to flowering plant morphology* (1st ed.). New York: Oxford University Press.

Bomblies, K., Wang, R. L., Ambrose, B. A., Schmidt, R. J., Meeley, R. B., & Doebley, J. (2003). Duplicate FLORICAULA/LEAFY homologs zfl1 and zfl2 control inflorescence architecture and flower patterning in maize. *Development, 130*, 2385–2395.

Bortiri, E., Chuck, G., Vollbrecht, E., Rocheford, T., Martienssen, R., & Hake, S. (2006). ramosa2 encodes a LATERAL ORGAN BOUNDARY domain protein that determines the fate of stem cells in branch meristems of maize. *Plant Cell, 18*, 574–585.

Brambilla, V., & Fornara, F. (2013). Molecular control of flowering in response to day length in rice. *Journal of Integrative Plant Biology, 55*, 410–418.

Chuck, G., Cigan, A. M., Saeteurn, K., & Hake, S. (2007). The heterochronic maize mutant Corngrass1 results from overexpression of a tandem microRNA. *Nature Genetics, 39*, 544–549.

Chuck, G., Whipple, C., Jackson, D., & Hake, S. (2010). The maize SBP-box transcription factor encoded by tasselsheath4 regulates bract development and the establishment of meristem boundaries. *Development, 137*, 1243–1250.

Clark, L. G., & Pohl, R. W. (1997). *The structure of grasses explained for beginners* (4th ed.). Smithsonian Institution Press.

Danilevskaya, O. N., Meng, X., Selinger, D. A., Deschamps, S., Hermon, P., Vansant, G., et al. (August 2008). Involvement of the MADS-box gene ZMM4 in floral induction and inflorescence development in maize. *Plant Physiology, 147*(4), 2054–2069.

Doebley, J., Stec, A., & Hubbard, L. (1997). The evolution of apical dominance in maize. *Nature, 386*, 485–488.

Doust, A. N., Devos, K. M., Gadberry, M. D., Gale, M. D., & Kellogg, E. A. (2005). The genetic basis for inflorescence variation between foxtail and green millet (poaceae). *Genetics, 169*, 1659–1672.

Fornara, F., Gregis, V., Pelucchi, N., Colombo, L., & Kater, M. (2008). The rice StMADS11-like genes OsMADS22 and OsMADS47 cause floral reversions in *Arabidopsis* without complementing the svp and agl24 mutants. *Journal of Experimental Botany, 59*, 2181–2190.

Furutani, I., Sukegawa, S., & Kyozuka, J. (2006). Genome-wide analysis of spatial and temporal gene expression in rice panicle development. *Plant Journal, 46*, 503–511.

Gallavotti, A., Long, J. A., Stanfield, S., Yang, X., Jackson, D., Vollbrecht, E., et al. (2010). The control of axillary meristem fate in the maize ramosa pathway. *Development, 137*, 2849–2856.

Gallavotti, A., Malcomber, S., Gaines, C., Stanfield, S., Whipple, C., Kellogg, E., et al. (2011). BARREN STALK FASTIGIATE1 is an AT-hook protein required for the formation of maize ears. *Plant Cell, 23*, 1756–1771.

Gallavotti, A., Zhao, Q., Kyozuka, J., Meeley, R. B., Ritter, M. K., Doebley, J. F., et al. (2004). The role of barren stalk1 in the architecture of maize. *Nature, 432*, 630–635.

Gao, X., Liang, W., Yin, C., Ji, S., Wang, H., Su, X., et al. (2010). The SEPALLATA-like gene OsMADS34 is required for rice inflorescence and spikelet development. *Plant Physiology, 153*, 728–740.

Gao, Z. Y., Zhao, S. C., He, W. M., Guo, L. B., Peng, Y. L., Wang, J. J., et al. (2013). Dissecting yield-associated loci in super hybrid rice by resequencing recombinant inbred lines and improving parental genome sequences. *Proceedings of the National Academy of Sciences of the United States of America, 110*, 14492–14497.

Gibson, D. J. (2009). *Grasses and grassland ecology*. New York: Oxford University Press.

Gocal, G. F., King, R. W., Blundell, C. A., Schwartz, O. M., Andersen, C. H., & Weigel, D. (April 2001). Evolution of floral meristem identity genes. Analysis of *Lolium temulentum* genes related to APETALA1 and LEAFY of *Arabidopsis*. *Plant Physiology, 125*(4), 1788–1801.

Greb, T., Clarenz, O., Schafer, E., Muller, D., Herrero, R., Schmitz, G., et al. (2003). Molecular analysis of the LATERAL SUPPRESSOR gene in *Arabidopsis* reveals a conserved control mechanism for axillary meristem formation. *Genes & Development, 17*, 1175–1187.

Hake, S. (2008). Inflorescence architecture: The transition from branches to flowers. *Current Biology, 18*, 23–24.

Han, J. J., Jackson, D., & Martienssen, R. (2012). Pod corn is caused by rearrangement at the tunicate1 locus. *Plant Cell, 24*, 2733–2744.

Hartmann, U., Höhmann, S., Nettesheim, K., Wisman, E., Saedler, H., & Huijser, P. (2000). Molecular cloning of SVP: a negative regulator of the floral transition in *Arabidopsis*. *Plant Journal, 21*, 351–360.

Hibara, K., Karim, M. R., Takada, S., Taoka, K., Furutani, M., Aida, M., et al. (2006). *Arabidopsis* CUP-SHAPED COTYLEDON3 regulates postembryonic shoot meristem and organ boundary formation. *Plant Cell, 18*, 2946–2957.

Hibara, K., Takada, S., & Tasaka, M. (2003). CUC1 gene activates the expression of SAM-related genes to induce adventitious shoot formation. *Plant Journal, 36*, 687–696.

Huang, X., Qian, Q., Liu, Z., Sun, H., He, S., Luo, D., et al. (2009). Natural variation at the DEP1 locus enhances grain yield in rice. *Nature Genetics, 41*, 494–497.

Huang, X. Y., Chao, D. Y., Gao, J. P., Zhu, M. Z., Shi, M., & Lin, H. X. (2009). A previously unknown zinc finger protein, DST, regulates drought and salt tolerance in rice via stomatal aperture control. *Genes & Development, 23*, 1805–1817.

Ikeda, K., Ito, M., Nagasawa, N., Kyozuka, J., & Nagato, Y. (2007). Rice ABERRANT PANICLE ORGANIZATION 1, encoding an F-box protein, regulates meristem fate. *Plant Journal, 51*, 1030–1040.

Ikeda, K., Nagasawa, N., & Nagato, Y. (2005). ABERRANT PANICLE ORGANIZATION 1 temporally regulates meristem identity in rice. *Developmental Biology, 282*, 349–360.

Ikeda-Kawakatsu, K., Maekawa, M., Izawa, T., Itoh, J. I., & Nagato, Y. (2011). ABERRANT PANICLE ORGANIZATION 2/RFL, the rice ortholog of *Arabidopsis* LEAFY, suppresses the transition from inflorescence meristem to floral meristem through interaction with APO1. *Plant Journal, 69*, 168–180.

Ikeda-Kawakatsu, K., Yasuno, N., Oikawa, T., Iida, S., Nagato, Y., Maekawa, M., et al. (2009). Expression level of ABERRANT PANICLE ORGANIZATION1 determines rice inflorescence form through control of cell proliferation in the meristem. *Plant Physiology, 150*, 736–747.

Imai, K., Murai, M., Hao, Y., Chiba, Y., Chiba, A., & Ishikawa, R. (2009). Mapping of rice Ur1 (Undulated rachis-1) gene with effect on increasing spikelet number per panicle and sink size, and development of selection markers for the breeding by the use of Ur1. *Hereditas, 146*, 260–268.
Irish, V. F. (2010). The flowering of *Arabidopsis* flower development. *Plant Journal, 61*, 1014–1028.
Iyer, L. M., & Aravind, L. (2012). ALOG domains: provenance of plant homeotic and developmental regulators from the DNA-binding domain of a novel class of DIRS1-type retroposons. *Biology Direct, 7*, 39.
Izawa, T., Oikawa, T., Sugiyama, N., Tanisaka, T., Yano, M., & Shimamoto, K. (2002). Phytochrome mediates the external light signal to repress FT orthologs in photoperiodic flowering of rice. *Genes & Development, 16*, 2006–2020.
Jiao, Y., Wang, Y., Xue, D., Wang, J., Yan, M., Liu, G., et al. (2010). Regulation of OsSPL14 by OsmiR156 defines ideal plant architecture in rice. *Nature Genetics, 42*, 541–544.
Kaur, P., Larson, S. R., Shaun Bushman, S. B., Wang, R. R., Mott, I. W., Hole, D., et al. (2008). Genes controlling plant growth habit in *Leymus* (Triticeae): maize barren stalk1 (ba1), rice lax panicle, and wheat tiller inhibition (tin3) genes as possible candidates. *Functional & Integrative Genomics, 8*, 375–386.
Kellogg, E. A., & Buell, C. R. (2009). Splendor in the grasses. *Plant Physiology, 149*, 1–3.
Kellogg, E. A., Camara, P. E., Rudall, P. J., Ladd, P., Malcomber, S. T., Whipple, C. J., et al. (2013). Early inflorescence development in the grasses (Poaceae). *Frontiers in Plant Science, 4*, 250.
Khush, G. S. (2001). Green revolution: the way forward. *Nature Reviews Genetics, 2*, 815–822.
Klein, J., Saedler, H., & Huijser, P. (1996). A new family of DNA binding proteins includes putative transcriptional regulators of the *Antirrhinum majus* floral meristem identity gene SQUAMOSA. *Molecular & General Genetics, 250*, 7–16.
Kobayashi, K., Maekawa, M., Miyao, A., Hirochika, H., & Kyozuka, J. (2010). PANICLE PHYTOMER2 (PAP2), encoding a SEPALLATA subfamily MADS-box protein, positively controls spikelet meristem identity in rice. *Plant Cell Physiology, 51*, 47–57.
Kobayashi, K., Yasuno, N., Sato, Y., Yoda, M., Yamazaki, R., Kimizu, M., et al. (2012). Inflorescence meristem identity in rice is specified by overlapping functions of three AP1/FUL-like MADS box genes and PAP2, a SEPALLATA MADS box gene. *Plant Cell, 24*, 1848–1859.
Komatsu, K., Maekawa, M., Ujiie, S., Satake, Y., Furutani, I., Okamoto, H., et al. (2003). LAX and SPA: major regulators of shoot branching in rice. *Proceedings of the National Academy of Sciences of the United States of America, 100*, 11765–11770.
Komatsu, M., Maekawa, M., Shimamoto, K., & Kyozuka, J. (2001). The LAX1 and FRIZZY PANICLE 2 genes determine the inflorescence architecture of rice by controlling rachis-branch and spikelet development. *Developmental Biology, 15*, 364–373.
Komatsuda, T., Pourkheirandish, M., He, C., Azhaguvel, P., Kanamori, H., Perovic, D., et al. (2007). Six-rowed barley originated from a mutation in a homeodomain-leucine zipper I-class homeobox gene. *Proceedings of the National Academy of Sciences of the United States of America, 104*, 1424–1429.
Komiya, R., Ikegami, A., Tamaki, S., Yokoi, S., & Shimamoto, K. (2008). Hd3a and RFT1 are essential for flowering in rice. *Development, 135*, 767–774.
Komiya, R., Yokoi, S., & Shimamoto, K. (2009). A gene network for long-day flowering activates RFT1 encoding a mobile flowering signal in rice. *Development, 136*, 3443–3450.
Koppolu, R., Anwar, N., Sakuma, S., Tagiri, A., Lundqvist, U., Pourkheirandish, M., et al. (August 6, 2013). Six-rowed spike4 (Vrs4) controls spikelet determinacy and row-type in barley. *Proceedings of the National Academy of Sciences of the United States of America, 110*(32), 13198–13203.

Kyozuka, J., Konishi, S., Nemoto, K., Izawa, T., & Shimamoto, K. (1998). Down-regulation of RFL, the FLO/LFY homolog of rice, accompanied with panicle branch initiation. *Proceedings of the National Academy of Sciences of the United States of America, 95*, 1979–1982.

Lee, J. H., Yoo, S. J., Park, S. H., Hwang, I., Lee, J. S., & Ahn, J. H. (2007). Role of SVP in the control of flowering time by ambient temperature in *Arabidopsis*. *Genes & Development, 21*, 397–402.

Lee, S., Choi, S. C., & An, G. (2008). Rice SVP-group MADS-box proteins, OsMADS22 and OsMADS55, are negative regulators of brassinosteroid responses. *Plant Journal, 54*, 93–105.

Li, S., Zhao, B., Yuan, D., Duan, M., Qian, Q., Tang, L., et al. (2013). Rice zinc finger protein DST enhances grain production through controlling Gn1a/OsCKX2 expression. *Proceedings of the National Academy of Sciences of the United States of America, 110*, 3167–3172.

Li, X., Qian, Q., Fu, Z., Wang, Y., Xiong, G., Zeng, D., et al. (2003). Control of tillering in rice. *Nature, 422*, 618–621.

Lippman, Z.B., Cohen, O., Alvarez, J.P., Abu-Abied, M., Pekker, I., Paran, I., et al. (2008). The making of a compound inflorescence in tomato and related nightshades. *PLoS Biology, 18*, e288.

Liu, C., Teo, Z. W., Bi, Y., Song, S., Xi, W., Yang, X., et al. (2013). A conserved genetic pathway determines inflorescence architecture in *Arabidopsis* and rice. *Developmental Cell, 25*, 612–622.

Lu, F., Ammiraju, J. S. S., Sanyal, A., Zhang, S., Song, R., Chen, J., et al. (2009). Comparative sequence analysis of MONOCULM1-orthologous regions in 14 *Oryza* genomes. *Proceedings of the National Academy of Sciences of the United States of America, 106*, 2071–2076.

Luo, L., Li, W., Miura, K., Ashikari, M., & Kyozuka, J. (2012). Control of tiller growth of rice by OsSPL14 and Strigolactones, which work in two independent pathways. *Plant Cell Physiology, 53*, 1793–1801.

MacAlister, C. A., Park, S. J., Jiang, K., Marcel, F., Bendahmane, A., Izkovich, Y., et al. (2012). Synchronization of the flowering transition by the tomato TERMINATING FLOWER gene. *Nature Genetics, 44*, 1393–1398.

McSteen, P., & Leyser, O. (2005). Shoot branching. *Annual Review of Plant Biology, 56*, 353–374.

Miura, K., Ikeda, M., Matsubara, A., Song, X.J., Ito, M., Asano, K., et al. (2010). OsSPL14 promotes panicle branching and higher grain productivity in rice. *Nature Genetics, 42*, 545–549.

Molinero-Rosales, N., Jamilena, M., Zurita, S., Gómez, P., Capel, J., & Lozano, R. (1999). FALSIFLORA, the tomato orthologue of FLORICAULA and LEAFY, controls flowering time and floral meristem identity. *Plant Journal, 20*, 685–693.

Moyroud, E., Kusters, E., Monniaux, M., Koes, R., & Parcy, F. (2010). LEAFY blossoms. *Trends in Plant Science, 15*, 346–352.

Murai, M., & Iizawa, M. (1994). Effects of major genes controlling morphology of panicle in rice. *Breeding Science, 44*, 247–255.

Nakagawa, M., Shimamoto, K., & Kyozuka, J. (2002). Overexpression of RCN1 and RCN2, rice TERMINAL FLOWER 1/CENTRORADIALIS homologs, confers delay of phase transition and altered panicle morphology in rice. *Plant Journal, 29*, 743–750.

Oikawa, T., & Kyozuka, J. (2009). Two-step regulation of LAX PANICLE1 protein accumulation in axillary meristem formation in rice. *Plant Cell, 21*, 1095–1108.

Ookawa, T., Hobo, T., Yano, M., Murata, K., Ando, T., Miura, H., et al. (2010). New approach for rice improvement using a pleiotropic QTL gene for lodging resistance and yield. *Nature Communications, 1*, 132.

Palmer, S. A., Moore, J. D., Clapham, A. J., Rose, P., & Allaby, R. G. (2009). Archaeogenetic evidence of ancient nubian barley evolution from six to two-row indicates local adaptation. *PLoS One, 22*, e6301.

Pin, P. A., & Nilsson, O. (2012). The multifaceted roles of FLOWERING LOCUS T in plant development. *Plant Cell and Environment, 35*, 1742–1755.
Poethig, R. S. (2013). Vegetative phase change and shoot maturation in plants. *Current Topics in Developmental Biology, 105*, 125–152.
Preston, J. C., & Hileman, L. C. (2013). Functional evolution in the plant SQUAMOSA-PROMOTER BINDING PROTEIN-LIKE (SPL) gene family. *Frontiers in Plant Science, 4*, 80.
Prusinkiewicz, P., Erasmus, Y., Lane, B., Harder, L.D., & Coen, E. (2007). Evolution and development of inflorescence architectures. *Science, 316*, 1452–1456.
Ramsay, L., Comadran, J., Druka, A., Marshall, D. F., Thomas, W.T., Macaulay, M., et al. (2011). INTERMEDIUM-C, a modifier of lateral spikelet fertility in barley, is an ortholog of the maize domestication gene TEOSINTE BRANCHED 1. *Nature Genetics, 43*, 169–172.
Rao, N. N., Prasad, K., Kumar, P. R., & Vijayraghavan, U. (2008). Distinct regulatory role for RFL, the rice LFY homolog, in determining flowering time and plant architecture. *Proceedings of the National Academy of Sciences of the United States of America, 105*, 3646–3651.
Ritter, M. K., Padilla, C. M., & Schmidt, R. J. (2002). The role of auxin transport during inflorescence development in maize (*Zea mays*, Poaceae). *American Journal of Botany, 89*, 203–210.
Sakuma, S., Pourkheirandish, M., Hensel, G., Kumlehn, J., Stein, N., Tagiri, A., et al. (2013). Divergence of expression pattern contributed to neofunctionalization of duplicated HD-Zip I transcription factor in barley. *The New Phytologist, 197*, 939–948.
Sakuma, S., Pourkheirandish, M., Matsumoto, T., Koba, T., & Komatsuda, T. (2010). Duplication of a well-conserved homeodomain-leucine zipper transcription factor gene in barley generates a copy with more specific functions. *Functional & Integrative Genomics, 10*, 123–133.
Sakuma, S., Salomon, B., & Komatsuda, T. (2011). The domestication syndrome genes responsible for the major changes in plant form in the Triticeae crops. *Plant Cell Physiology, 52*, 738–749.
Satoh-Nagasawa, N., Nagasawa, N., Malcomber, S., Sakai, H., & Jackson, D. (2006). A trehalose metabolic enzyme controls inflorescence architecture in maize. *Nature, 441*, 227–230.
Schwab, R., Palatnik, J. F., Riester, M., Schommer, C., Schmid, M., & Weigel, D. (2005). Specific effects of microRNAs on the plant transcriptome. *Developmental Cell, 8*, 517–527.
Schwarz, S., Grande, A. V., Bujdoso, N., Saedler, H., & Huijser, P. (2008). The microRNA regulated SBP-box genes SPL9 and SPL15 control shoot maturation in *Arabidopsis*. *Plant Molecular Biology, 67*, 183–195.
Sentoku, N., Kato, H., Kitano, H., & Imai, R. (2005). OsMADS22, an STMADS11-like MADS-box gene of rice, is expressed in non-vegetative tissues and its ectopic expression induces spikelet meristem indeterminacy. *Molecular Genetics and Genomics, 273*, 1–9.
Shitsukawa, N., Takagishi, A., Ikari, C., Takumi, S., & Murai, K. (2006). WFL, a wheat FLORICAULA/LEAFY ortholog, is associated with spikelet formation as lateral branch of the inflorescence meristem. *Genes & Genetic Systems, 81*, 13–20.
Skirpan, A., Wu, X., & McSteen, P. (2008). Genetic and physical interaction suggest that BARREN STALK 1 is a target of BARREN INFLORESCENCE2 in maize inflorescence development. *Plant Journal, 55*, 787–797.
Souer, E., Rebocho, A. B., Bliek, M., Kusters, E., de Bruin, R. A., & Koes, R. (2008). Patterning of inflorescences and flowers by the F-Box protein DOUBLE TOP and the LEAFY homolog ABERRANT LEAF AND FLOWER of petunia. *Plant Cell, 20*, 2033–2048.
Spanudakis, E., & Jackson, S. (2014). The role of microRNAs in the control of flowering time. *Journal of Experimental Botany, 65*, 365–380.
Srikanth, A., & Schmid, M. (2011). Regulation of flowering time: all roads lead to Roma. *Cellular and Molecular Life Sciences, 68*, 2013–2037.

Sussex, I. M., & Kerk, N. M. (2002). The organization and function of plant meristems. In M. T. McManus, & B. E. Veit (Eds.), *Meristematic tissue in plant growth and development*. Shefield: Schefield Academic Press.
Tabuchi, H., Zhang, Y., Hattori, S., Omae, M., Shimizu-Sato, S., Oikawa, T., et al. (2011). LAX PANICLE2 of rice encodes a novel nuclear protein and regulates the formation of axillary meristems. *Plant Cell, 23*, 3276–3287.
Takacs, E. M., Li, J., Du, C., Ponnala, L., Janick-Buckner, D., Yu, J., et al. (2012). Ontogeny of the maize shoot apical meristem. *Plant Cell, 24*, 3219–3234.
Takeda, S., Hanano, K., Kariya, A., Shimizu, S., Zhao, L., Matsui, M., et al. (2011). CUP-SHAPED COTYLEDON1 transcription factor activates the expression of LSH4 and LSH3, two members of the ALOG gene family, in shoot organ boundary cells. *Plant Journal, 66*, 1066–1077.
Tamaki, S., Matsuo, S., Wong, H. L., Yokoi, S., & Shimamoto, K. (May 18, 2007). Hd3a protein is a mobile flowering signal in rice. *Science, 316*(5827), 1033–1036.
Tanahashi, T., Sumikawa, N., Kato, M., & Hasebe, M. (2005). Diversification of gene function: homologs of the floral regulator FLO/LFY control the first zygotic cell division in the moss *Physcomitrella patens*. *Development, 132*, 1727–1736.
Taoka, K., Ohki, I., Tsuji, H., Furuita, K., Hayashi, K., Yanase, T., et al. (2011). 14-3-3 proteins act as intracellular receptors for rice Hd3a florigen. *Nature, 476*, 332–335.
Terao, T., Nagata, K., Morino, K., & Hirose, T. (2010). A gene controlling the number of primary rachis branches also controls the vascular bundle formation and hence is responsible to increase the harvest index and grain yield in rice. *Theoretical and Applied Genetics, 120*, 875–893.
Trevaskis, B., Tadege, M., Hemming, M. N., Peacock, W. J., Dennis, E. S., & Sheldon, C. (2007). Short vegetative phase-like MADS-box genes inhibit floral meristem identity in barley. *Plant Physiology, 143*, 225–235.
Vollbrecht, E., Springer, P. S., Goh, L., Buckler, E. S., & Martienssen, R. (2005). Architecture of floral branch systems in maize and related grasses. *Nature, 436*, 1119–1126.
Wang, J. W., Czech, B., & Weigel, D. (2009). miR156-regulated SPL transcription factors define an endogenous flowering pathway in *Arabidopsis thaliana*. *Cell, 138*, 738–749.
Wang, J. W., Schwab, R., Czech, B., Mica, E., & Weigel, D. (2008). Dual effects of miR156-targeted SPL genes and CYP78A5/KLUH on plastochron length and organ size in *Arabidopsis thaliana*. *Plant Cell, 20*, 1231–1243.
Wang, Y., & Li, J. (2011). Branching in rice. *Current Opinion in Plant Biology, 14*, 94–99.
Weberling, F. (1989). *Morphology of flowers and inflorescences*. Cambridge: Cambridge University Press.
Wingen, L. U., Münster, T., Faigl, W., Deleu, W., Sommer, H., Saedler, H., et al. (2012). Molecular genetic basis of pod corn (Tunicate maize). *Proceedings of the National Academy of Sciences of the United States of America, 109*, 7115–7120.
Woods, D. P., Hope, C. L., & Malcomber, S. T. (2011). Phylogenomic analyses of the BARREN STALK1/LAX PANICLE1 (BA1/LAX1) genes and evidence for their roles during axillary meristem development. *Molecular Biology and Evolution, 28*, 2147–2159.
Wu, G., & Poethig, R. S. (2006). Temporal regulation of shoot development in *Arabidopsis thaliana* by miR156 and its target SPL3. *Development, 133*, 3539–3547.
Wu, X., & McSteen, P. (2007). Genetic and physical interaction suggest that BARREN STALK 1 is a target of BARREN INFLORESCENCE2 in maize inflorescence development. *American Journal of Botany, 94*, 1745–1755.
Xue, W., Xing, Y., Weng, X., Zhao, Y., Tang, W., Wang, L., et al. (2008). Natural variation in Ghd7 is an important regulator of heading date and yield potential in rice. *Nature Genetics, 8*(40), 761–767.
Yamaguchi, A., Wu, M. F., Yang, L., Wu, G., Poethig, R. S., & Wagner, D. (2009). The microRNA-regulated SBP-Box transcription factor SPL3 is a direct upstream activator of LEAFY, FRUITFULL, and APETALA1. *Developmental Cell, 17*, 268–278.

Yan, L., Loukoianov, A., Tranquilli, G., Helguera, M., Fahima, T., & Dubcovsky, J. (2003). Positional cloning of the wheat vernalization gene VRN1. *Proceedings of the National Academy of Sciences of the United States of America, 100*, 6263–6268.

Yan, W. H., Wang, P., Chen, H. X., Zhou, H. J., Li, Q. P., Wang, et al. (2011). A major QTL, Ghd8, plays pleiotropic roles in regulating grain productivity, plant height, and heading date in rice. *Molecular Plant, 4*, 319–330.

Yang, F., Wang, Q., Schmitz, G., Müller, D., & Theres, K. (2012). The bHLH protein ROX acts in concert with RAX1 and LAS to modulate axillary meristem formation in *Arabidopsis*. *Plant Journal, 71*, 61–70.

Yang, Z., Wang, X., Gu, S., Hu, Z., Xu, H., & Xu, C. (2007). Comparative study of SBP-box gene family in *Arabidopsis* and rice. *Gene, 407*, 1–11.

Yoshida, A., Sasao, M., Yasuno, N., Takagi, K., Daimon, Y., Chen, R., et al. (2013). TAWAWA1, a regulator of rice inflorescence architecture, functions through the suppression of meristem phase transition. *Proceedings of the National Academy of Sciences of the United States of America, 110*, 767–772.

Yoshida, A., Suzaki, T., Tanaka, W., & Hirano, H. (2009). The homeotic gene long sterile lemma (G1) specifies sterile lemma identity in the rice spikelet. *Proceedings of the National Academy of Sciences of the United States of America, 106*, 20103–20108.

Zhao, L., Nakazawa, M., Takase, T., Manabe, K., Kobayashi, M., Seki, M., et al. (2004). Overexpression of LSH1, a member of an uncharacterised gene family, causes enhanced light regulation of seedling development. *Plant Journal, 37*, 694–706.

CHAPTER EIGHT

Flower Development in Rice

Wakana Tanaka, Taiyo Toriba and Hiro-Yuki Hirano[1]
Department of Biological Sciences, Graduate School of Science, The University of Tokyo, Tokyo, Japan
[1]Corresponding author: e-mail address: hyhirano@biol.s.u-tokyo.ac.jp

Contents

Abstract

In rice, the flower consisting of lodicules, stamens and carpels is enclosed by the lemma and palea to form the floret, which together with sterile lemmas and rudimentary glumes constitutes the spikelet. Thus, the flower and the inflorescence units of rice are distinct from those of eudicots. The ABC model, which explains the genetic mechanism underlying floral organ specification in eudicots, is largely applicable to the

Advances in Botanical Research, Volume 72
ISSN 0065-2296
http://dx.doi.org/10.1016/B978-0-12-417162-6.00008-0

specialized flowers of rice. For instance, the function of class B genes is conserved to specify the lodicule (a petal homologue) and the stamen. Two class C genes are functionally diversified in rice: one specifies stamen identity together with class B genes, whereas the other is mainly responsible for the determinacy of the flower meristem. By contrast, carpel specification in rice is regulated by a *YABBY* gene, *DROOPING LEAF* (*DL*). Homeotic transformation of the stamen or carpel in loss-of-function mutants of class B genes or *DL* reveals a mutual repression mode of action for these genes. Additional genes responsible for the development of spikelet organs, such as the lemma, palea, sterile lemma and rudimentary glume, have been identified in rice. Mutations in some of these genes affect the development of only spikelet organs, whereas mutations in others affect the development of both flower and non-floral spikelet organs. In this review, we describe the genetic mechanism underlying flower and spikelet development in rice, and discuss the regulation of maintenance and fate of reproductive meristems, the activity of which is closely associated with flower and spikelet development.

8.1 INTRODUCTION

Angiosperms generate a diverse range of flower types. Flowers produced by plants of the rose or orchid families are conspicuous and beautiful, catching the eye and inspiring the mind. These flowers also attract insects to serve as pollinators, and flowering plants together with their insect pollinators have been shaped by tens of millions years of coevolution. Other plants produce inconspicuous flowers without obvious petals and such flowers are mainly pollinated by wind. Rice, like other grasses such as maize and wheat, is an example of the latter.

In the broad sense, rice encompasses all species in the genus *Oryza* of the *Poaceae* (grass) family. The *Oryza* genus comprises 21 wild species and 2 domesticated species, *Oryza sativa* and *Oryza glaberrima*. *O. sativa* contains two subspecies, *japonica* and *indica*, both of which are derived from the wild rice, *Oryza rufipogon* (Hirano, Hirai, Sano, & Sasaki, 2008; Huang et al., 2012; Oka, 1988). Rice flowers, which are made up of carpels, stamens and lodicules, are distinct from those of eudicots (Figure 8.1) (Hirano, Tanaka, & Toriba, 2014; Tanaka, Pautler, Jackson, & Hirano, 2013; Yoshida & Nagato, 2011). This rice flower is enclosed in a palea and a lemma, two organs that along with the lodicule are unique to grasses. Lodicules open the lemma and palea to ensure wind-mediated pollination when flowers reach maturity. Although their morphologies and functions are different from each other, the lodicule and the petals are homologous organs, sharing the same evolutionarily origin (Hoshikawa, 1989; Troll, 2000). Deciphering the conservation and specification of genes involved in the development of lodicules and petals is a matter of much interest. A long needlelike

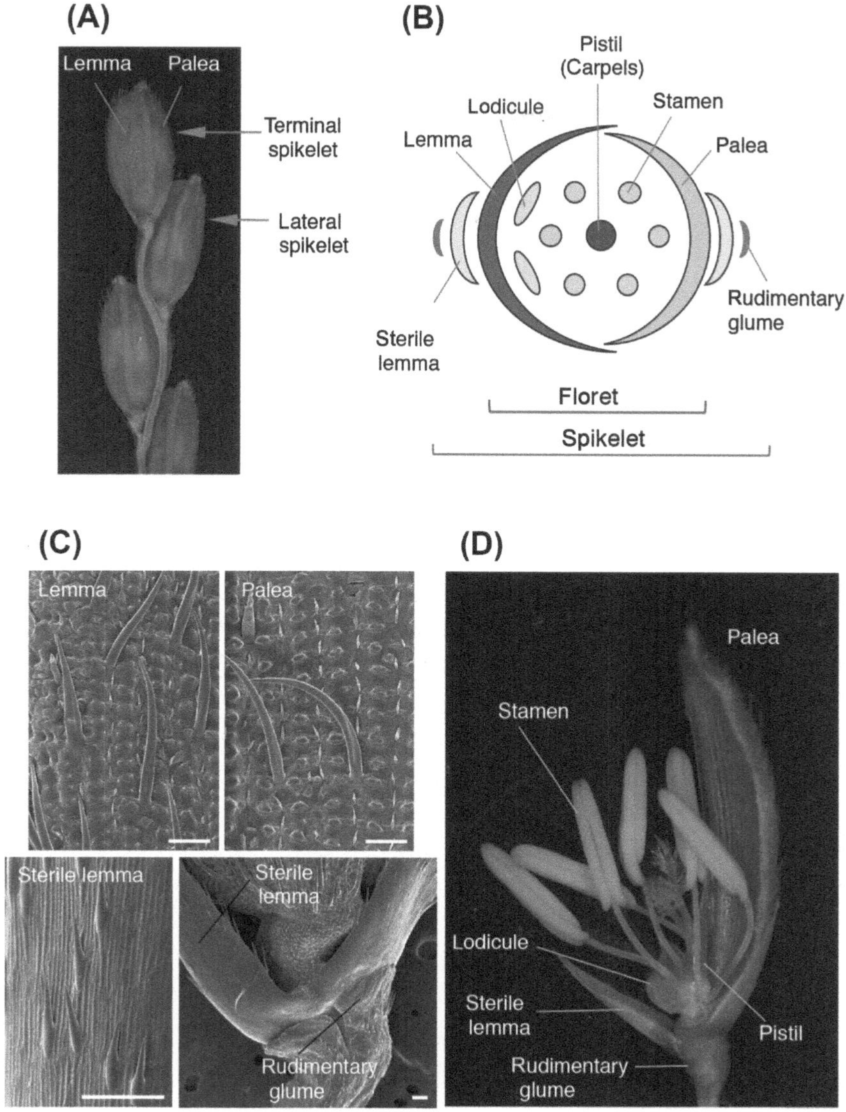

Figure 8.1 Spikelet and flower structure of rice. (A) An apical region of the primary branch producing the terminal and lateral spikelets. (B) Schematic representation of the rice spikelet. (C) Epidermal abaxial surface of the lemma, palea and sterile lemma, and the basal region of the spikelet. The lemma and palea have a rough abaxial surface with many tubercles, whereas the abaxial surface of the sterile lemma is smooth. (D) Flower and spikelet organs. The lemma and a part of the palea are removed. Scale bars: 100 μm. (See the colour plate.)

structure called the awn is formed on top of the lemma in most wild rice species (Luo et al., 2013; Toriba & Hirano, 2014). By contrast, awn formation is suppressed in most strains of *Oryza sativa*. Thus, studies that focus on awn development and suppression are likely to provide insight into the function of genes involved in rapid morphological changes during domestication (i.e. microevolution).

The molecular mechanisms underlying flower development are well understood in model eudicots such as *Arabidopsis thaliana* and *Antirrhinum majus* (Jack, 2004; Lohmann & Weigel, 2002; Prunet & Jack, 2014). The ABC model, which was proposed in the early 1990s, represents a milestone in our understanding of flower development, and subsequent molecular genetic discoveries, including the identification of floral homeotic genes, have contributed to our understanding of this process. By the end of 1990s, the basic framework of flower development in *Arabidopsis* had been established. This advance prompted rice researchers to ask whether the ABC model was directly applicable to the grass flowers. In addition, it raised the question of what kinds of genes are required for the development of spikelet organs that are not present in eudicot flowers.

In the early 2000s, a few important genes that regulate flower and spikelet development began to be isolated in rice (Jeon et al., 2000; Nagasawa et al., 2003; Yamaguchi et al., 2004). Since then, numerous genes responsible for floral homeotic mutations and abnormal morphologies of spikelet organs have been identified and their molecular functions have been elucidated (Hirano et al., 2014; Tanaka et al., 2013; Yoshida & Nagato, 2011). We have found that the ABC model is indeed applicable in grass flowers, but in a modified form.

In this review, we describe the genes required for the specification and differentiation of the organs in the flower and spikelet in rice, and present a modified version of the ABC model that explains rice flower development. Like other developmental processes in plants, the regulation of flower development cannot be separated from the function of the meristem. Therefore, we describe the regulation of the reproductive meristems by focussing on fate, maintenance and determinacy. We will also discuss the conserved and specific functions of developmental genes by comparing rice and *Arabidopsis*. The rice genes that regulate flower and spikelet development are summarized in Table 8.1, together with their orthologues or related genes in *Arabidopsis*. Due to the abundance of primary research papers on flower development in *Arabidopsis*, we mainly cite review articles for the *Arabidopsis* research. Studies on flower/inflorescence development and meristem

Table 8.1 Genes Responsible for Rice Development and Their Related Genes in *Arabidopsis*

Rice	*Arabidopsis*	Protein
Regulation of Meristem Fate and Maintenance		
FLORAL ORGAN NUMBER1 (FON1)	*CLAVATA1 (CLV1)*	LRR-receptor kinase
FLORAL ORGAN NUMBER2 (FON2)	*CLAVATA3 (CLV3)*	CLE domain
FON2-LIKE CLE PROTEIN1 (FCP1), FCP2	*CLAVATA3 (CLV3)*	CLE domain
FON2 SPARE1 (FOS1)	*CLE8*	CLE domain
OSH1	*SHOOTMERISTEMLESS (STM), KNAT1*	Homeo domain (class I KNOX)
OSH15	*SHOOTMERISTEMLESS (STM), KNAT1*	Homeo domain (class I KNOX)
OsWOX4	*WUSCHEL-RELATED HOMEOBOX4 (WOX4)*	Homeo domain (WOX)
CYTOKININ OXIDASE2 (CKX2)	*CYTOKININ OXIDASE*	Cytokinin oxidase
LONELY GUY (LOG)	*AtLOGs*	Cytokininriboside 5′-monophosphate phosphoribohydrolase
FRIZZY PANICLE (FZP)	*PUCHI*	AP2/ERF domain
SUPERNUMERARY BRACT (SNB)	*TARGET OF EAT1 (TOE1), TOE2, TOE3*	AP2/ERF domain
OsIDS1	*TARGET OF EAT1 (TOE1), TOE2, TOE3*	AP2/ERF domain
miR172b	*miR172*	MicroRNA
LEAFY HULL STERILE1 (LHS1)/ OsMADS1	*SEP-like*	MADS domain
ABERRANT SPIKELET AND PANICLE1 (ASP1)	*TOPLESS (TPL), TOPLESS-RELATED1 (TPR1), TPR2, TPR3, TPR4*	Transcriptional corepressor
TONGARI-BOUSHI1 (TOB1)	*FILAMENTOUS FLOWER (FIL), YABBY3*	YABBY domain

Continued

Table 8.1 Genes Responsible for Rice Development and Their Related Genes in *Arabidopsis*—cont'd

Rice	*Arabidopsis*	Protein
Flower and Spikelet Development		
SPW1/OsMADS16	*APETALA3 (AP3)*	MADS domain
OsMADS2	*PISTILLATA (PI)*	MADS domain
OsMADS4	*PISTILLATA (PI)*	MADS domain
OsMADS3	*AGAMOUS (AG)*	MADS domain
OsMADS58	*AGAMOUS (AG)*	MADS domain
OsMADS13	*SEEDSTIK (STK)*	MADS domain
DROOPING LEAF (DL)	*CRABS CLAW (CRC)*	YABBY domain
RFL/ABBERANT PANICLE ORGANIZATION2 (APO2)	*LEAFY (LFY)*	Plant-specific transcription factor
ABBERANT PANICLE ORGANIZATION1 (APO1)	*UNUSUAL FLORAL ORGANS (UFO)*	F-box protein
OPEN BEAK (OPB)/STAMENLESS1 (SL1)	*JAGGED (JAG)/NUBBIN (NUB)*	Zinc finger
OsETTIN2	*ETTIN/ARF3*	
OsETTIN3	*ETTIN/ARF3*	Auxin response factor
OsPHB3	*PHABULOSA (PHB), PHAVULOTA (PHV)*	HD-ZIPIII motif
DEPRESSED PALEA1 (DP1)	*AT-HOOK MOTIF NUCLEAR-LOCALIZED PROTEIN18 (AHL18), AHL22*	AT-hook motif
TRIANGULAR HULL1 (TH1)/BEAK LIKE SPIKELET1 (BLS1)	*LIGHT-DEPENDENT SHORT HYPOCOTYL1 (LSH1), LSH2, LSH3, LSH4*	ALOG domain
MOSAIC FLORAL ORGAN1/OsMADS6	*AGL6, AGL13*	MADS domain
RETARDED PALEA1 (REP1)	*TCP1, TCP12, TCP18*	TCP domain
MULTI-FLORET SPIKELET1 (MFS1)	*ERF12*	AP2/ERF domain
CHIMERIC FLORAL ORGANS1 (CFO1)/OsMADS32	None	MADS domain

DWARF AND DEFORMED1 (DDF1)	*FLAVIN-BINDING-KELCH REPEAT-F-BOX1 (FKF1), SLEEPY1 (SLY1), SNEEZY (SNE)*	F-box protein
LONG STERILE LEMMA/GLUME1 (G1)	None	ALOG domain
PANICLE PHYTOMER2 (PAP2)/ OSMADS34	*SEPALLATA1 (SEP1), SEP2, SEP4*	MADS domain
AN-1	None	bHLH domain
Anther Differentiation		
MICROSPORELESS1 (MIL1)	*ROXY1, ROXY2*	CC-type glutaredoxin
MULTIPLE SPOROCYTE1 (MSP1)	*EXCESS MICROSPOROCYTES1 (EMS1)/ EXTRA SPOROGENOUS CELLS (EXS)*	LRR-receptor kinase
MICROSPORELESS2 (MIL2)/OsTDL1A	*TAPETUM DETERMINANT1 (TPD1)*	
DEFECTIVE TAPETUM AND MEIOCYTES1 (DTM1)	None	
UNDEVELOPED TAPETUM1 (UDT1)	*DYSFUNCTIONAL TAPETUM1 (DYT1)*	bHLH domain
TAPETUM DEGENERATION RETARDATION (TDR)	*ABORTEDMICROSPORE (AMS)*	bHLH domain
ETERNAL TAPETUM1 (EAT1)	*AtbHLH010, AtbHLH089*	bHLH domain
PERSISTENT TAPETAL CELL1 (PTC1)	*MALE STERILITY1 (MS1)*	PHD finger motif
OsGAMYB	*AtMYB33, AtMYB65*	MYB domain
CYP703A3	*CYP703A2*	Cytochrome P450
OsC6	None	Lipid transfer protein
MICROSPORE AND TAPETUM REGULATOR1 (MTR1)	*FASCICLIN-LIKE ARABINOGALACTAN PROTEIN20 (FLA20)*	Fasciclin glycoprotein

KNOX, Knotted1-like homeobox; LRR, leucine-rich repeat.

regulation are also active in maize, another grass species. Comparing the functions of developmental genes between closely related species provides insight into the morphological evolution and diversification of gene function. Although we focus on rice development in this review, several recent reviews have summarized current understanding of flower and inflorescence development in both rice and maize (Hirano et al., 2014; Pautler, Tanaka, Hirano, & Jackson, 2013; Tanaka et al., 2013).

8.2 STRUCTURE OF THE FLOWER AND SPIKELET IN RICE

Rice belongs to the *Poaceae* (grass) family, which is widely distributed throughout the world. Grasses form a complex inflorescence, consisting of a main axis, branches, spikelets and florets (Bortiri & Hake, 2007; Hirano et al., 2014; Itoh et al., 2005; Tanaka et al., 2013). The spikelet is a unit of the inflorescence in grasses and consists of the glumes and one or several florets, which contain the floral organs.

Rice generates a raceme-type inflorescence made up of long branches called 'panicles'. Spikelets are formed on the primary and secondary branches. In contrast to the maize tassel or wheat spike, no spikelets are formed on the main axis (rachis) of the rice panicle. Each rice spikelet comprises a single floret, a pair of sterile lemmas and rudimentary glumes (Figure 8.1) (Hirano et al., 2014; Itoh et al., 2005; Tanaka et al., 2013; Yoshida & Nagato, 2011).

The floret contains the floral organs, including a pistil, stamens and lodicules, enclosed by the lemma and palea, organs that are specific to grasses (Figure 8.1(B) and (D)). The pistil is formed from three congenitally fused carpels, and a single ovule develops within the pistil. The apical region of the pistil bifurcates into two styles, on top of which the stigma hairs differentiate. Similar to stamens in other angiosperms, the stamen of rice consists of an anther and a filament. The lodicules are small semitransparent organs (Figure 8.1(D)). Classical morphological studies demonstrate that lodicules are homologues of petals (Hoshikawa, 1989; Troll, 2000). Two lodicules are asymmetrically formed at the lemma side of the floret (Figure 8.1(B) and (D)), and act as the driving force of anthesis by pushing the lemma and palea apart to promote outcrosses. The rice flower lacks organs corresponding to sepals. Thus, the carpel, stamen and lodicule are the floral organs in rice. Despite the lack of sepals, the regions where the lodicule, stamen and carpel develop are referred to as whorls 2, 3 and 4, respectively, as in general eudicot flowers, to avoid confusion.

The lemma is a bract that subtends the floral branchlet, which in turn consists of the palea and floral organs, and the palea is a modified prophyll (Clayton, 1990; Clifford, 1987; Troll, 2000). Although the lodicules are asymmetrically distributed, the floral organs, including the lodicules and stamens, are formed in a whorled phyllotaxy (Figure 8.1(B)). By contrast, the lemma and the palea initiate alternately, suggesting that the initiation of the lemma/palea differs from that of floral organs. Two tiny white leaf-like organs, the sterile lemmas, are formed outside of the floret (Figure 8.1 (B)–(D)). Although the naming of this organ is a matter of some confusion in the literature, a recent molecular genetic study confirmed the classic idea that these tiny organs are remnants of the lemmas of two degenerated florets (Arber, 1934; Hirano et al., 2014; Yoshida & Nagato, 2011). Thus, 'sterile lemma' is the most appropriate name for this organ. No such organs are observed in other grasses. In general, grass spikelets contain several florets enclosed within a pair of glumes (Bortiri & Hake, 2007; Hirano et al., 2014; Tanaka et al., 2013). In rice, however, the glumes are highly degenerated and appear as tiny flat protrusions, which are therefore called rudimentary glumes (Figure 8.1(C) and (D)).

Several types of meristems are involved in inflorescence and flower development (Figure 8.2(B)) (Hirano et al., 2014; Tanaka et al., 2013). Inflorescence development starts with the transition of a vegetative shoot apical meristem (SAM) into an inflorescence meristem (IM). The IM generates primary branch meristems (pBMs) as axillary meristems and supplies cells to form the rachis beneath. The pBMs generate secondary branch meristems (sBMs) and the spikelet meristem (SM), by forming long branches. Thus, higher-order branches are derived from the IM, pBMs and sBMs, and these meristems are indeterminate.

There are two types of spikelets, which are distinguished based on their developmental origin. Lateral spikelets develop from SMs that are produced from BMs as axillary meristems, whereas terminal spikelets are derived from SMs that are converted from BMs. In the case of spikelets that produce more than two florets, such as maize male spikelets, the transition from the SM to the flower/floret meristem (FM) can be clearly observed by scanning electron microscopy (Chuck, Meeley, & Hake, 1998). In rice, however, it is difficult to distinguish the FM from the SM because the SM forms a single FM. Although the sterile lemmas may have originated from the FM of the lateral floret during rice evolution, the sterile lemmas initiate as if they were lateral organs of the SM. In this article, our description of spikelet development is based on the assumption that the SM initiates the rudimentary

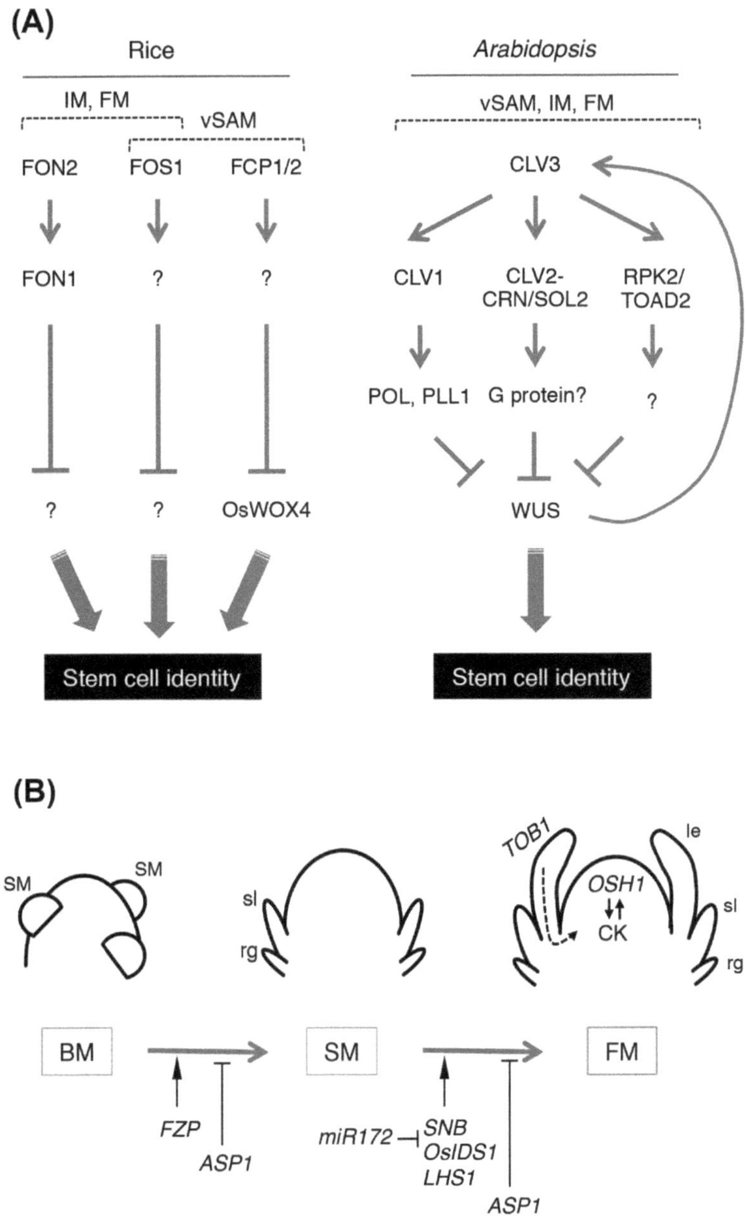

Figure 8.2 Regulation of stem cell maintenance and meristem fate. (A) Genes that regulate stem cell identity in rice and *Arabidopsis*. Question marks indicate unidentified factors. (B) Transition of the reproductive meristems and genes required for meristem fate in rice. The BM generates the SM, and the SM then initiates primordia of the rudimentary glume and sterile lemma. The floret/floral meristem initiates primordia of the lemma and the palea, and those of floral organs. BM, branch meristem; FM, floral meristem; IM, inflorescence meristem; le, lemma; sl, sterile lemma; SM, spikelet meristem; rg, rudimentary glume; vSAM, vegetative shoot apical meristem.

glumes and sterile lemmas, and the FM initiates the lemma/palea and the floral organs. Although the floret meristem is strictly speaking different from the flower meristem, we do not distinguish between these two meristems here.

8.3 MERISTEM REGULATION

8.3.1 Maintenance of Stem Cells and Undifferentiated Cells

Stem cells are located in the uppermost central region of all aerial meristems. They continue to produce daughter cells for both self-renewal and the formation of lateral organs. Therefore, stem cells are the ultimate source of the plant body, and knowledge of the genetic mechanisms underlying stem cell maintenance is important for understanding plant development. In *Arabidopsis*, the CLAVATA (CLV)–WUSCHEL (WUS) negative feedback loop is a major genetic pathway that maintains the stem cell population in meristems (Figure 8.2(A)) (Aichinger, Kornet, Friedrich, & Laux, 2012; Ha, Jun, & Fletcher, 2010; Miyawaki, Tabata, & Sawa, 2013). *WUS* promotes stem cell identity while the *CLV* genes negatively regulate stem cell proliferation by restricting the expression of *WUS*. Mutations in *CLV* genes result in enlarged meristems due to the overaccumulation of stem cells, which leads to the production of supernumerary organs in the flower. *CLV3* is expressed in the stem cell region, and encodes a small secreted protein, the CLE domain of which is processed into a small CLE peptide. *CLV1* encodes a leucine-rich repeat (LRR) receptor kinase that transduces the external signal mediated by the CLE peptide and ultimately represses *WUS* expression. Thus, CLV signalling facilitates communication between the stem cell region and the organizing centre, where *WUS* is expressed.

Rice *floral organ number1* (*fon1*) and *fon2*/*fon4* have larger FMs, which lead to an increased number of floral organs (Chu et al., 2006; Nagasawa, Miyoshi, Kitano, Satoh, & Nagato, 1996; Suzaki et al., 2004; Suzaki et al., 2006). The phenotype of the *fon1 fon2* double mutant resembles that of *fon1* or *fon2* single mutants, indicating that both *FON1* and *FON2* function in the same genetic pathway (Suzaki et al., 2004; Suzaki et al., 2006). *FON1* encodes an LRR-receptor kinase that is orthologous to *Arabidopsis* CLV1 (Suzaki et al., 2004), whereas *FON2* encodes a small CLE peptide similar to *Arabidopsis* CLV3 (Chu et al., 2006; Suzaki et al., 2006). Therefore, the function of *CLV* genes as negative regulators of stem cell proliferation is probably conserved in rice (Figure 8.2(A)).

FON2 SPARE1 (*FOS1*) was identified as a suppressor of the *fon2* mutation in *indica* (Suzaki, T., Ohneda, Toriba, Yoshida, & Hirano, 2009). *FOS1* encodes a protein in which the CLE domain is closely related to that of FON2. *FON2* and *FOS1* likely act redundantly to maintain the stem cell population in the FM of *indica* and wild rice species (Figure 8.2(A)). Interestingly, all cultivars of *japonica* rice have a mutation at the signal peptide processing site of FOS1, suggesting that *FOS1* function is defective in *japonica* and that the mutation is associated with the domestication of *japonica* (Suzaki et al., 2009).

CLV2 constitutes a receptor complex that functions independently of the CLV1 homodimer; however, the signalling mechanism downstream of CLV2 has not yet been elucidated in *Arabidopsis* (Müller, Bleckmann, & Simon, 2008; Nimchuk, Tarr, & Meyerowitz, 2011). A recent study shed light on the CLV2 signalling pathway in maize (Bommert, Je, Goldshmidt, & Jackson, 2013). Maize FASCIATED EAR2 (FEA2), which is closely related to CLV2, physically interacts with COMPACT PLANT2 (CT2), and both *fea2* and *ct2* act in the same genetic pathway to regulate IM maintenance in maize (Bommert et al., 2013; Taguchi-Shiobara, Yuan, Hake, & Jackson, 2001). The *ct2* encodes the α-subunit (Gα) of a heterotrimeric GTP-binding protein, suggesting that heterotrimeric G protein-mediated signalling is involved in meristem maintenance (Bommert et al., 2013). In rice, *DWARF1* (*D1*) encodes Gα, and its loss-of-function mutation results in a phenotype similar to that of *ct2*, including compact panicles and dwarfism (Ashikari, Wu, Yano, Sasaki, & Yoshimura, 1999; Fujisawa et al., 1999). This raises the questions of whether *D1* is involved in FM maintenance in rice and how it interacts with the *FON* genes.

FON2-LIKE CLE PROTEIN1 (*FCP1*) and *FOS1* likely act as negative regulators of stem cell maintenance in the vegetative SAM (Suzaki et al., 2009; Suzaki, Yoshida, & Hirano, 2008). This was initially inferred from the observation that overexpression of either *FCP1* or *FOS1* results in SAM termination in shoots regenerated from calli (Suzaki et al., 2009; Suzaki et al., 2008;). A recent study using an inducible knockdown system further elucidated the function of *FCP1* (Ohmori, Tanaka, Kojima, Sakakibara, & Hirano, 2013). The expression domain of *FON2*, which marks the stem cell region, expands in inducible knockdown lines of both *FCP1* and *FCP2*. Whereas *OSH1*, which marks undifferentiated cells, is downregulated at the leaf initiation site in wild type, this downregulation is inhibited by inducible knockdown of *FCP1*. Thus, *FCP1* and *FCP2* seem to function in SAM maintenance by regulating stem cell identity in the central region and promoting cell differentiation in the peripheral region. In addition, constitutive overexpression of *FCP1* and *FOS2* results in premature termination of the SAM in shoots regenerated from *fon1* calli, indicating that both factors might act through a receptor other than FON1 (Suzaki et al., 2009; Suzaki et al., 2008;). Therefore, it is likely that independent signalling pathways regulate stem cell maintenance, depending on the type of meristem in rice (Figure 8.2(A)). This is in contrast to the situation in *Arabidopsis*, where CLV3 alone acts as the negative regulator in all aerial meristems (Figure 8.2(A)).

In rice, our understanding of positive regulators of stem cell maintenance is insufficient. Recently, it was reported that *OsWOX4*, a member of the *WUSCHEL-RELATED HOMEOBOX* (*WOX*) family, is a positive regulator of stem cell maintenance in the SAM (Figure 8.2(A)) (Ohmori et al., 2013). Inducible RNA silencing of *OsWOX4* causes the downregulation of both *FON2* and *OSH1*, and its constitutive silencing results in smaller or flattened SAMs in the regenerating shoots. These meristem defects are also observed when *FCP1* is overexpressed, as described above. Consistent with these findings, *OsWOX4* expression is repressed by *FCP1*. Thus, the FCP1–OsWOX4 pathway appears to resemble the *Arabidopsis* CLV–WUS pathway. However, there are several differences between the pathways in rice and *Arabidopsis*, as discussed in the original report (Ohmori et al., 2013).

The class I *Knotted1-like homeobox* (*KNOX*) gene is required for the maintenance of undifferentiated states of cells in the meristem. In the *osh1* mutant, the SAM fails to keep a dome shape in the seedling, suggesting that meristem maintenance is compromised. Although the *osh1* homozygous mutation is lethal at the seedling stage, plants generated from *osh1* homozygous calli can grow to the reproductive phase. Such *osh1* plants exhibit small panicles and abnormal spikelets with fewer floral organs (Tsuda, Ito, Sato, & Kurata, 2011). These observations suggest that *OSH1* is required for the maintenance of all aerial meristems, including the vegetative SAM, IM and FM. The double mutant of *OSH1* and its close paralogue *OSH15* fails to form a SAM both during embryogenesis and shoot regeneration, indicating that both genes are redundantly required for the initiation of the meristem, in addition to meristem maintenance (Tsuda et al., 2011). *OSH1* positively regulates its own expression as well as that of other *KNOX* genes, including *OSH15*. Thus, this positive autoregulatory mechanism of *KNOX* genes appears to be essential for meristem maintenance in rice.

Several lines of evidence suggest that the plant hormone cytokinin functions in meristem maintenance. The *lonely guy* (*log*) mutant exhibits a small panicle with a reduced number of branches, spikelets and floral organs (Kurakawa et al., 2007). These phenotypes appear to result from the weak activity of the reproductive meristems. *LOG* encodes an enzyme that catalyses the final step of cytokinin biosynthesis (Kurakawa et al., 2007). The identification of *LOG* in rice promotes a deeper understanding of stem cell maintenance in *Arabidopsis*: the active cytokinin generated by LOG4 acts as a positional cue for patterning the WUS domain in the organizing centre (Chickarmane, Gordon, Tarr, Heisler, & Meyerowitz, 2012). In contrast to *LOG*, mutation in a gene encoding *CYTOKININ OXIDASE2* (*CKX2*), which catalyses the degradation of active cytokinin, results in increased numbers of branches and spikelets (Ashikari et al., 2005). Overexpression of *OSH15* and *OsWOX4* accelerates cytokinin production, whereas *OSH1* and *OSH15* are upregulated by exogenous cytokinin application (Figure 8.2(B)) (Ohmori et al., 2013; Sakamoto et al., 2006; Tsuda et al., 2011). Phyllotaxy is also associated with cytokinin action through meristem activity. Leaves are initiated alternately in grasses, but maize *abphyl1 (abph1)* and rice *decussate* (*dec*) mutants have enlarged SAMs, resulting in decussate phyllotaxy (Giulini, Wang, & Jackson, 2004; Itoh, Hibara, Kojima, Sakakibara, & Nagato, 2012). *abph1* encodes a type A response regulator that negatively regulates cytokinin action (Giulini et al., 2004). Although the biochemical function of DEC is unknown, circumstantial evidence suggests that DEC function is associated with cytokinin signalling (Itoh et al., 2012).

8.3.2 Transition of Meristem Fate

Rice exhibits complex inflorescence architecture, as described above, and inflorescence and spikelet development largely depends on the fate of the reproductive meristems. Several genes regulating meristem fate have been identified.

FRIZZY PANICLE (*FZP*) regulates the transition from the BM to the SM (Komatsu, Chujo, Nagato, Shimamoto, & Kyozuka, 2003). In the *fzp* mutant, small incomplete branches are repetitively produced in the axil of rudimentary glumes, resulting in a highly branched inflorescence that lacks spikelets. This phenotype suggests that *FZP* specifies SM identity by repressing indeterminate BM fate (Figure 8.2(B)). *FZP* encodes a transcription factor that contains an AP2/ERF (APETALA2/ethylene responsive factor) domain. The *branched silkless1* (*bd1*) gene, an orthologue of rice *FZP*, has the same function in maize (Chuck, Muszynski, Kellogg, Hake, & Schmidt, 2002).

SUPERNUMERARY BRACT (*SNB*) and *OsIDS1* appear to regulate the timing of the transition from the SM to the FM (Lee & An, 2012; Lee,

Lee, Moon, Park, & An, 2007). In the *snb* mutant, the spikelet forms a number of rudimentary glumes and exhibits delayed formation of the floret (Lee et al., 2007). In severe cases, spikelets completely lack floral organs after generating excess rudimentary glumes. Only a weak phenotype is observed in a mutant of *OsIDS1*, which is the closest paralogue of *SNB*. However, the *osids1 snb* double mutant produces a larger number of rudimentary glumes in a spikelet than does the *snb* single mutant (Lee & An, 2012). Thus, both *SNB* and *OsIDS1* are redundantly required for the acquisition of FM fate and the determinacy of the SM (Figure 8.2(B)) (Lee & An, 2012; Lee et al., 2007). *SNB* and *OsIDS1* encode members of a class of AP2/ERF transcription factors closely related to maize *INDETERMINATE SPIKELET1* (*IDS1*) and *SISTER OF INDETERMINATE SPIKELET1* (*SID1*) (Chuck et al., 1998; Chuck, Meeley, & Hake, 2008). Maize *IDS1* and *SID1* also regulate the determinacy of the SM by controlling the transition from SM to FM. Both *IDS1* and *SID1* are negatively regulated by microRNA *miR172* (Chuck, Meeley, Irish, Sakai, & Hake, 2007). Consistent with this, the overexpression of *miR172b* mimics the *snb osids1* double mutant in rice (Zhu, Upadhyaya, Gubler, & Helliwell, 2009).

Another regulator of FM identity is *LEAFY HULL STERILE1* (*LHS1*)/*OsMADS1* (Agrawal, Abe, Yamazaki, Miyao, & Hirochika, 2005; Jeon et al., 2000). The loss-of-function *lhs* mutant exhibits pleiotropic phenotypes ranging from mild to severe (Agrawal et al., 2005; Hirano et al., 2014; Jeon et al., 2000; Kinoshita, Hidano, & Takahashi, 1977). In mild cases, the lemma and palea are transformed into leaflike organs, suggesting that *LHS1* is required for lemma and palea identity, as described below (Section 8.7.1). In severe cases, the spikelet repetitively produces leaflike organs and rarely produces inner floral organs. This phenotype is probably due to the loss of determinacy of the SM and failure of the SM in the transition to the FM (Figure 8.2(B)) (Hirano et al., 2014; Jeon et al., 2000; Yoshida & Nagato, 2011). *LHS1* is expressed in the meristem and in the primordia of the lemma and palea, but not in those of the floral organs, such as lodicules and stamens (Prasad, Parameswaran, & Vijayraghavan, 2005; Prasad, Sriram, Kumar, Kushalappa, & Vijayraghavan, 2001). *LHS1*/*OsMADS1* is a member of the class E gene family of MADS-box transcription factors and is similar to *Arabidopsis SEPALLATA* (*SEP*) genes. In contrast to rice *LHS1*/*OsMADS1*, *Arabidopsis SEP* genes are expressed in the floral organs, such as the petals, stamens and carpels, and are involved in floral organ specification together with the ABC-floral homeotic genes (Jack, 2004; Lohmann & Weigel, 2002). Thus, the function of class E genes appears not to be conserved in rice and *Arabidopsis*, unlike previously suggested (Agrawal et al., 2005; Cui et al., 2010; Prasad et al., 2005).

8.3.3 Meristem Maintenance and Spikelet Organ Development

Mutations in *ABERRANT SPIKELET AND PANICLE1* (*ASP1*) and *TONGARI-BOUSHI1* (*TOB1*) result in pleiotropic phenotypes affecting inflorescence and spikelet traits, and these phenotypes are closely related to both the maintenance and fate of meristems (Tanaka, Toriba, Ohomori, & Hirano, 2012a; Tanaka et al., 2012b). The *asp1* mutants generate small panicles with short branches, including some that degenerate during early developmental stages, and a reduced number of spikelets, with aberrant morphology (Yoshida et al., 2012). These abnormalities are probably due to precocious transition from BM to SM and failure to maintain the BM and SM (Figure 8.2(B)). In addition, disturbed phyllotaxy in both leaf and branch initiation and derepression of the axillary meristem are observed, suggesting that *ASP1* is involved in auxin signalling. *ASP1* encodes a transcriptional corepressor, similar to *Arabidopsis* TOPLESS (TPL) (Long, Ohno, Smith, & Meyerowitz, 2006; Yoshida et al., 2012). Therefore, it is likely that the pleiotropic defects in *asp1* are due to the derepression of multiple genes related to meristem function. No phenotypes are observed even in quadruple recessive mutants of *TPL* and its paralogues (*TOPLESS-RELATED)* in *Arabidopsis*, whereas a single mutation in *ASP1* causes various developmental defects in rice (Long et al., 2006; Yoshida et al., 2012). Therefore, the rice *ASP1* gene may provide an opportunity to clarify the function of a TPL-like corepressor in plant development.

The *tob1* mutant shows pleiotropic defects in spikelets, such as awn elongation, reduced growth of the lemma/palea, formation of a cone-shaped organ and premature termination of the SM (Tanaka et al., 2012a; Tanaka et al., 2012b). Cone-shaped organs are formed instead of the lemma and palea, and appear to result from the disorganization of the FM. Thus, *tob1* has defects not only in lateral organ development, but also in meristem function. *TOB1* encodes a member of the *YABBY* transcription factors, and is closely related to *Arabidopsis FILAMENTOUS FLOWER* (*FIL*) and *YABBY3* (Siegfried, Eshed, tsuga, Drews, & Bowman, 1999; Sawa et al., 1999). *TOB1* is expressed in all lateral organ primordia of the spikelet, but not in the meristem (Tanaka et al., 2012b). Therefore, the meristem defect in *tob1* is a consequence of a failure in the non-cell autonomous action of *TOB1*, which is expressed in the lateral organs

(Figure 8.2(B)). There are several differences in *YABBY* activity between rice and *Arabidopsis*: *YABBY* genes like *FIL* show abaxially-localized expression and regulate leaf characteristics, including adaxial–abaxial establishment (Sarojam et al., 2010; Sawa et al., 1999; Siegfried et al., 1999), whereas rice *YABBY* genes do not show such localized expression patterns and are unrelated to adaxial–abaxial polarity (Tanaka et al., 2012b; Toriba et al., 2007). However, *FIL* and its related *YABBY* genes also maintain meristem activity in a non-cell autonomous manner (Goldshmidt, Alvarez, Bowman, & Eshed, 2008). Therefore, the role of YABBY in facilitating communication between lateral organs and the meristem to promote meristem activity is likely conserved in rice and *Arabidopsis*, and this may be the ancestral function of *YABBY* genes in this clade in the angiosperms. Since neither the mRNA nor protein of *FIL* is mobile (Goldshmidt et al., 2008), the *YABBY* genes may act through an unknown signal (Goldshmidt et al., 2008; Tanaka et al., 2012b). Identification of this signal and clarification of which genes are involved in the *YABBY* signalling pathway await further study.

8.4 FLORAL ORGAN SPECIFICATION—FUNCTION OF FLORAL HOMEOTIC GENES

The identification and characterization of a number of mutants and their causative genes in *Arabidopsis* and *Antirrhinum* during the past two decades have greatly advanced our understanding of floral organ development. In particular, the ABC model explains how the activity of just three classes of genes can specify four floral organs (Coen & Meyerowitz, 1991; Jack, 2004; Lohmann & Weigel, 2002; Prunet & Jack, 2014): the ABC genes act in two adjacent whorls, alone or in combination, to specify the fate of floral organs (Figure 8.3(A)).

In the past decade, our understanding of the genes that regulate floral organ specification in rice has also progressed.

8.4.1 Lodicule Differentiation

In *Arabidopsis*, the petals are specified by a combination of class A and class B genes (Coen & Meyerowitz, 1991; Jack, 2004; Lohmann & Weigel, 2002; Prunet & Jack, 2014). Whereas it is not clear whether class A genes have a role in floral organ specification in rice, class B genes are known to be involved in the specification of the lodicule and stamen (Hirano et al., 2014; Yoshida & Nagato, 2011). Rice has three class B MADS-box genes: *SUPERWOMAN1* (*SPW1*), an orthologue of *Arabidopsis APETALA3* (*AP3*), and

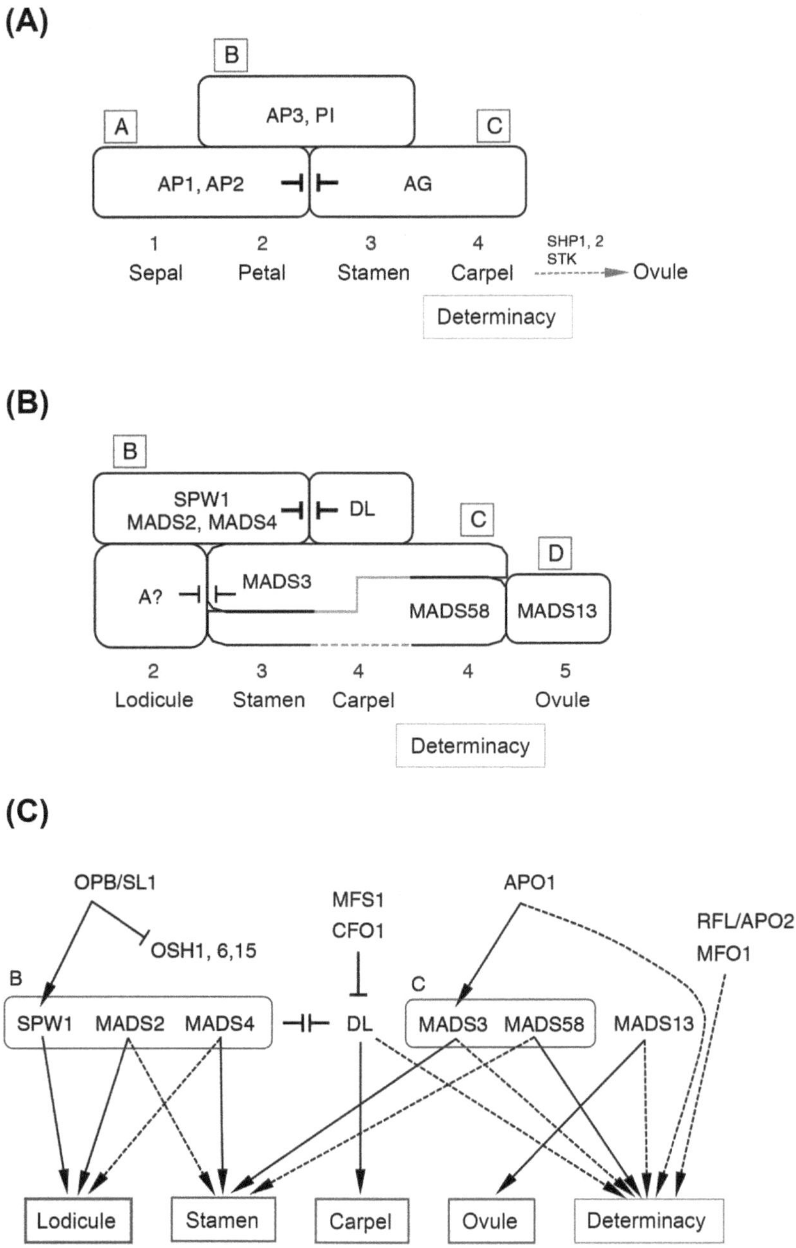

Figure 8.3 Mechanism of floral organ specification. (A) The ABC model of flower development in *Arabidopsis*. (B) A modified ABC model of flower development in rice. (C) Genetic networks of genes involved in floral organ development. Solid and broken lines indicate strong and moderate contribution to the specification of each organ, respectively.

OsMADS2 and *OsMADS4*, co-orthologues of *Arabidopsis PISTILLATA* (*PI*). These rice genes are expressed specifically in the presumptive region where the lodicule and stamen initiate in the FM (Kyozuka, Kobayashi, Morita, & Shimamoto, 2000; Nagasawa et al., 2003; Yadav, Prasad, & Vijayraghavan, 2007; Yoshida et al., 2007).

Loss of function of *SPW1* causes a complete homeotic transformation of lodicules into glume/leaf-like organs (Nagasawa et al., 2003). RNA silencing of *OsMADS2* results in mild phenotypes, whereas silencing of *OsMADS4* causes no visible abnormality in the lodicule (Yoshida et al., 2007). However, double knockdown of these two *PI*-like genes leads to complete homeotic transformation of lodicules into glume/leaf-like organs, suggesting that both *PI*-like genes act redundantly to specify the lodicule, but that their contributions differ (Yao, Ohmori, Kimizu, & Yoshida, 2008). Taken together, both *AP3*- and *PI*-like class B genes are required for lodicule specification in rice (Figure 8.3(B) and (C)).

As described above, lodicules are homologous to petals (Hoshikawa, 1989; Troll, 2000). Thus, the finding that class B genes specify the lodicule confirmed the idea proposed by classical comparative morphological studies. The lodicules and petals are morphologically and functionally different from each other. Therefore, target genes under the regulation of class B genes that encode transcription factors may differ in rice and *Arabidopsis*. It will be intriguing to identify the genes that regulate lodicule morphology and function.

A weak *spw1* mutant (*spw1-cls*) exhibits a developmental defect in the lodicule, but not in the stamen (Yoshida et al., 2007). Although the florets of *spw1-cls* do not open, this mutant is fertile and produces seeds by self-pollination. This demonstrates that the genetic manipulation of the activity of class B genes is an efficient method of artificially inducing cleistogamy; an important agronomic trait that prevents gene flow from genetically modified crops.

8.4.2 Stamen Specification

Stamens are specified by class B and C genes in *Arabidopsis* (Coen & Meyerowitz, 1991; Jack, 2004; Lohmann & Weigel, 2002; Prunet & Jack, 2014). Similar to lodicule specification, class B genes are required for stamen specification in rice. Loss of function of *SPW1* or double knockdown of *OsMADS2* and *OsMADS4* causes a homeotic transformation of stamens into carpels (Figure 8.3(B) and (C)) (Nagasawa et al., 2003; Yao et al., 2008). Single knockdown of *OsMADS2* or *OsMADS4* does not result in an obvious phenotype, suggesting that both genes function redundantly in specifying the stamen (Yao et al., 2008; Yoshida et al., 2007).

The biochemical properties of class B proteins are conserved. Like *Arabidopsis* AP3 and PI, SPW1 and OsMADS2 or OsMADS4 physically interact with each other, forming heterodimers in rice (Yao et al., 2008). This interaction is conserved in maize as well (Whipple et al., 2004).

There are two class C genes, *OsMADS3* and *OsMADS58,* in rice. Similar to the *Arabidopsis* class C gene *AGAMOUS* (*AG*), these two rice genes are expressed in both whorl 3 and whorl 4 (Dreni et al., 2011; Kyozuka & Shimamoto, 2002; Yamaguchi et al., 2006). A severe mutation in *OsMADS3* results in homeotic transformation of stamens into lodicules (Dreni et al., 2011; Yamaguchi et al., 2006). Conversely, the overexpression of *OsMADS3* causes homeotic transformation of lodicules into stamens (Kyozuka & Shimamoto, 2002). Although knockdown or knockout of *OsMADS58* does not result in severe defects in stamen development, combination of *OsMADS58* knockdown or knockout mutant with *osmads3* mutants causes complete loss of stamen identity (Dreni et al., 2011; Yamaguchi et al., 2006). Therefore, *OsMADS3* and *OsMADS58* redundantly regulate stamen specification but *OsMADS3* plays a more crucial role than *OsMADS58* (Figure 8.3(B) and (C)).

Taken together, these results indicate that rice class B and C genes are involved in the specification of stamen identity, as is the case in *Arabidopsis* (Figure 8.3(A)–(C)).

8.4.3 Carpel Specification

Whereas the ABC model of eudicot flower development can be applied to explain the specification of the stamen and lodicule in rice, it does not fit carpel specification perfectly. Carpel identity in *Arabidopsis* is specified by a class C MADS-box gene, *AG* (Coen & Meyerowitz, 1991; Jack, 2004; Lohmann & Weigel, 2002; Prunet & Jack, 2014). By contrast, carpel specification in rice is regulated by *DROOPING LEAF* (*DL*), which encodes a YABBY transcription factor (Figure 8.3(B) and (C)) (Yamaguchi et al., 2004). A mutation in *DL* causes homeotic transformation of carpels into stamens, which are generated indeterminately (Nagasawa et al., 2003; Yamaguchi et al., 2004). *DL* is expressed in the presumptive region of carpel initiation in the meristem and developing carpels (Yamaguchi et al., 2004). Its expression in the meristem implies that *DL* is involved in determining the fate of undifferentiated cells. Similar expression patterns were observed for *DL* orthologues in maize, wheat and barley, suggesting that the functions of *DL* orthologues are conserved in grasses (Ishikawa et al., 2009).

DL is an orthologue of *CRABS CLAW* (*CRC*) of *Arabidopsis*, which is partially responsible for carpel identity (Bowman & Smyth, 1999). *CRC* is expressed in the abaxial region of carpel primordia, but not in the meristem (Bowman & Smyth, 1999). Thus, *DL* orthologues have gained a crucial function in carpel specification during grass evolution.

In *Arabidopsis*, the activities of class A and C genes are mutually antagonistic, and this antagonistic interaction is a tenet of the ABC model (Figure 8.3(A)) (Coen & Meyerowitz, 1991). A similar antagonistic interaction exists between the class B genes and the *DL* gene in rice (Nagasawa et al., 2003; Yamaguchi et al., 2004). Carpels are formed in whorl 3 in the *spw1* mutant or the *OsMADS2/OsMADS4* double knockdown lines, because *DL* is ectopically expressed in this region (Nagasawa et al., 2003; Yamaguchi et al., 2004; Yao et al., 2008). This supports the idea that *DL* is negatively regulated by the function of class B genes in the wild type. Consistent with this idea, overexpression of *SPW1* by a constitutive promoter results in homeotic conversion of carpels into stamens, suggesting that the expression of endogenous *DL* is repressed (Lee et al., 2003). Conversely, class B genes appear to be repressed by *DL* in whorl 4, because stamens develop in whorl 4 in the *dl* mutant (Yamaguchi et al., 2004). Thus, the class B genes and *DL* antagonistically regulate each other (Figure 8.3(B) and (C)).

OPEN BEAK/STAMENLESS1 is a positive regulator of class B genes (see below in detail) (Horigome et al., 2009; Xiao et al., 2009). In the *opb/sl1* mutant, carpels develop instead of stamens in whorl 3, and *DL* is ectopically expressed in this region. This observation confirms the idea that *DL* specifies carpel identity and is negatively regulated by class B genes.

8.4.4 Ovule Differentiation

In *Arabidopsis*, ovules arise from the placental tissues after carpels develop (Colombo, Battaglia, & Kater, 2008). Three MADS-box genes, *SEEDSTICK* (*STK*), *SHATTERPROOF1* (*SHP1*) and *SHP2*, play essential roles in ovule specification (Pinyopich et al., 2003). By contrast, in rice, the ovule is formed directly from the FM immediately after carpel initiation (Dreni et al., 2007; Yamaki, Nagato, Kurata, & Nonomura, 2011). Despite their different origins, ovule identity is determined by *OsMADS13*, an orthologue of *STK*, in rice, as demonstrated by the finding that ovule development is defective in the *osmads13* mutant (Figure 8.3(B) and (C)) (Dreni et al., 2007; Yamaki et al., 2011).

Reiterative formation of the carpel implies that meristem determinacy is lost in *osmads13* (Dreni et al., 2007; Yamaki et al., 2011). This loss of determinacy of the meristem is likely to be associated with the fact that the ovule is the final organ differentiated from the FM in rice, a process that completely consumes the indeterminate cells. Conversely, a defect in FM activity affects ovule development. A weak mutation in *LOG* (*log-3*), which is required for meristem maintenance, as described above (see Section 8.3.1), inhibits ovule formation (Yamaki et al., 2011). This ovuleless phenotype of *log-3* is rescued by combination with the *fon1* mutation, which results in an excess number of stem cells in the FM (Yamaki et al., 2011). Thus, ovule development is closely associated with FM activity.

Although the placental tissues are differentiated from the carpel margin in *Arabidopsis*, these tissues appear to have an indeterminate nature. *SHOOT MERISTEMLESS* and *KNAT2* genes, which mark indeterminate tissues, are expressed in the placenta (Long, Moan, Medford, & Barton, 1996; Pautot et al., 2001). Thus, the function of the MADS-box genes in promoting ovule development from indeterminate tissues may be conserved in rice and *Arabidopsis*.

8.5 GENETIC INTERACTIONS AND FLOWER MERISTEM DETERMINACY

8.5.1 Genes that Interact with Floral Homeotic Genes

LEAFY (*LFY*) acts as a central regulator of flower development in *Arabidopsis* by promoting FM identity and the expression of floral homeotic genes (Lohmann & Weigel, 2002; Jack, 2004; Prunet & Jack, 2014). Downregulation of the *LFY/FLO* orthologue of rice (*RFL*) (Kyozuka, Konishi, Nemoto, Izawa, & Shimamoto, 1998) results in pleiotropic phenotypes, such as reduced plant height, a delay in flowering and small panicles (Rao, Prasad, Kumar, & Vijayraghavan, 2008). The recent isolation of a mutant of *RFL*, named *aberrant panicle organization2* (*apo2*), revealed a more detailed function of the *LFY/FLO* orthologue in rice (Ikeda-Kawakatsu, Maekawa, Izawa, Itoh, & Nagato, 2012). The IM size is reduced in *apo2*, and the small panicle phenotype is likely to result from precocious transition of the IM or BM into the SM. The *apo2* also exhibits floral organ defects, such as partial homeotic transformation of the lodicule, reduction in stamen number and the loss of FM determinacy. The floral organ defects observed in *apo2* are much weaker than those observed in *Arabidopsis lfy* mutants. Therefore, unlike *LFY*, *RFL/APO2* appears not to play a crucial role in regulating floral homeotic genes, and other key factors would be required for the activation of these genes in rice. Although there are inconsistencies in the reported expression patterns of *RFL/APO2* (Kyozuka et al., 1998; Rao et al., 2008), this gene appears to be expressed in all aerial meristems, including the vegetative SAM (Ikeda-Kawakatsu et al., 2012). Interestingly, *RFL/LFY* expression transiently disappears when the BM is converted into the SM. Thus, the pleiotropic phenotypes of *apo2* described above are likely to be closely associated with the activity of this gene in the meristem.

Flower and inflorescence development is also compromised in the *apo1* mutant. In the flower, the stamens are homeotically transformed into lodicules, and the carpels are formed iteratively. In situ analysis showed that *OsMADS3* expression is reduced in *apo1*, suggesting that *APO1* promotes *OsMADS3* expression (Figure 8.3(C)). Accordingly, the homeotic transformation of stamens in *apo1* is likely due to the reduced ability of class C genes to specify the stamen. Iterative formation of the carpel implies a reduction in FM determinacy. It will be interesting to examine whether the expression of *OsMADS58* is affected in *apo1*, because this class C gene is strongly associated with FM determinacy (see below). *APO1* encodes a protein similar to the *Arabidopsis* UNUSUAL FLORAL ORGANS (UFO) F-box protein. UFO is a positive regulator of class B genes in *Arabidopsis*, and is expressed in whorl 2 and 3 of the FM. By contrast, *APO1* is expressed throughout the meristem and is involved in the regulation of the class C gene *OsMADS3*. Thus, studies of *APO1* and *APO2* have revealed that the functions of *LFY* and *UFO* homologues are likely not conserved in rice.

The lodicule and stamen are partially transformed into lemma/palea-like organs and carpels in the *open beak* (*opb*)/*stamenless* (*sl1*) mutant (Horigome et al., 2009; Xiao et al., 2009). In severe cases, stamens are completely transformed into carpels. *SPW1* expression is highly downregulated in *opb/sl1*. Therefore, *OPB/SL1* is involved in both lodicule and stamen development through the positive regulation of the B class gene *SPW1* (Figure 8.3(C)). In addition, *OPB/SL1* represses class I *KNOX* genes, such as *OSH1* and *OSH15*. Ectopic expression of the class I *KNOX* genes in *opb/sl1* results in the formation of a cell mass in the carpel by abnormal proliferation (Horigome et al., 2009). *OPB/SL1* encodes a protein with a C_2H_2 zinc finger domain, and is co-orthologous to *Arabidopsis JAGGED* and *NUBBIN*, which are responsible for morphogenesis of the stamen and carpel (Ohno et al., 2004; Dinneny et al., 2006).

8.5.2 Flower Meristem Determinacy

Whereas stem cells are maintained in indeterminate meristems, such as the vegetative SAM, IM and BM, they are consumed by the formation of the final lateral organs differentiated from the FM. Thus, the FM is determinate. In wild-type *Arabidopsis*, the stem cells disappear after carpel development due to the activity of *AG*. In the *ag* mutant, by contrast, a set of floral organs (sepals followed by two whorls of petals) is iteratively formed from the remaining stem cells (Lohmann & Weigel, 2002; Jack, 2004; Prunet & Jack, 2014). The continuous generation of stem cells in *ag* is due to prolonged *WUS* activity, which is repressed by *AG* in a later developmental stage of wild-type flowers. Thus, *AG* is a key gene regulating FM determinacy (Figure 8.3(A)).

In rice, one of the *AG* orthologues, *OsMADS58*, plays a critical role in the determinacy of the FM, as first evidenced by the finding that complete knockdown of this gene results in flowers consisting of reiterating whorls of lodicules, stamens and carpels (Yamaguchi et al., 2006). By contrast, mutation in *OsMADS3* results in weak loss of determinacy: a few carpels are repeatedly formed. Thus, the contribution to FM determinacy differs between the two class C genes, *OsMADS58* and *OsMADS3*, in rice (Figure 8.3(B) and (C)).

Because the ovule is the final organ to be differentiated from the FM, failure in ovule formation disturbs FM determinacy (Yamaki et al., 2011), as described above (Section 8.4.4). Combination of *osmads13* and *osmads3* was reported to enhance the mild phenotype of FM indeterminacy found in each single mutant (Dreni et al., 2011). In the same work, an *OsMADS58* knockout line was found not to exhibit an obvious phenotype (Dreni et al., 2011), in contrast to the previous study (Yamaguchi et al., 2006), possibly due to differences in genetic background. However, when the *osmads58* mutation was combined with *osmads3* or *osmads13*, the indeterminate FM phenotype of each single mutant was strongly enhanced, confirming that *OsMADS58* is required for the control of FM determinacy (Dreni et al., 2011). Further studies are required to elucidate the function of *OsMADS58*.

The carpel is specified by *DL* in rice. Severe *dl* mutations produce a variable number of stamens instead of the carpel, suggesting that *DL* also functions in FM determinacy (Figure 8.3(C)) (Yamaguchi et al., 2006). Consistent with this, the *dl* mutation enhances the indeterminate phenotypes of the FM in *osmads3*, *apo1* and *mfo1* (Ikeda et al., 2007; Li et al., 2011c; Li et al., 2011a; Ohmori et al., 2009). Mutation in *MOSAIC FLORAL ORGAN1 (MFO1)/OsMADS6,* a member of the AGL6 subfamily, causes pleiotropic flower phenotypes (Li et al., 2010; Ohmori et al., 2009). Meristem determinacy is compromised in *mfo1* and the defect in SM determinacy in the *lhs1* mutant is enhanced in the *mfo1 lhs1* double mutant (Li et al., 2010; Ohmori et al., 2009). *MFO1/OsMADS6* is also involved in the development of spikelet/flower organs, such as the palea, lodicule and carpel, as described below (Li et al., 2010; Ohmori et al., 2009).

Thus, a number of genes are involved in the determinacy of the FM. However, the genes that promote stem cell proliferation in the FM, such as *WUS* in *Arabidopsis*, remain to be identified in rice.

8.6 ORGAN DEVELOPMENT AND DIFFERENTIATION

8.6.1 Establishment of Adaxial–Abaxial Polarity

8.6.1.1 Lemma

Lateral organs are formed from founder cells in the peripheral region of the apical meristem. The founder cells differentiate into distinct tissues characteristic along the adaxial–abaxial axis. Several types of transcription factors and the small RNAs that regulate some of them are key determinants in establishing adaxial–abaxial polarity (Kidner & Timmermans, 2010). The establishment of this polarity is closely associated with planar growth of the lamina, and failure to establish polarity results in the formation of radialized cylindrical organs that lack lamina outgrowth (Kidner & Timmermans, 2010).

Several rice mutants are associated with the failure to establish adaxial–abaxial identity during flower/spikelet development (Abe et al., 2010; Itoh et al., 2000; Itoh et al., 2008; Liu et al., 2007; Song et al., 2012; Toriba et al., 2010). In these mutants, the lemma is radially symmetrical (because of the absence of lateral outgrowth) and forms a needlelike structure together with an elongated awn. A study of one of these mutants, *rodlike lemma* (*rol*), revealed that the radialized lemma is fully abaxialized (Toriba et al., 2010). In agreement with this, an abaxial determinant, *OsETTIN3* (*OsETT3*) (a homologue of *Arabidopsis ETTIN/ARF3*), is expressed throughout the epidermal and subepidermal tissues, whereas an adaxial determinant, *OsPHB3* (an HD-ZIPIII gene) is not expressed in these tissues (Toriba et al., 2010).

All of the mutants exhibiting radial lemmas have defects in the biosynthesis of *trans*-acting siRNAs (ta-siRNAs) (Abe et al., 2010; Liu et al., 2007; Nagasaki et al., 2007; Song et al., 2012; Toriba et al., 2010). Because the targets of the ta-siRNAs are *OsETT* genes in rice, the abaxialised lemma appears to result from misregulation of *OsETTs*. Several lines of evidence support this idea. Firstly, *OsETT* genes are upregulated or ectopically expressed in these mutants (Abe et al., 2010; Itoh et al., 2008; Liu et al., 2007; Nagasaki et al., 2007; Song et al., 2012; Toriba et al., 2010). Secondly, RNA silencing of

OsETT3 rescues the abaxialised phenotypes in one of the mutants (Toriba & Hirano, 2014). Lastly, overexpression of a modified *OsETT1* gene, which is resistant to ta-siRNAs, causes abaxialisation of the lemma (Song et al., 2012).

8.6.1.2 Stamens

The ABC model reveals that floral organs are modified leaves, because all floral organs are transformed into leaves in an *abc* triple mutant (Coen & Meyerowitz, 1991). Unlike petals and sepals, which are flat organs like leaves, the stamen is composed of the anther, which is a set of cylindrical structures, and the filament, which is a rodlike structure. What mechanisms underlie the formation of these structures? How is the establishment of adaxial–abaxial polarity associated with stamen development? A study based on the rice *rol* mutant proposed a unique model to answer these questions (Toriba et al., 2010).

In general, the anther is composed of two thecae that each consist of two pollen sacs and are joined to each other by connective tissue (Goldberg et al., 1993; Hufford, 1996). In the *rol* mutant, the anther frequently lacks one or both thecae, and anthers with one or three pollen sacs are not observed, suggesting that the theca is the developmental unit of anther patterning (Toriba et al., 2010). RNA in situ analysis using adaxial and abaxial determinants, such as *OsPHB3* and *OsETT1*, revealed that the adaxial–abaxial axis is duplicated and rearranged during anther development (Figure 8.4). After the rearrangement, a new axis is established within each theca primordium. Failure to establish this axis appears to result in loss of theca production in the *rol* mutant. The region between the adaxial and abaxial domains subsequently protrudes and differentiates to form pollen sacs (Figure 8.4). This protrusion resembles the planar outgrowth observed at the boundary between the adaxial and abaxial region during leaf development. Thus, anther patterning appears to rely on mechanisms that establish adaxial–abaxial polarity. The anther is supported by the filament, a radially symmetrical rodlike structure. Spatial expression patterns of both adaxial and abaxial determinants show that the filament is completely abaxialised in wild type (Figure 8.4) (Toriba et al., 2010). Thus, morphogenesis of the stamen is closely associated with regulation of adaxial–abaxial polarity in rice.

The above model seems to be widely applicable to stamen development in angiosperms. In *Arabidopsis*, the expression of *PHB* and *ETT* is rearranged in a similar manner as that observed in rice stamen development, and some mutants that have defects in adaxial–abaxial polarity establishment lack thecae like the rice *rol* mutant (see references in Toriba et al., 2010).

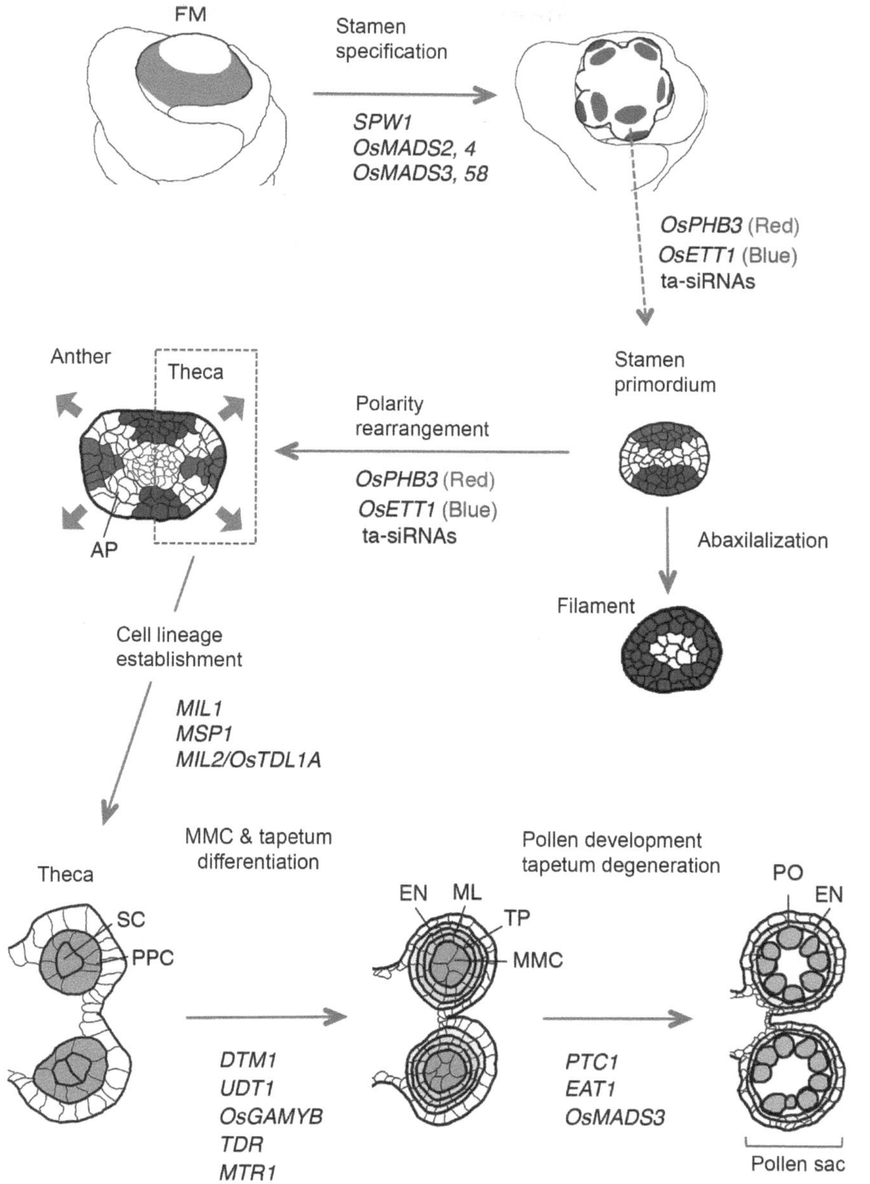

Figure 8.4 Stamen development and anther differentiation. Stamen initiation is specified by the combination of class B genes (*SPW1, OsMADS2, OsMADS4*) and class C genes (*OsMADS3, OsMADS58*). The adaxial–abaxial polarity is initially established in the stamen primordia in a similar way as in the leaf primordia, and then the polarity is rearranged and re-established in the theca as a developmental unit. The cells in the area between the regions of adaxial and abaxial identity (marked by expression of *OsPHB3* and *OsETT1*, respectively) proliferate and differentiate into pollen sacs. AP, archesporial cells; EN, endothecium; FM, flower meristem; ML, middle layer; MMC, microspore mother cell; PO developing pollen; PPC, primary parietal cell; SC, sporogenous cell; TP, tapetum. (See the colour plate.)

8.6.2 Anther Differentiation

In rice anther development, the subepidermal cells at the four corners of the anther primordium give rise to archesporial cells, which differentiate into two cell types, primary parietal and sporogenous cells (Figure 8.4) (Guo & Liu, 2012; Hong et al., 2012b; Nonomura et al., 2003; Zhang et al., 2011). In the somatic cell lineage, the primary parietal cells generate three cell layers, which differentiate into the tapetum, the middle layer and the endothecium. The tapetum supports microspore development by serving as a source of nutrients, and degenerates as the pollen matures. In the reproductive cell lineage, the sporogenous cells differentiate into microspore mother cells, which enter meiosis. Genetic studies on male-sterile mutants have provided much insight into the mechanisms underlying anther differentiation in rice.

MICROSPORELESS1 (*MIL1*) is required for both reproductive and somatic cell lineages (Figure 8.4) (Hong et al., 2012a). In *mil1*, the archesporial cells apparently generate both sporogenous and primary parietal cells, but the progeny of the archesporial cells are not differentiated into the typical cell types. *MIL1* encodes a glutaredoxin, a protein involved in redox regulation by mediating the reduction of protein disulfide bonds. A recent paper demonstrated that a low redox potential induces differentiation of the reproductive and surrounding somatic cells in maize (Kelliher & Walbot, 2012). Thus, *MIL1*-dependent redox control may promote cell fate determination to establish the two cell lineages in rice. Consistent with this notion, *MIL1* is expressed in the archesporial cells and their descendant cells in both lineages (Hong et al., 2012a). MIL1 physically interacts with a TGA transcription factor with a bZIP motif, suggesting that TGA is involved in anther development in rice as well as in *Arabidopsis* (Murmu et al., 2010; Hong et al., 2012a).

Loss-of-function mutants of *MULTIPLE SPOROCYTE1* (*MSP1*) and *MICROSPORELESS2* (*MIL2*)/*OsTDL1A* show almost identical phenotypes, i.e. the formation of excess sporogenous cells and a lack of somatic cell layers such as the tapetum and middle layer (Hong et al., 2012b; Nonomura et al., 2003; Zhao et al., 2008). *MSP1* and *MIL2/OsTDL1A* encode an LRR-receptor-like kinase and a putative ligand, respectively. MSP1 and MIL2/OsTDL1A proteins physically interact with each other (Zhao et al., 2008). *MSP1* is expressed in the primary parietal cells and their progeny (Nonomura et al., 2003), suggesting that MSP1-MIL2/OsTDL1A signalling is involved in determining the fate of these cells (Figure 8.4). *MSP1* is not expressed in the sporogenous cells. The excess sporogenous cells produced in these mutants may be a side effect of the failure to develop a somatic cell lineage (Hong et al., 2012b; Nonomura et al., 2003). A signalling pathway similar to MSP1-MIL2/OsTDL1A is involved in anther differentiation in both *Arabidopsis* and maize (Zhang & Yang, 2014).

An endoplasmic reticulum (ER) protein named DEFECTIVE TAPETUM AND MEIOCYTES1 (DTM1) is important for early tapetum development (Figure 8.4) (Yi et al., 2012). Although the biochemical function of DTM1 is unclear, it seems to be involved in ER development.

Several transcription factors are essential for both the development and subsequent degeneration of the tapetum. *UNDEVELOPED TAPETUM1* (*UDT1*) and *TAPETUM DEGENERATION RETARDATION* (*TDR*), both of which encode putative basic helix-loop-helix (bHLH) transcriptional factors, function in tapetum development (Figure 8.4) (Jung et al., 2005; Li et al., 2006; Zhang et al., 2008). The *udt1* mutant exhibits abnormalities in tapetal cells from an earlier stage than does the *tdr* mutant. In addition, *TDR* expression is reduced in the *udt1* mutant, suggesting that this gene acts downstream of *UDT1*. In *Arabidopsis*, bHLH transcription factors similar to TDR and UDT1 are also involved in tapetum development (Sorensen et al., 2003; Zhang et al., 2006). Another rice bHLH transcription factor, ETERNAL TAPETUM1 (EAT1), is involved in the programmed cell death that induces tapetum degeneration (Niu et al., 2013). *EAT1* directly regulates the genes encoding aspartic proteases that act in programmed cell death. EAT1 may act together with TDR, because the two proteins physically interact with each other (Niu et al., 2013).

Tapetum degeneration is also regulated by *PERSISTENT TAPETAL CELL1* (*PTC1*), *OsMADS3* and *OsGAMYB* (Figure 8.4) (Aya et al., 2009; Hu et al., 2011; Liu et al., 2010; Li et al., 2011b; Kaneko et al., 2004). *PTC1* encodes a PHD finger protein related to *Arabidopsis* MALE STERILITY1 (Li et al., 2011b). Although *OsMADS3* is required for stamen specification, as described above (Yamaguchi et al., 2006), this gene is also required for tapetum degeneration by controlling the homeostasis of reactive oxygen species (Hu et al., 2011). *OsGAMYB*, a member of the MYB transcription factor family, regulates tapetum degeneration by modulating gibberellin signalling (Aya et al., 2009; Liu et al., 2010; Kaneko et al., 2004). *OsGAMYB* is additionally involved in tapetum differentiation by directly activating several genes, including *CYP703A3* and *OsC6* (Aya et al., 2009). *CYP703A3* encodes a cytochrome P450 hydroxylase involved in the formation of Ubisch bodies, which develop on tapetal cell walls and function in the transfer of proteins and lipids from tapetal cells to microspores (Aya et al., 2009; Jung et al., 2006; Wang et al., 2003). *OsC6* encodes a lipid transfer protein that acts in the tapetum to support microspore development (Zhang et al., 2010). Intriguingly, *OsC6* is a direct target of TDR (Li et al., 2006), suggesting that the MYB and bHLH transcription factors function cooperatively.

A study of *MICROSPORE AND TAPETUM REGULATOR1* (*MTR1*) suggests the existence of putative reverse developmental regulation of tapetal cells by microspore mother cells (Tan et al., 2012). Although *MTR1* is expressed in microspore mother cells and their progeny, the *mtr1* mutant exhibits serious defects not only in the microspore, but also in the tapetum. *MTR1* encodes a fasciclin glycoprotein localized to the plasma membrane. It is suggested that *MTR1* promotes tapetum development by mediating cell–cell adhesion or signalling from microspore mother cells to the tapetal cells (Figure 8.4).

8.7 DEVELOPMENT OF NONFLORAL SPIKELET ORGANS

8.7.1 Lemma and Palea

Rice flowers lack sepals, and floral organs are enclosed by the lemma and the palea. Because the glumes are highly degenerated, the lemma and palea are the most prominent organs in the rice spikelet (Hirano et al., 2014; Itoh et al., 2005; Tanaka et al., 2013; Yoshida and Nagato, 2011). Several genes affecting the development and morphology of both of these organs have been reported.

Loss-of-function or dominant negative mutants of *LHS1* fail to form a normal lemma and palea, and, instead, form repeated whorls of leaflike organs (Agrawal et al., 2005; Jeon et al., 2000). The latter phenotype is associated with loss of determinacy of the SM, as described above (Section 8.3.2). The lemma and palea have a rough abaxial surface, which is covered with a number of small convex structures, called tubercles (Figure 8.1(C)) (Terrell et al., 2001). RNA silencing of *LHS1* reduces the number of tubercles present on the lemma and palea, resulting in relatively smooth abaxial surfaces (Prasad et al., 2005). Conversely, overexpression of *LHS1* promotes the formation of tubercles on the sterile lemma, the surface of which is smooth in the wild type (Figure 8.1(C)). Thus, *LHS1* regulates the morphological identity of the lemma and palea. In a double mutant of *lhs1* and *osmads15* (a class A gene mutant), the leaflike character of the *lhs1* spikelet is strongly enhanced (Wang et al., 2010). The simultaneous down-regulation of class E genes, such as *OsMADS5, 7* and *8*, has a similar effect on the spikelet (Cui et al., 2010).

Mutation of *TOB1* disturbs the development of the lemma and palea, resulting in pleiotropic phenotypes (Tanaka et al., 2012b). For example, the two organs are reduced in size or replaced with a cone-shaped organ. These developmental defects are also closely associated with meristem activity, as described above. Despite these morphological defects, the surfaces of *tob1* mutants are unchanged. The differentiation of the lateral region of both the lemma and palea is suppressed in the *opb/sl1* mutant, resulting in the formation of a beaklike lemma/palea that fails to enclose the inner organs (Horigome et al., 2009; Xiao et al., 2009). Lateral growth of the lemma and palea is weakly affected in *triangular hull1* (*th1*)/*beak like spikelet1* (*bls1*) mutants, which have a defect in a gene encoding a protein with an ALOG (*Arabidopsis* LSH1 and *Oryza* G1) domain (Li et al., 2012; Ma et al., 2013).

Mutations in *DEPRESSED PALEA1* (*DP1*) affect the morphology and growth of the palea; in particular, the central region of the palea is deleted, and two marginal regions of the palea are formed separately (Jin et al., 2011). Because the lemma is largely normal, it appears that *DP1* predominantly plays roles in palea development. Similarly, the growth of the central region of the palea is repressed in *retarded palea1* (*rep1*) and *multi-floret spikelet1* (*mfs1*) (Ren et al., 2013; Yuan et al., 2009). *REP1* and *MFS1* encode TCP (TEOSINTE BRANCHED1/CYCLOIDEA/PCF) and AP2/ERF transcription factors, respectively. *REP1* expression is reduced in the *dp1* mutant, suggesting that *DP1* acts upstream of *REP1* (Jin et al., 2011). In contrast to these mutants, the marginal regions of the palea are morphologically changed in *mfo1/osmads6*, and are similar to those of the lemma (Li et al., 2010; Ohmori et al., 2009). Close examination of internal tissue differentiation suggests that the palea is partially transformed into the lemma in *mfo1* (Ohmori et al., 2009). In *chimeric floral organs1* (*cfo1*), the palea margins are wider than those of the wild type and the abaxial epidermis is rough with tubercles (Sang et al., 2012). In addition, various chimeric floral organs showing partial carpel characteristics are formed in *cfo1*, and *DL* is ectopically expressed in these abnormal organs, suggesting that *DL* is involved in the abnormalities observed in *cfo1* spikelets. *cfo1* is mutated in *OsMADS32*, which belongs to a monocot-specific clade of the MADS-box gene family.

The development of the lemma and palea is strongly compromised in the *degenerated hull1* (*dh1*) mutant; in severe cases, both lemma and palea are absent (Li et al., 2008). *DH1* encodes a protein with an LOB (LATERAL ORGAN BOUNDARY) domain and although detailed developmental analysis has not yet been conducted, it will be intriguing to determine the function of this LOB domain protein in lemma/palea development. Lemma and palea development are also defective in the mutant of *DWARF AND DEFORMED1* (*DDF1*), which encodes an F-box protein (Duan et al., 2012). By contrast, extra or ectopic lemma/palea-like structures are formed in mutants such as *extra glume1* (*eg1*) and *mfs1* (Li et al., 2009; Ren et al., 2013). *EG1* encodes a protein similar to phospholipase.

8.7.2 Sterile Lemma and Rudimentary Glume

A pair of sterile lemmas is formed between the rudimentary glumes and the floret. This organ is not found in the spikelets of other grasses, such as maize and wheat. Over 75 years ago, Arber (Arber, 1934) proposed that the sterile lemmas are derived from the lemmas of two lateral florets, which degenerated during the evolution of ancestral species of the *Oryza* genus.

A classical mutant, *long sterile lemma/glume1* (*g1*), generates sterile lemmas that are enlarged like a lemma or a palea, and has been used as a morphological marker to construct the rice genetic map (Kinoshita, 1984; Nagato & Yoshimura, 1998). A close examination by Yoshida et al. revealed that this large organ has properties of the lemma, such as a rough abaxial surface with many tubercles and lemma-like marginal structures (Yoshida et al., 2009). This suggests that the sterile lemma is homeotically transformed into a lemma in the *g1* mutant. *G1* encodes a nuclear-localized protein that bears an ALOG domain, and is expressed predominantly in sterile lemma primordia. Therefore, *G1* is likely to repress lemma characteristics to specify the sterile lemma identity in wild type. It is, therefore, an attractive hypothesis that *G1* was recruited to repress lemma identity during the evolution of the *Oryza* genus. The notion that *G1* is essential for sterile lemma identity was confirmed by analysis of the *msf1* mutant by another group (Ren et al., 2013). The sterile lemmas of *msf1* have rough surfaces instead of smooth, suggesting that the sterile lemma identity is partially compromised in this mutant. Consistently, *G1* expression is reduced in this mutant.

Large sterile lemmas are generated in the *panicle phytomer2* (*pap2*)/*osmads34* mutant, which has a defect in meristem fate (Gao et al., 2010; Kobayashi et al., 2010). In contrast to *g1*, the epidermal surfaces of the sterile lemmas in *pap2*/*osmads34* are like those of the wild type.

The development of the sterile lemmas and rudimentary glumes are affected in the *asp1* mutant (Yoshida et al., 2012). In severe cases, the initiation of both organs is compromised, resulting in a reduction in their numbers. This may be associated with a defect in the meristem activity of *asp1*, as described above. In addition, the identities of both organs are compromised. For example, the sterile lemmas exhibit a lemma-like epidermal surface or chimeric surface that displays properties of both the lemma and sterile lemma, and the rudimentary glume exhibits a sterile lemma-like surface. These abnormalities suggest that the developmental programmes of these organs are disturbed in the *asp1* mutant.

8.7.3 Awn

An awn is a needlelike appendage formed on the lemma (Vaughan, 1989). As described in the Introduction, the ability to form an awn has been reduced during the domestication of most *japonica* and some *indica* strains.

Several studies have been performed to identify the gene responsible for awn formation by quantitative trait locus (QTL) analysis using awned wild rice species and awnless domesticated rice, and more than 10 loci have been genetically identified (Cai & Morishima, 2002; Gu et al., 2005; Kubo et al., 2001). Recently, *AN-1*, which promotes both the initiation and elongation of the awn, was molecularly identified by QTL analysis and map-based cloning and found to encode a bHLH transcription factor (Luo et al., 2013). The top of the lemma primordium is larger in an awned floret than in an awnless one, suggesting that a putative awn primordium differentiates before awn development (Luo et al., 2013; Toriba & Hirano, 2014). *AN-1* is strongly expressed in the awn primordium, where it is thought to promote cell division for awn formation (Luo et al., 2013). Nucleotide variations in *AN-1* that resulted in awn suppression may have accumulated during domestication.

In some morphological mutants, the awn elongates even in *japonica*, in which awn formation is largely suppressed (Abe et al., 2010; Itoh et al., 2000; Itoh et al., 2008; Liu et al., 2007; Song et al., 2012; Tanaka et al., 2012b; Toriba et al., 2010). This suggests that the *japonica* strains still have the potential to form awns. Focussing on a couple of developmental mutants of *japonica*, Toriba and Hirano (2014) revealed that *DL* and *OsETT2* play crucial roles in awn formation. Whereas *OsETT2* is expressed in the awn primordium and developing awn, *DL* is expressed beneath the awn primordium, but not in the primordium itself. Thus, *DL* appears to act in a non-cell autonomous manner to form the awn. Genetic analysis indicates that *DL* and *OsETT2* function independently in awn development. *DL* is expressed in awnless strains in a pattern similar to that observed in awned strains whereas *OsETT2* expression is markedly downregulated in the putative awn primordium in awnless strains (Toriba & Hirano, 2014). Therefore, downregulation in *OsETT2* expression is likely essential for the repression of awn development in awnless *japonica* strains, and this downregulation may have been subjected to artificial selection during the domestication of *japonica* rice.

8.8 CONCLUDING REMARKS

In the past decade, much progress has been made in understanding meristem regulation and flower/spikelet development in rice. Several lines of evidence demonstrate that the ABC model of flower development in eudicots is applicable to rice in large part. However, carpel specification in rice does not comply with the ABC model, as it involves the *YABBY* gene *DL* instead of a class C gene. In addition, genes in the *AG* and *PI* clade are duplicated and functionally diverged in rice. A similar mechanism to the CLV–WUS negative feedback loop appears to be partially conserved in rice for meristem maintenance, although multiple CLE peptides are involved in SAM maintenance in rice. There has also been rapid progress in our understanding of the genes required for the development and morphogenesis of spikelet organs, such as lemmas, paleae and sterile lemmas. All of these advances have been primarily achieved through genetic studies, in which genes responsible for developmental mutants are first isolated and then their functions are analysed. Reverse genetic approaches have been successful in *Arabidopsis* research, because it is relatively easy to identify knockout mutants by screening a large number of T-DNA tagged lines. By contrast, it is challenging to identify rice mutants disrupted in a gene of interest because of the small size of available knockout libraries. It is now possible to use new genome editing technologies, such as transcription activator-like effector nuclease (TALEN) and CRISPR-CAS9, to disrupt a specific gene of interest (Ran et al., 2013; Voytas, 2013). The application of these technologies promises to advance our understanding of the developmental mechanisms in rice at the molecular level.

REFERENCES

Abe, M., Yoshikawa, T., Nosaka, M., Sakakibara, H., Sato, Y., Nagato, Y., et al. (2010). *WAVY LEAF1*, an ortholog of *Arabidopsis HEN1*, regulates shoot development by maintaining microRNA and trans-acting small interfering RNA accumulation in rice. *Plant Physiology*, *154*, 1335–1346.

Agrawal, G. K., Abe, K., Yamazaki, M., Miyao, A., & Hirochika, H. (2005). Conservation of the E-function for floral organ identity in rice revealed by the analysis of tissue culture-induced loss-of-function mutants of the *OsMADS1* gene. *Plant Molecular Biology*, *59*, 125–135.

Aichinger, E., Kornet, N., Friedrich, T., & Laux, T. (2012). Plant stem cell niches. *Annual Review of Plant Biology*, *63*, 615–636.

Arber, A. (1934). *The Gramineae: A Study of Cereal, Bamboo, and Grasses*. Cambridge: University Press.

Ashikari, M., Wu, J., Yano, M., Sasaki, T., & Yoshimura, A. (1999). Rice gibberellin-insensitive dwarf mutant gene *Dwarf 1* encodes the alpha-subunit of GTP-binding protein. *Proceedings of the National Academy of Sciences of the United States of America, 96*, 10284–10289.

Ashikari, M., Sakakibara, H., Lin, S., Yamamoto, T., Takashi, T., Nishimura, A., et al. (2005). Cytokinin oxidase regulates rice grain production. *Science, 309*, 741–745.

Aya, K., Ueguchi-Tanaka, M., Kondo, M., Hamada, K., Yano, K., Nishimura, M., et al. (2009). Gibberellin modulates anther development in rice via the transcriptional regulation of GAMYB. *Plant Cell, 21*, 1453–1472.

Bommert, P., Je, B. I., Goldshmidt, A., & Jackson, D. (2013). The maize Gα gene *COMPACT PLANT2* functions in CLAVATA signalling to control shoot meristem size. *Nature, 502*, 555–558.

Bortiri, E., & Hake, S. (2007). Flowering and determinacy in maize. *Journal of Experimental Botany, 58*, 909–916.

Bowman, J. L., & Smyth, D. R. (1999). *CRABS CLAW*, a gene that regulates carpel and nectary development in *Arabidopsis*, encodes a novel protein with zinc finger and helix-loop-helix domains. *Development, 126*, 2387–2396.

Cai, W., & Morishima, H. (2002). QTL clusters reflect character associations in wild and cultivated rice. *Theoretical and Applied Genetics, 104*, 1217–1228.

Chickarmane, V. S., Gordon, S. P., Tarr, P. T., Heisler, M. G., & Meyerowitz, E. M. (2012). Cytokinin signaling as a positional cue for patterning the apical-basal axis of the growing *Arabidopsis* shoot meristem. *Proceedings of the National Academy of Sciences of the United States of America, 109*, 4002–4007.

Chu, H., Qian, Q., Liang, W., Yin, C., Tan, H., & Yao, X. (2006). The *FLORAL ORGAN NUMBER4* gene encoding a putative ortholog of *Arabidopsis* CLAVATA3 regulates apical meristem size in rice. *Plant Physiology, 142*, 1039–1052.

Chuck, G., Meeley, R. B., & Hake, S. (1998). The control of maize spikelet meristem fate by the APETALA2-like gene *indeterminate spikelet1*. *Genes & Development, 12*, 1145–1154.

Chuck, G., Meeley, R., & Hake, S. (2008). Floral meristem initiation and meristem cell fate are regulated by the maize *AP2* genes *ids1* and *sid1*. *Development, 135*, 3013–3019.

Chuck, G., Muszynski, M., Kellogg, E., Hake, S., & Schmidt, R. J. (2002). The control of spikelet meristem identity by the branched silkless1 gene in maize. *Science, 298*, 1238–1241.

Chuck, G., Meeley, R., Irish, E., Sakai, H., & Hake, S. (2007). The maize *tasselseed4* microRNA controls sex determination and meristem cell fate by targeting *Tasselseed6/indeterminate spikelet1*. *Nature Genetics, 39*, 1517–1521.

Clayton, W. D. (1990). The spikelet. In G. P. Chapman (Ed.), *Reproductive versatility in the grasses* (pp. 22–51). Cambridge: Cambridge University Press.

Clifford, H. T. (1987). Spikelet and floral morphology. In T. R. Soderstrom, K. W. Hilu, C. S. Campbell, & M. E. Barkworth (Eds.), *Grass Systematics and Evolution* (pp. 21–30). Washington, DC: Smithsonian Institution Press.

Coen, E. S., & Meyerowitz, E. M. (1991). The war of the whorls: genetic interactions controlling flower development. *Nature, 353*, 31–37.

Colombo, L., Battaglia, R., & Kater, M. M. (2008). *Arabidopsis* ovule development and its evolutionary conservation. *Trends in Plant Science, 13*, 444–450.

Cui, R., Han, J., Zhao, S., Su, K., Wu, F., Du, X., et al. (2010). Functional conservation and diversification of class E floral homeotic genes in rice (*Oryza sativa*). *Plant Journal, 61*, 767–781.

Dinneny, J. R., Weigel, D., & Yanofsky, M. F. (2006). *NUBBIN* and *JAGGED* define stamen and carpel shape in *Arabidopsis*. *Development, 133*, 1645–1655.

Dreni, L., Jacchia, S., Fornara, F., Fornari, M., Ouwerkerk, P. B. F., An, G., et al. (2007). The D-lineage MADS-box gene *OsMADS13* controls ovule identity in rice. *Plant Journal, 52*, 690–699.

Dreni, L., Pilatone, A., Yun, D., Erreni, S., Pajoro, A., Caporali, E., et al. (2011). Functional analysis of all AGAMOUS subfamily members in rice reveals their roles in reproductive organ identity determination and meristem determinacy. *Plant Cell, 23*, 2850–2863.

Duan, Y., Li, S., Chen, Z., Zheng, L., Diao, Z., Zhou, Y., et al. (2012). *Dwarf and deformed flower 1*, encoding an F-box protein, is critical for vegetative and floral development in rice (*Oryza sativa* L.). *Plant Journal, 72*, 829–842.

Fujisawa, Y., Kato, T., Ohki, S., Ishikawa, A., Kitano, H., Sasaki, T., et al. (1999). Suppression of the heterotrimeric G protein causes abnormal morphology, including dwarfism, in rice. *Proceedings of the National Academy of Sciences of the United States of America, 96*, 7575–7580.

Gao, X., Liang, W., Yin, C., Ji, S., Wang, H., & Su, X. (2010). The *SEPALLATA*-like gene *OsMADS34* is required for rice inflorescence and spikelet development. *Plant Physiology, 153*, 728–740.

Giulini, A., Wang, J., & Jackson, D. (2004). Control of phyllotaxy by the cytokinin-inducible response regulator homologue *ABPHYL1*. *Nature, 430*, 1031–1034.

Goldberg, R. B., Beals, T. P., & Sanders, P. M. (1993). Anther development: basic principles and practical applications. *Plant Cell, 5*, 1217–1229.

Goldshmidt, A., Alvarez, J. P., Bowman, J. L., & Eshed, Y. (2008). Signals derived from *YABBY* gene activities in organ primordia regulate growth and partitioning of *Arabidopsis* shoot apical meristems. *Plant Cell, 20*, 1217–1230.

Gu, X. Y., Kianian, S. F., Hareland, G. A., Hoffer, B. L., & Foley, M. E. (2005). Genetic analysis of adaptive syndromes interrelated with seed dormancy in weedy rice (*Oryza sativa*). *Theoretical and Applied Genetics, 110*, 1108–1118.

Guo, J. X., & Liu, Y. G. (2012). Molecular control of male reproductive development and pollen fertility in rice. *Journal of Integrative Plant Biology, 54*, 967–978.

Ha, C. M., Jun, J. H., & Fletcher, J. C. (2010). Shoot apical meristem form and function. *Current Topics in Developmental Biology, 91*, 103–140.

Hirano, H.-Y., Tanaka, W., & Toriba, T. (2014). Grass flower development. In J. L. Riechmann, & F. Wellmer (Eds.), *Flower development - Methods and protocols* (pp. 57–84). New York: Springer.

Hirano, H.-Y., Hirai, A., Sano, Y., & Sasaki, T. (2008). *Rice Biology in the Genomics Era*. Heidelberg.

Hong, L., Tang, D., Zhu, K., Wang, K., Li, M., & Cheng, Z. (2012a). Somatic and reproductive cell development in rice anther is regulated by a putative glutaredoxin. *Plant Cell, 24*, 577–588.

Hong, L., Tang, D., Shen, Y., Hu, Q., Wang, K., Li, M., et al. (2012b). MIL2 (MICROSPORELESS2) regulates early cell differentiation in the rice anther. *New Phytologist, 196*, 402–413.

Horigome, A., Nagasawa, N., Ikeda, K., Ito, M., Itoh, J.-I., & Nagato, Y. (2009). Rice *OPEN BEAK* is a negative regulator of class 1 *knox* genes and a positive regulator of class B floral homeotic gene. *Plant Journal, 58*, 724–736.

Hoshikawa, K. (1989). *The Growing Rice Plant: An Anatomical Monograph*. Tokyo: Nobunkyo.

Hu, L., Liang, W., Yin, C., Cui, X., Zong, J., Wang, X., et al. (2011). Rice MADS3 regulates ROS homeostasis during late anther development. *Plant Cell, 23*, 515–533.

Huang, X., Kurata, N., Wei, X., Wang, Z. X., Wang, A., Zhao, Q., et al. (2012). A map of rice genome variation reveals the origin of cultivated rice. *Nature, 490*, 497–501.

Hufford, L. (1996). Origin and early evolution of angiosperm stamens. In W. G. D'Arcy, & R. C. Keating (Eds.), *The Anther: Form, Function, and Phylogeny*. Cambridge: Cambridge University Press.

Ikeda, K., Ito, M., Nagasawa, N., Kyozuka, J., & Nagato, Y. (2007). Rice *ABERRANT PANICLE ORGANIZATION 1*, encoding an F-box protein, regulates meristem fate. *Plant Journal, 51*, 1030–1040.

Ikeda-Kawakatsu, K., Maekawa, M., Izawa, T., Itoh, J.-I., & Nagato, Y. (2012). *ABERRANT PANICLE ORGANIZATION 2/RFL*, the rice ortholog of *Arabidopsis LEAFY*, suppresses the transition from inflorescence meristem to floral meristem through interaction with *APO1*. *Plant Journal*, *69*, 168–180.

Ishikawa, M., Ohmori, Y., Tanaka, W., Hirabayashi, C., Murai, K., Ogihara, Y., et al. (2009). The spatial expression patterns of *DROOPING LEAF* orthologs suggest a conserved function in grasses. *Genes and Genetic Systems*, *84*, 137–146.

Itoh, J., Sato, Y., & Nagato, Y. (2008). The *SHOOT ORGANIZATION2* Gene Coordinates Leaf Domain Development Along the Central-Marginal Axis in Rice. *Plant and Cell Physiology*, *49*, 1226–1236.

Itoh, J.-I., Kitano, H., Matsuoka, M., & Nagato, Y. (2000). *SHOOT ORGANIZATION* genes regulate shoot apical meristem organization and the pattern of leaf primordium initiation in rice. *Plant Cell*, *12*, 2161–2174.

Itoh, J.-I., Nonomura, K.-I., Ikeda, K., Yamaki, S., Inukai, Y., Yamagishi, H., et al. (2005). Rice plant development: from zygote to spikelet. *Plant and Cell Physiology*, *46*, 23–47.

Itoh, J. I., Hibara, K. I., Kojima, M., Sakakibara, H., & Nagato, Y. (2012). Rice *DECUSSATE* controls phyllotaxy by affecting the cytokinin signaling pathway. *Plant Journal*, *72*, 869–881.

Jack, T. (2004). Molecular and genetic mechanisms of floral control. *Plant Cell*, *16*, S1–S17.

Jeon, J. S., Jang, S., Lee, S., Nam, J., Kim, C., Lee, S. H., et al. (2000). *leafy hull sterile1* is a homeotic mutation in a rice MADS box gene affecting rice flower development. *Plant Cell*, *12*, 871–884.

Jin, Y., Luo, Q., Tong, H., Wang, A., Cheng, Z., Tang, J., et al. (2011). An AT-hook gene is required for palea formation and floral organ number control in rice. *Developmental Biology*, *359*, 277–288.

Jung, K. H., Han, M. J., Lee, Y. S., Kim, Y. W., Hwang, I., Kim, M. J., et al. (2005). Rice *Undeveloped Tapetum1* is a major regulator of early tapetum development. *Plant Cell*, *17*, 2705–2722.

Jung, K. H., Han, M. J., Lee, D. Y., Lee, Y. S., Schreiber, L., Franke, R., et al. (2006). *Wax-deficient anther1* is involved in cuticle and wax production in rice anther walls and is required for pollen development. *Plant Cell*, *18*, 3015–3032.

Kaneko, M., Inukai, Y., Ueguchi-Tanaka, M., Itoh, H., Izawa, T., Kobayashi, Y., et al. (2004). Loss-of-function mutations of the rice *GAMYB* gene impair α-amylase expression in aleurone and flower development. *Plant Cell*, *16*, 33–44.

Kelliher, T., & Walbot, V. (2012). Hypoxia triggers meiotic fate acquisition in maize. *Science*, *337*, 345–348.

Kidner, C. A., & Timmermans, M. C. P. (2010). Signaling sides: adaxial-abaxial patterning in leaves. *Current Topics in Developmental Biology*, *91*, 141–168.

Kinoshita, T. (1984). Gene analysis and linkage map. In S. Tsunoda, & N. Takahashi (Eds.), *Biology of Rice*. Tokyo/Amsterdam: Japan Sci. Soc. Press/Elsevier.

Kinoshita, T., Hidano, Y., & Takahashi, M. (1977). A mutant 'long hull sterile' found out in the rice variety, 'Sorachi': Genetic studies on rice plant. LXVII. *Journal of the Faculty of Agriculture, Hokkaido University*, *10*, 247–268.

Kobayashi, K., Maekawa, M., Miyao, A., Hirochika, H., & Kyozuka, J. (2010). *PANICLE PHYTOMER2 (PAP2)*, encoding a SEPALLATA subfamily MADS-box protein, positively controls spikelet meristem identity in rice. *Plant and Cell Physiology*, *51*, 47–57.

Komatsu, M., Chujo, A., Nagato, Y., Shimamoto, K., & Kyozuka, J. (2003). *FRIZZY PANICLE* is required to prevent the formation of axillary meristems and to establish floral meristem identity in rice spikelets. *Development*, *130*, 3841–3850.

Kubo, T., Takano-Kai, N., & Yoshimura, A. (2001). RFLP mapping of genes for long kernel and awn on chromosome 3 in rice. *Rice Genetics Newsletter*, *18*, 26–28.

Kurakawa, T., Ueda, N., Maekawa, M., Kobayashi, K., Kojima, M., Nagato, Y., et al. (2007). Direct control of shoot meristem activity by a cytokinin-activating enzyme. *Nature*, *445*, 652–655.

Kyozuka, J., & Shimamoto, K. (2002). Ectopic expression of *OsMADS3*, a rice ortholog of *AGAMOUS*, caused a homeotic transformation of lodicules to stamens in transgenic rice plants. *Plant and Cell Physiology*, *43*, 130–135.

Kyozuka, J., Kobayashi, T., Morita, M., & Shimamoto, K. (2000). Spatially and temporally regulated expression of rice MADS box genes with similarity to *Arabidopsis* Class A, B and C Genes. *Plant and Cell Physiology*, *41*, 710–718.

Kyozuka, J., Konishi, S., Nemoto, K., Izawa, T., & Shimamoto, K. (1998). Down-regulation of *RFL*, the *FLO/LFY* homolog of rice, accompanied with panicle branch initiation. *Proceedings of the National Academy of Sciences of the United States of America*, *95*, 1979–1982.

Lee, D.-Y., Lee, J., Moon, S., Park, S. Y., & An, G. (2007). The rice heterochronic gene *SUPERNUMERARY BRACT* regulates the transition from spikelet meristem to floral meristem. *Plant Journal*, *49*, 64–78.

Lee, D.Y., & An, G. (2012). Two AP2 family genes, *SUPERNUMERARY BRACT (SNB)* and *OsINDETERMINATE SPIKELET 1 (OsIDS1)*, synergistically control inflorescence architecture and floral meristem establishment in rice. *Plant Journal*, *69*, 445–461.

Lee, S., Jeon, J.-S., An, K., Moon, Y.-H., Lee, S., Chung, Y.-Y., et al. (2003). Alteration of floral organ identity in rice through ectopic expression of OsMADS16. *Planta*, *217*, 904–911.

Li, A., Zhang, Y., Wu, X., Tang, W., Wu, R., Dai, Z., et al. (2008). DH1, a LOB domain-like protein required for glume formation in rice. *Plant Molecular Biology*, *66*, 491–502.

Li, H., Liang, W., Yin, C., Zhu, L., & Zhang, D. (2011a). Genetic interaction of *OsMADS3*, *DROOPING LEAF*, and *OsMADS13* in specifying rice floral organ identities and meristem determinacy. *Plant Physiology*, *156*, 263–274.

Li, H., Liang, W., Jia, R., Yin, C., Zong, J., Kong, H., et al. (2010). The *AGL6*-like gene *OsMADS6* regulates floral organ and meristem identities in rice. *Cell Research*, *20*, 299–313.

Li, H., Yuan, Z., Vizcay-Barrena, G., Yang, C., Liang, W., Zong, J., et al. (2011b). *PERSISTENT TAPETAL CELL1* encodes a PHD-finger protein that is required for tapetal cell death and pollen development in rice. *Plant Physiology*, *156*, 615–630.

Li, H., Xue, D., Gao, Z., Yan, M., Xu, W., Xing, Z., et al. (2009). A putative lipase gene *EXTRA GLUME1* regulates both empty-glume fate and spikelet development in rice. *Plant Journal*, *57*, 593–605.

Li, H., Liang, W., Hu, Y., Zhu, L., Yin, C., Xu, J., et al. (2011c). Rice *MADS6* interacts with the floral homeotic genes *SUPERWOMAN1, MADS3, MADS58, MADS13*, and *DROOPING LEAF* in specifying floral organ identities and meristem fate. *Plant Cell*, *23*, 2536–2552.

Li, N., Zhang, D. S., Liu, H. S., Yin, C. S., Li, X. X., Liang, W. Q., et al. (2006). The rice *Tapetum Degeneration Retardation* gene is required for tapetum degradation and anther development. *Plant Cell*, *18*, 2999–3014.

Li, X., Sun, L., Tan, L., Liu, F., Zhu, Z., Fu, Y., et al. (2012). *TH1*, a DUF640 domain-like gene controls lemma and palea development in rice. *Plant Molecular Biology*, *78*, 351–359.

Liu, B., Chen, Z., Song, X., Liu, C., Cui, X., Zhao, X., et al. (2007). *Oryza sativa dicer-like4* reveals a key role for small interfering RNA silencing in plant development. *Plant Cell*, *19*, 2705–2718.

Liu, Z., Bao, W., Liang, W., Yin, J., & Zhang, D. (2010). Identification of *gamyb-4* and analysis of the regulatory role of *GAMYB* in rice anther development. *Journal of Integrative Plant Biology*, *52*, 670–678.

Lohmann, J. U., & Weigel, D. (2002). Building beauty: The genetic control of floral patterning. *Developmental Cell*, *2*, 135–142.

Long, J. A., Moan, E. I., Medford, J. I., & Barton, M. K. (1996). A member of the KNOTTED class of homeodomain proteins encoded by the *STM* gene of *Arabidopsis*. *Nature*, *379*, 66–69.

Long, J. A., Ohno, C., Smith, Z. R., & Meyerowitz, E. M. (2006). TOPLESS regulates apical embryonic fate in *Arabidopsis*. *Science*, *312*, 1520–1523.

Luo, J., Liu, H., Zhou, T., Gu, B., Huang, X., Shangguan, Y., et al. (2013). *An-1* encodes a basic helix-loop-helix protein that regulates awn development, grain size, and grain number in rice. *Plant Cell*, *25*, 3360–3376.

Ma, X. D., Cheng, Z. J., Wu, F. Q., Jin, M. N., Zhang, L. G., Zhou, F., et al. (2013). *BEAK LIKE SPIKELET1* is required for lateral development of lemma and palea in rice. *Plant Molecular Biology Reporter*, *31*, 98–108.

Miyawaki, K., Tabata, R., & Sawa, S. (2013). Evolutionarily conserved CLE peptide signaling in plant development, symbiosis, and parasitism. *Current Opinion in Plant Biology*, *16*, 598–606.

Müller, R., Bleckmann, A., & Simon, R. (2008). The receptor kinase CORYNE of *Arabidopsis* transmits the stem cell-limiting signal CLAVATA3 independently of CLAVATA1. *Plant Cell*, *20*, 934–946.

Murmu, J., Bush, M. J., DeLong, C., Li, S., Xu, M., Khan, M., et al. (2010). *Arabidopsis* basic leucine-zipper transcription factors TGA9 and TGA10 interact with floral glutaredoxins ROXY1 and ROXY2 and are redundantly required for anther development. *Plant Physiology*, *154*, 1492–1504.

Nagasaki, H., Itoh, J., Hayashi, K., Hibara, K., Satoh-Nagasawa, N., Nosaka, M., et al. (2007). The small interfering RNA production pathway is required for shoot meristem initiation in rice. *Proceedings of the National Academy of Sciences of the United States of America*, *104*, 14867–14871.

Nagasawa, N., Miyoshi, M., Kitano, H., Satoh, H., & Nagato, Y. (1996). Mutations associated with floral organ number in rice. *Planta*, *198*, 627–633.

Nagasawa, N., Miyoshi, M., Sano, Y., Satoh, H., Hirano, H.-Y., Sakai, H., et al. (2003). SUPERWOMAN 1 and DROOPING LEAF genes control floral organ identity in rice. *Development*, *130*, 705–718.

Nagato, Y., & Yoshimura, A. (1998). Report of the committee on gene symbolization, nomenclature and linkage groups. *Rice Genetics Newsletter*, *15*, 13–74.

Nimchuk, Z. L., Tarr, P. T., & Meyerowitz, E. M. (2011). An evolutionarily conserved pseudokinase mediates stem cell production in plants. *Plant Cell*, *23*, 851–854.

Niu, N., Liang, W., Yang, X., Jin, W., Wilson, Z. A., Hu, J., et al. (2013). EAT1 promotes tapetal cell death by regulating aspartic proteases during male reproductive development in rice. *Nature Communications*, *4*, 1445.

Nonomura, K., Miyoshi, K., Eiguchi, M., Suzuki, T., Miyao, A., Hirochika, H., et al. (2003). The *MSP1* gene is necessary to restrict the number of cells entering into male and female sporogenesis and to initiate anther wall formation in rice. *Plant Cell*, *15*, 1728–1739.

Ohmori, S., Kimizu, M., Sugita, M., Miyao, A., Hirochika, H., Uchida, E., et al. (2009). *MOSAIC FLORAL ORGANS1*, an *AGL6*-like MADS box gene, regulates floral organ identity and meristem fate in rice. *Plant Cell*, *21*, 3008–3025.

Ohmori, Y., Tanaka, W., Kojima, M., Sakakibara, H., & Hirano, H.-Y. (2013). *WUSCHEL-RELATED HOMEOBOX4* is involved in meristem maintenance and is negatively regulated by the CLE gene *FCP1* in rice. *Plant Cell*, *25*, 229–241.

Ohno, C. K., Reddy, G. V., Heisler, M. G. B., & Meyerowitz, E. M. (2004). The *Arabidopsis JAGGED* gene encodes a zinc finger protein that promotes leaf tissue development. *Development*, *131*, 1111–1122.

Oka, H. I. (1988). *Origin of Cultivated Rice*. Tokyo: Japan Sci. Soc. Press.

Pautler, M., Tanaka, W., Hirano, H.-Y., & Jackson, D. (2013). Grass meristem I – Shoot apical meristem maintenance, axillary meristem determinacy, and the floral transition. *Plant and Cell Physiology*, *54*, 302–312.

Pautot, V., Dockx, J., Hamant, O., Kronenberger, J., Grandjean, O., Jublot, D., et al. (2001). *KNAT2*: evidence for a link between knotted-like genes and carpel development. *Plant Cell*, *13*, 1719–1734.

Pinyopich, A., Ditta, G. S., Savidge, B., Liljegren, S. J., Baumann, E., Wisman, E., et al. (2003). Assessing the redundancy of MADS-box genes during carpel and ovule development. *Nature, 424*, 85–88.

Prasad, K., Parameswaran, S., & Vijayraghavan, U. (2005). *OsMADS1*, a rice MADS-box factor, controls differentiation of specific cell types in the lemma and palea and is an early-acting regulator of inner floral organs. *Plant Journal, 43*, 915–928.

Prasad, K., Sriram, P., Kumar, C. S., Kushalappa, K., & Vijayraghavan, U. (2001). Ectopic expression of rice *OsMADS1* reveals a role in specifying the lemma and palea, grass floral organs analogous to sepals. *Development Genes and Evolution, 211*, 281–290.

Prunet, N., & Jack, T. P. (2014). Flower development in *Arabidopsis*: there is more to it than learning your ABCs. In J. L. Riechmann, & F. Wellmer (Eds.), *Flower development - methods and protocols* (pp. 3–33). New York: Springer.

Ran, F. A., Hsu, P. D., Wright, J., Agarwala, V., Scott, D. A., & Zhang, F. (2013). Genome engineering using the CRISPR-Cas9 system. *Nature Protocols, 8*, 2281–2308.

Rao, N. N., Prasad, K., Kumar, P. R., & Vijayraghavan, U. (2008). Distinct regulatory role for *RFL*, the rice *LFY* homolog, in determining flowering time and plant architecture. *Proceedings of the National Academy of Sciences of the United States of America, 105*, 3646–3651.

Ren, D., Li, Y., Zhao, F., Sang, X., Shi, J., Wang, N., et al. (2013). *MULTI-FLORET SPIKELET1*, which encodes an AP2/ERF protein, determines spikelet meristem fate and sterile lemma identity in rice. *Plant Physiology, 162*, 872–884.

Sakamoto, T., Sakakibara, H., Kojima, M., Yamamoto, Y., Nagasaki, H., Inukai, Y., et al. (2006). Ectopic expression of KNOTTED1-like homeobox protein induces expression of cytokinin biosynthesis genes in rice. *Plant Physiology, 142*, 54–62.

Sang, X., Li, Y., Luo, Z., Ren, D., Fang, L., Wang, N., et al. (2012). *CHIMERIC FLORAL ORGANS1*, encoding a monocot-specific MADS box protein, regulates floral organ identity in rice. *Plant Physiology, 160*, 788–807.

Sarojam, R., Sappl, P. G., Goldshmidt, A., Efroni, I., Floyd, S. K., Eshed, Y., et al. (2010). Differentiating *Arabidopsis* shoots from leaves by combined YABBY activities. *Plant Cell, 22*, 2113–2130.

Sawa, S., Watanabe, K., Goto, K., Liu, Y. G., Shibata, D., Kanaya, E., et al. (1999). *FILAMENTOUS FLOWER*, a meristem and organ identity gene of *Arabidopsis*, encodes a protein with a zinc finger and HMG-related domains. *Genes & Development, 13*, 1079–1088.

Siegfried, K. R., Eshed, Y., Baum, S. F., Otsuga, D., Drews, G. N., & Bowman, J. L. (1999). Members of the *YABBY* gene family specify abaxial cell fate in *Arabidopsis*. *Development, 126*, 4117–4128.

Song, X., Wang, D., Ma, L., Chen, Z., Li, P., Cui, X., et al. (2012). Rice RNA-dependent RNA polymerase 6 acts in small RNA biogenesis and spikelet development. *Plant Journal, 71*, 378–389.

Sorensen, A. M., Krober, S., Unte, U. S., Huijser, P., Dekker, K., & Saedler, H. (2003). The *Arabidopsis ABORTED MICROSPORES (AMS)* gene encodes a MYC class transcription factor. *Plant Journal, 33*, 413–423.

Suzaki, T., Yoshida, A., & Hirano, H.-Y. (2008). Functional diversification of CLAVATA3-related CLE proteins in meristem maintenance in rice. *Plant Cell, 20*, 2049–2058.

Suzaki, T., Ohneda, M., Toriba, T., Yoshida, A., & Hirano, H.-Y. (2009). *FON2 SPARE1* redundantly regulates floral meristem maintenance with *FLORAL ORGAN NUMBER2* in rice. *PLoS Genetics, 5*, e1000693.

Suzaki, T., Sato, M., Ashikari, M., Miyoshi, M., Nagato, Y., & Hirano, H.-Y. (2004). The gene *FLORAL ORGAN NUMBER1* regulates floral meristem size in rice and encodes a leucine-rich repeat receptor kinase orthologous to *Arabidopsis* CLAVATA1. *Development, 131*, 5649–5657.

Suzaki, T., Toriba, T., Fujimoto, M., Tsutsumi, N., Kitano, H., & Hirano, H.-Y. (2006). Conservation and diversification of meristem maintenance mechanism in *Oryza sativa*: function of the *FLORAL ORGAN NUMBER2* gene. *Plant and Cell Physiology, 47*, 1591–1602.

Taguchi-Shiobara, F., Yuan, Z., Hake, S., & Jackson, D. (2001). The *fasciated ear2* gene encodes a leucine-rich repeat receptor-like protein that regulates shoot meristem proliferation in maize. *Genes & Development, 15*, 2755–2766.
Tan, H., Liang, W., Hu, J., & Zhang, D. (2012). *MTR1* encodes a secretory fasciclin glycoprotein required for male reproductive development in rice. *Developmental Cell, 22*, 1127–1137.
Tanaka, W., Toriba, T., Ohomori, Y., & Hirano, H.-Y. (2012a). Formation of two florets within a single spikelet in the rice *tongari-boushi1* mutant. *Plant Signaling and Behavior, 7*, 793–795.
Tanaka, W., Pautler, M., Jackson, D., & Hirano, H.-Y. (2013). Grass meristem II – Flower development and meristem fate. *Plant and Cell Physiology, 54*, 313–324.
Tanaka, W., Toriba, T., Ohmori, Y., Yoshida, A., Kawai, A., Mayama-Tsuchida, T., et al. (2012b). The *YABBY* gene *TONGARI-BOUSHI1* is involved in lateral organ development and maintenance of meristem organization in the rice spikelet. *Plant Cell, 24*, 80–95.
Terrell, E. E., Peterson, P. M., & Wergin, W. P. (2001). Epidermal features and spikelet micromorphology in Oryza and related genera (Poaceae: Oryzeae). *Smithsonian Contributions to Botany, 91*, 1–50.
Toriba, T., & Hirano, H.-Y. (2014). The *DROOPING LEAF* and *OsETTIN2* genes promote awn development in rice. *Plant Journal, 77*, 616–626.
Toriba, T., Suzaki, T., Yamaguchi, T., Ohmori, Y., Tsukaya, H., & Hirano, H.-Y. (2010). Distinct regulation of adaxial-abaxial polarity in anther patterning in rice. *Plant Cell, 22*, 1452–1462.
Toriba, T., Harada, K., Takamura, A., Nakamura, H., Ichikawa, H., Suzaki, T., et al. (2007). Molecular characterization the *YABBY* gene family in *Oryza sativa* and expression analysis of *OsYABBY1*. *Molecular Genetics and Genomics, 277*, 457–468.
Troll, W. (2000). *Praktische Einführung in die Pflanzenmorphologie*. Heiderlberg: Akademischer Verlag.
Tsuda, K., Ito, Y., Sato, Y., & Kurata, N. (2011). Positive autoregulation of a *KNOX* gene is essential for shoot apical meristem maintenance in rice. *Plant Cell, 23*, 4368–4381.
Vaughan, D. A. (1989). The genus *Oryza* L.: current status of taxonomy. *IRRI Research Paper Series, 138*, 1–21.
Voytas, D. F. (2013). Plant genome engineering with sequence-specific nucleases. *Annual Review of Plant Biology, 64*, 327–350.
Wang, A., Xia, Q., Xie, W., Datla, R., & Selvaraj, G. (2003). The classical Ubisch bodies carry a sporophytically produced structural protein (RAFTIN) that is essential for pollen development. *Proceedings of the National Academy of Sciences of the United States of America, 100*, 14487–14492.
Wang, K., Tang, D., Hong, L., Xu, W., Huang, J., Li, M., et al. (2010). *DEP* and *AFO* regulate reproductive habit in rice. *PLoS Genetics, 6*, e1000818.
Whipple, C. J., Ciceri, P., Padilla, C. M., Ambrose, B. A., Bandong, S. L., & Schmidt, R. J. (2004). Conservation of B-class floral homeotic gene function between maize and *Arabidopsis*. *Development, 131*, 6083–6091.
Xiao, H., Tang, J., Li, Y., Wang, W., Li, X., Jin, L., et al. (2009). *STAMENLESS 1*, encoding a single C2H2 zinc finger protein, regulates floral organ identity in rice. *Plant Journal, 59*, 789–801.
Yadav, S. R., Prasad, K., & Vijayraghavan, U. (2007). Divergent regulatory *OsMADS2* functions control size, shape and differentiation of the highly derived rice floret second-whorl organ. *Genetics, 176*, 283–294.
Yamaguchi, T., Nagasawa, N., Kawasaki, S., Matsuoka, M., Nagato, Y., & Hirano, H.-Y. (2004). The *YABBY* gene *DROOPING LEAF* regulates carpel specification and midrib development in *Oryza sativa*. *Plant Cell, 16*, 500–509.
Yamaguchi, T., Lee, Y. D., Miyao, A., Hirochika, H., An, G., & Hirano, H.-Y. (2006). Functional diversification of the two C-class genes, *OSMADS3* and *OSMADS58*, in *Oryza sativa*. *Plant Cell, 18*, 15–28.

Yamaki, S., Nagato, Y., Kurata, N., & Nonomura, K.-I. (2011). Ovule is a lateral organ finally differentiated from the terminating floral meristem in rice. *Developmental Biology, 351*, 208–216.

Yao, S.-G., Ohmori, S., Kimizu, M., & Yoshida, H. (2008). Unequal genetic redundancy of rice *PISTILLATA* orthologs, *OsMADS2* and *OsMADS4*, in lodicule and stamen development. *Plant and Cell Physiology, 49*, 853–857.

Yi, J., Kim, S. R., Lee, D. Y., Moon, S., Lee, Y. S., Jung, K. H., et al. (2012). The rice gene *DEFECTIVE TAPETUM AND MEIOCYTES 1 (DTM1)* is required for early tapetum development and meiosis. *Plant Journal, 70*, 256–270.

Yoshida, A., Suzaki, T., Tanaka, W., & Hirano, H.-Y. (2009). The homeotic gene *LONG STERILE LEMMA (G1)* specifies sterile lemma identity in the rice spikelet. *Proceedings of the National Academy of Sciences of the United States of America, 106*, 20103–20108.

Yoshida, A., Ohmori, Y., Kitano, H., Taguchi-Shiobara, F., & Hirano, H.-Y. (2012). *ABERRANT SPIKELET AND PANICLE1*, encoding a TOPLESS-related transcriptional co-repressor, is involved in the regulation of meristem fate in rice. *Plant Journal, 70*, 327–339.

Yoshida, H., & Nagato, Y. (2011). Flower development in rice. *Journal of Experimental Botany, 62*, 4719–4730.

Yoshida, H., Itoh, J.-I., Ohmori, S., Miyoshi, K., Horigome, A., Uchida, E., et al. (2007). *superwoman1-cleistogamy*, a hopeful allele for gene containment in GM rice. *Plant Biotechnology Journal, 5*, 835–846.

Yuan, Z., Gao, S., Xue, D.-W., Luo, D., Li, L.-T., Ding, S.-Y., et al. (2009). *RETARDED PALEA1* controls palea development and floral zygomorphy in rice. *Plant Physiology, 149*, 235–244.

Zhang, D., & Yang, L. (2014). Specification of tapetum and microsporocyte cells within the anther. *Current Opinion in Plant Biology, 17*, 49–55.

Zhang, D., Luo, X., & Zhu, L. (2011). Cytological analysis and genetic control of rice anther development. *Journal of Genetics and Genomics, 38*, 379–390.

Zhang, D., Liang, W., Yin, C., Zong, J., Gu, F., & Zhang, D. (2010). *OsC6*, encoding a lipid transfer protein, is required for postmeiotic anther development in rice. *Plant Physiology, 154*, 149–162.

Zhang, D. S., Liang, W. Q., Yuan, Z., Li, N., Shi, J., Wang, J., et al. (2008). Tapetum Degeneration Retardation is critical for aliphatic metabolism and gene regulation during rice pollen development. *Molecular Plant, 1*, 599–610.

Zhang, W., Sun, Y., Timofejeva, L., Chen, C., Grossniklaus, U., & Ma, H. (2006). Regulation of *Arabidopsis* tapetum development and function by *DYSFUNCTIONAL TAPETUM1 (DYT1)* encoding a putative bHLH transcription factor. *Development, 133*, 3085–3095.

Zhao, X., de Palma, J., Oane, R., Gamuyao, R., Luo, M., Chaudhury, A., et al. (2008). OsTDL1A binds to the LRR domain of rice receptor kinase MSP1, and is required to limit sporocyte numbers. *Plant Journal, 54*, 375–387.

Zhu, Q.-H., Upadhyaya, N. M., Gubler, F., & Helliwell, C. A. (2009). Over-expression of miR172 causes loss of spikelet determinacy and floral organ abnormalities in rice (*Oryza sativa*). *BMC Plant Biology, 9*, 149.

CHAPTER NINE

Genetic and Hormonal Regulation of Maize Inflorescence Development

Beth Thompson[1]
Biology Department, East Carolina University, Greenville, NC, USA
[1]E-mail address: thompsonb@ecu.edu

Contents

Abstract

Plant development is determined by the activity of meristems, pools of undifferentiated cells that generate organs throughout the life of the plant. Inflorescence morphology is determined by the position and activity of meristems that form during reproductive development. Grass inflorescences are complex, highly branched structures that initiate higher order meristems before the floral meristems, which ultimately produce the floral organs. Maize produces two inflorescences: the tassel is formed at the apex of the plant and bears male flowers and the ear is formed in the axil of a leaf and bears female flowers. Despite their distinct morphologies at maturity, the tassel and ear arise from strikingly similar inflorescence primordia and their patterning depends largely on the same set of developmental regulators. Many of these developmental regulators have been identified through mutant analysis

Advances in Botanical Research, Volume 72
ISSN 0065-2296
http://dx.doi.org/10.1016/B978-0-12-417162-6.00009-2

and cloning efforts, and include multiple transcription factors, microRNAs and plant hormones. This chapter focuses on the molecular mechanisms that regulate the maintenance, identity, and activity of meristems in maize inflorescences, as well as floral development and sex determination.

9.1 INTRODUCTION

Meristems are pools of stems cells that are required for plant growth and development. Unlike animals, in which the basic body plan and organs are formed during embryogenesis, plants continuously initiate organs throughout their lives. Meristems are essential for both generating differentiated cells that form plant organs as well as maintaining a pool of stem cells. The shoot apical meristem (SAM) is formed during embryogenesis and gives rise to all of the aerial structures of the plant either by directly initiating organ primordia or by first initiating additional meristems. The identity and activity of meristems governs growth and differentiation in plants and ultimately determines plant architecture. Inflorescence architecture determines the number and position of flowers and since flowers produce the seeds, inflorescence architecture is a major determinate of yield. Inflorescence architecture varies greatly among angiosperms and the final architecture is determined by the degree of branching and the position of flowers along these branches.

Grass inflorescence architecture is of particular interest due to the economic importance of the cereal crops, including maize, rice, wheat, barley and sorghum. Furthermore, the variation in inflorescence architectures among the grasses, combined with emerging molecular and genomic tools in diverse model grass species makes the grasses an excellent system to investigate the evolution of inflorescence morphologies. This chapter focuses on the genetic regulation of maize inflorescence architecture, however the reader is also directed to related chapters in this volume describing other systems (Chapters 7, 8 and 10).

In addition to its obvious economic importance, maize has many advantages as a model grass to dissect the molecular regulation of inflorescence architecture as well as other aspects of development. The long, rich history of maize genetics has produced a large and ever growing mutant collection that includes spontaneous mutants as well as mutants from transposon and chemical mutagenesis screens. The completion of the maize B73 reference genome has made positional cloning of maize mutants routine and many classical mutants that affect inflorescence architecture have now been cloned. Many of these mutants affect specific stages of inflorescence development and cloning of these mutants has identified genes that function at discrete developmental stages. In addition, the increasing numbers of reverse genetic resources and molecular tools to investigate gene function position maize as a premier model species within the grasses. Maize is also a unique evolutionary model given its relatively recent domestication from teosinte and several of the genes responsible for the dramatic morphological changes between maize and teosinte have been identified (Clark, Wagler, Quijada, & Doebley, 2006; Meyer & Purugganan, 2013; Wang et al., 2005; Wills et al., 2013). The significant genetic and morphological variation within maize, combined with modern genomic resources also make maize an excellent system to identify and understand the function of genes that control natural variation within a species.

9.2 A BRIEF PRIMER ON MAIZE INFLORESCENCE DEVELOPMENT

Maize produces two inflorescences, the tassel and ear. The tassel is formed at the apex of the plant when the SAM transitions to a reproductive or inflorescence meristem (IM) and bears the male flowers, whereas the ear is formed from an IM initiated in the axil of a leaf and bears the female flowers (Figure 9.1). Despite their distinct morphologies at maturity, the tassel and ear arise from strikingly similar inflorescence primordia (Figure 9.2) and are patterned largely by the same developmental programs. In the tassel, the first meristems initiated on the flanks of the IM are branch meristems (BM) that form long branches at the base of the tassel; ears lack BM and do not form long branches. The IM and BM initiate the first determinate meristems, the spikelet pair meristems (SPM), which in turn form two short branch, or spikelet meristems (SM). Spikelets are grass-specific structures that contain the flowers, called florets, the number of which can range from one to many, depending on the species. A defining feature of the *Andropogoneae*, a tribe of ~1000 grass species that includes maize and sorghum, is the presence of paired spikelets (Kellogg, 2000). Each spikelet pair consists of a pedicellate and sessile spikelet; the pedicellate spikelet is attached to a short branch, called the pedicel, while the sessile spikelet is a lateral spikelet that lacks a pedicel (Figure 9.1(D)). In maize, each SM initiates an upper- and lower floral meristem (UFM

Figure 9.1 Mature maize inflorescences. (A) Normal maize plant. The male inflorescence, the tassel, is formed at the apex of the plant, and the female inflorescence, the ear, is formed in the axil of a leaf. (B) Mature tassel. (C) Mature ear. (D) Mature spikelet pair from a tassel. (E) Tassel spikelet opened to expose two male florets (glumes have been removed). (F) Close-up of ear spikelets that each contain a single female floret. (See the colour plate.)

and LFM), which produce the floral organs. Grass florets are highly derived and contain grass-specific floral organs in addition to the stamens and carpels (the male and female sex organs), including palea, lemma and lodicules (Figures 9.1(E) and 9.3).

Maize is monoecious and male and female flowers are produced on distinct inflorescences; the tassel produces male (staminate) flowers and the ear produces female (pistillate) flowers. All flowers are initially bisexual and sex determination occurs through carpel abortion in the tassel and stamen arrest in the ear (Cheng, Greyson, & Walden, 1983). In addition, only the UFM fully develops in the ear and the lower floret aborts, resulting in ear spikelets that contain a single floret.

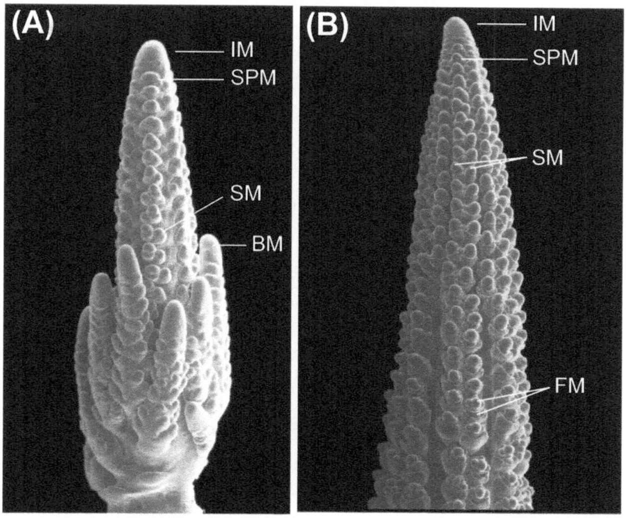

Figure 9.2 Tassels and ears develop from similar inflorescence primordia. SEMs of developing tassel (A) and ear (B) primordia. Development in the tassel and ear is similar except tassels develop long branches at the base of the inflorescence. IM, inflorescence meristem; BM, branch meristem; SPM, spikelet pair meristem; SM, spikelet meristem; FM, floral meristem.

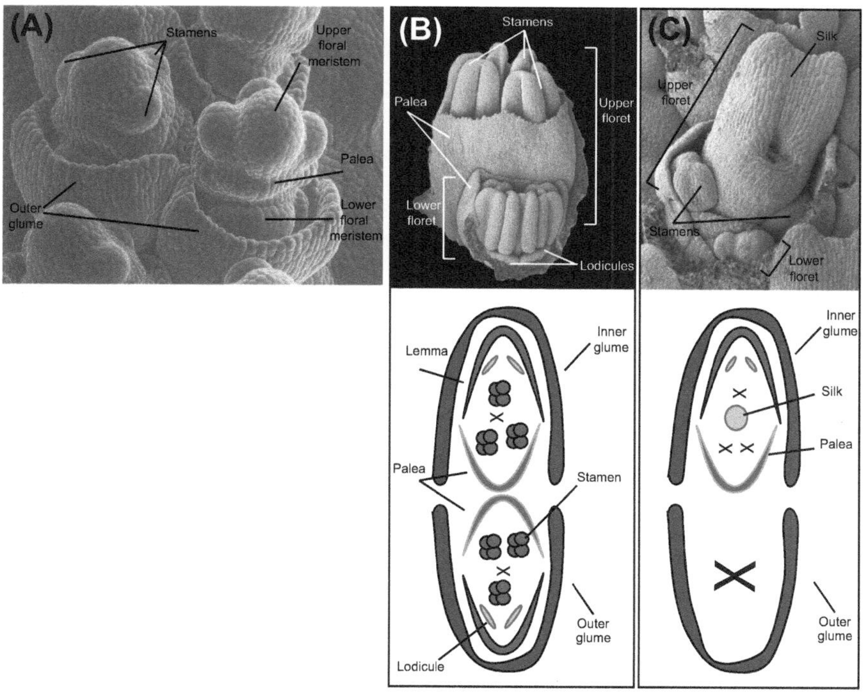

Figure 9.3 Maize floral development. (A) SEM of spikelet pair, each spikelet consists of an upper and lower floral meristem. (B) Late stage tassel spikelets contain two florets with stamens, but no carpels. (C) Late stage ear spikelets contains an upper floret with arrested stamens and carpels that have fused to form the silk. The lower floret aborts to produce a spikelet with a single floret.

9.3 STEM CELL MAINTENANCE AND HOMEOSTASIS IN INFLORESCENCE MERISTEMS

In response to intrinsic and environmental cues, the SAM ceases making vegetative structures (leaves) and transitions to a reproductive, or IM. The SAM at the apex of the plant, transitions to form the IM that will give rise to the tassel, while the ear is formed from a lateral IM in the axil of a leaf. The IM initiates meristems that form the long and short branches that ultimately form the florets. Like all meristems, the IM must carefully balance stem cell renewal in the central zone of the meristem with the generation of more differentiated cells at the periphery. Alteration in this balance of stem cell renewal and differentiation can change the size and shape of the IM and can influence inflorescence architecture. If a sufficient stem cell population is not maintained, stem cells are depleted and fewer lateral meristems are produced. In contrast, if the stem cell population is too large, meristem size increases and in extreme cases the IM becomes fasciated. Within a 'normal' window in which the IM maintains the proper domelike shape, IM size influences the number of lateral meristems produced. Therefore, regulation of IM size is critical to determine the number of seeds produced and has a direct impact on yield.

9.3.1 Meristem Maintenance by KNOX Transcription Factors

One of the first classes of stem cell regulators identified in plants is the class I KNOTTED1-like homeobox (KNOX) transcription factors (TFs). KNOX TFs are key regulators of stem cell fate in all angiosperms, including maize and other grasses (Hay & Tsiantis, 2010; Kerstetter, Laudencia-Chingcuanco, Smith, & Hake, 1997; Long, Moan, Medford, & Barton, 1996; Tsuda, Ito, Sato, & Kurata, 2011). The founding member of the KNOX family, *knotted1 (kn1)* is present in all vegetative and reproductive meristems, but absent from leaves and other differentiated tissues (Jackson, Veit, & Hake, 1994). *kn1* loss-of-function (lf) mutant phenotypes vary depending on inbred background, likely due to variations in meristem size and expression of other redundant *knox* genes that differ between inbreds. In the most restrictive inbred backgrounds, *kn1(lf)* mutants fail to maintain a SAM and exhibit a limited shoot phenotype (Vollbrecht, Reiser, & Hake, 2000), indicating that *kn1* is required for stem cell maintenance in the SAM. In permissive backgrounds, which have larger SAMs that likely allow *kn1(lf)* mutants to overcome these early vegetative defects, phenotypes are limited to the inflorescence. In these permissive backgrounds, *kn1(lf)* mutant inflorescences initiate fewer lateral meristems resulting in fewer long branches and spikelets (Kershstetter et al., 1997). IM size has not been directly measured in *kn1(lf)* mutants, but the reduced number of lateral meristems in *kn1(lf)* mutants is likely due to smaller IM size.

The KNOX TF family is comprised of a relatively small group of genes that fall into two phylogenetic classes and therefore referred to as class I and class II *KNOX* genes (Bharathan, Janssen, Kellogg, & Sinha, 1997; Bharathan, Janssen, Kellogg, & Sinha, 1999; Reiser, Sanchez-Baracaldo, & Hake, 2000). The class I *KNOX* genes, of which there are eight in maize, are primarily expressed in meristems where they promote meristem maintenance (McConnell & Barton, 1998; Millar & Gubler, 2005; Wu, Tian, & Reed, 2006). At least four class I *KNOX* genes are co-expressed in the inflorescence, including *kn1, rough sheath1 (rs1), knox3 (kn3)* and *knox8 (kn8)* and are thought to act at least partially redundantly (Jackson et al., 1994). Indeed, differential expression of *knox* genes between inbreds has been proposed to explain the inbred-specific phenotypes of *kn1(lf)* mutants (Vollbrecht et al., 2000). Recently, the *kn1* and *rs1 KNOX* genes have been shown to function partially redundantly. While *rs1(lf)* single mutants do not have a phenotype on their own, *rs1(lf)* mutants enhance *kn1(lf)* mutants in a dose-dependent manner, indicating the level of class I *KNOX* gene expression is critical in meristem maintenance (Bolduc, Tyers, Freeling, & Hake, 2014).

There has been considerable interest in understanding the downstream transcription targets of KN1 and other class I KNOX TFs that are critical for stem cell maintenance and identity. Chromatin immunoprecipitation (ChIP) experiments indicate that KN1 binds to thousands of sites in the maize genome, corresponding to nearly 5000 genes (Bolduc et al., 2012). Only ~600 of the genes bound by KN1 are also modulated in *kn1(lf)* or *kn1* gain-of-function (gf) mutants, suggesting that either only a subset of these genes are genuine targets of KN1 transcriptional regulation or that other KNOX TFs can compensate in the absence of KN1. Nevertheless, the identification of high-confidence KN1 target genes indicates that KN1 is positioned atop a transcriptional regulatory cascade and targets TF classes critical for inflorescence development, including other KNOX, MADS-box, and YABBY TFs (Bolduc et al., 2012). Furthermore, KN1 targets genes involved in multiple hormonal pathways, including the gibberellin (GA), brassinosteroid (BR), and auxin pathways. KN1 keeps GA levels low within the meristem through direct activation of *ga2ox1,* which encodes an enzyme involved in GA inactivation, as well as other enzymes involved in

GA response (Bolduc & Hake, 2009; Bolduc et al., 2012). KN1 also activates genes involved in auxin synthesis, transport and signalling, which are likely required for lateral meristem initiation (Bolduc et al., 2012). Thus, KN1 orchestrates a complex transcriptional and hormone signalling network to balance meristem maintenance and differentiation.

9.3.2 Regulation of Stem Cell Homeostasis by the CLAVATA Signalling Pathway

Stem cell homeostasis and meristem size are also regulated by a negative feedback loop involving the CLAVATA (CLV) signalling proteins and the WUSCHEL (WUS) TF. The main components of this negative feedback loop were first elucidated in the model plant *Arabidopsis*, and the core pathway appears largely conserved in maize. Briefly, *CLV3* encodes a small, secreted peptide expressed in the central stem cells of the meristem (Fletcher, Brand, Running, Simon, & Meyerowitz, 1999). The CLV3 ligand binds to receptor complexes composed of the leucine-rich repeat (LRR) receptor-like kinases CLV1, CLV2 and co-receptors CORYNE (CRN) and RPK2/TOAD2 in cells directly below the central stem cells (Betsuyaku et al., 2011; Clark, Williams, & Meyerowitz, 1997; Kinoshita et al., 2010; Miwa et al., 2008; Muller, Bleckmann, & Simon, 2008). Binding of the CLV3 ligand to its receptor complexes triggers a signalling cascade that ultimately restricts the expression of the homeodomain-containing TF, WUSCHEL (WUS) (Mayer et al., 1998; Schoof et al., 2000). WUS-expressing cells function as a stem cell niche and the size of the WUS-expression domain determines stem cell number and meristem size (Schoof et al., 2000). *CLV1, CLV2,* and *CLV3* are negative regulators of stem cell number and *clv* mutants have increased stem cell number and enlarged meristems, whereas *WUS* is a positive regulator of stem cell number and *wus* mutants fail to maintain stem cells within the meristem.

The maize mutants, *thick tassel dwarf1 (td1)* and *fasciated ear2 (fea2)* encode homologues of the CLV1 and CLV2 receptors, respectively (Bommert et al., 2005; Taguchi-Shiobara, Yuan, Hake, & Jackson, 2001). Similar to *clv1* and *clv2* mutants in *Arabidopsis,* both *td1* and *fea2* mutants have enlarged or fasciated meristems, indicating that the CLV signalling pathway also restricts stem cell number in maize. Florets in both *td1* and *fea2* mutants often make extra stamens, indicating that the CLV pathway also regulates FM size. In addition, the CLV signalling pathway likely controls natural variation in meristem size in maize. *td1* maps near a quantitative trait locus (QTL) that affects kernel row number and spikelet density (Bommert et al., 2005), and *fea2* maps near a QTL regulating kernel row number and meristem size (Bommert, Nagasawa, & Jackson, 2013). It is unknown if *td1* or a linked gene underlies the first QTL. *fea2* is thought to underlie the second QTL, however, since mild loss-of-function alleles of *fea2* do not have fasciated IM, but have larger IM that initiate more lateral meristems and thus increase kernel row number in the ear (Bommert, Nagasawa, et al., 2013). Therefore, *fea2* and potentially other members of the CLV-WUS signalling pathway control natural variation in meristem size. To date, neither the ligand(s) for the TD1 or FEA2 receptors, nor a WUS homologue that functions in regulation of stem cell homeostasis has been identified in maize, perhaps due to genetic redundancy.

The *compact plant2 (ct2)* mutant phenotypically resembles *td1* and *fea2* mutants, with fasciated ear IM and enlarged tassel IM, indicating that the *ct2* gene also regulates IM size (Bommert, Je, Goldshmidt, & Jackson, 2013). *Ct2* encodes a predicted α-subunit of a heterotrimeric GTP-binding protein, which physically associates with FEA2 and is required to transmit the CLV signal (Bommert, Je, et al., 2013). Thus *ct2* functions as part of the signalling pathway downstream of the TD1 and FEA2 receptors and suggests that G-proteins can function with LRR receptor complexes in addition to canonical seven transmembrane G-protein coupled receptors.

9.4 INITIATION AND DETERMINACY OF LATERAL MERISTEMS

9.4.1 Auxin Is Required to Initiate Lateral Primordia

The inflorescence is patterned by a series of branching events that require the initiation of lateral meristems on the flanks of the IM. The first lateral meristems formed are indeterminate BMs that develop into long branches at the base of the tassel (no BM are formed in ears). The IM and BM then initiate determinate meristems, SPM, which form short branches that bear a pair of spikelets (Figure 9.4). The initiation of all lateral meristems during vegetative and reproductive development depends on the plant hormone auxin. Disruption of auxin biosynthesis, transport or signalling results in naked, pin-like inflorescences that lack or have greatly reduced numbers of lateral meristems and organs. Incipient lateral primordia are marked by local auxin maxima that are formed by a combination of polar auxin transport and biosynthesis. Our knowledge of auxin biology and its role in initiating lateral primordia comes from a combination of mutant analysis and physiological studies that utilize inhibitors of auxin transport.

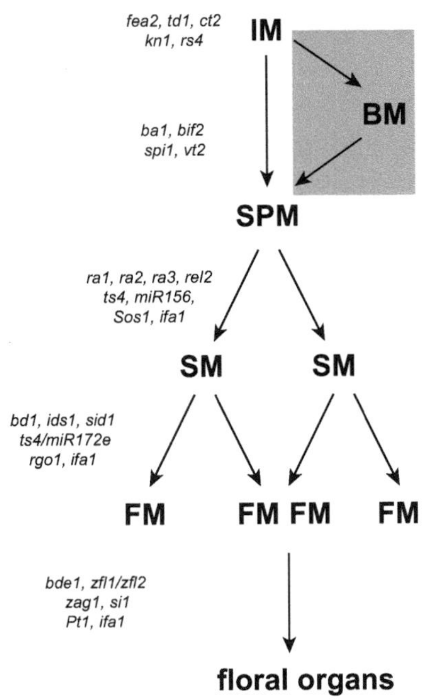

Figure 9.4 Schematic of maize inflorescence development and associated genes. Inflorescence meristems (IM) initiate spikelet pair meristems (SPM). Each SPM initiates two spikelet meristems (SM), which in turn initiate an upper and lower floral meristem (FM) that produce the floral organs. Tassel IM also initiate branch meristems (BM, gray box). Genes that regulate the identity or activity of given meristems are indicated. See text for details.

Polarized auxin transport is required to generate local auxin maxima and initiate lateral primordia. These local auxin maxima coincide with expression of the synthetic *DR5* reporter, which is a readout of auxin-induced transcription (Ulmasov, Murfett, Hagen, & Guilfoyle, 1997). In *Arabidopsis*, mutants in the auxin efflux carrier, *PINFORMED1 (PIN1)* form naked, pin-like inflorescences that lack lateral structures (Okada, Ueda, Komaki, Bell, & Shimura, 1991; Galweiler et al., 1998). In maize, no mutants in *pin* genes have been reported, likely due to functional redundancy. Indeed the maize genome contains at least 12 *pin*-like genes, four of which cluster in a single clade with *PIN1* (Forestan, Farinati, & Varotto, 2012). A translational fusion of *ZmPin1a* to YFP rescues *Arabidopsis pin1* mutants and in maize ZmPIN1a, is localized to lateral primordia in both vegetative and inflorescence meristems, suggesting that the function of *pin1* is conserved between *Arabidopsis* and maize (Gallavotti, Yang, Schmidt, & Jackson, 2008). In *Arabidopsis*, PIN1 localization is regulated by the serine/threonine kinase,

PINOID (PID). *pid* mutants phenotypically resemble *pin1* mutants and PIN1 is mislocalized in *pid* mutants (Bennett, Alvarez, Bossinger, & Smyth, 1995; Benjamins, Quint, Weijers, Hooykaas, & Offringa, 2001; Friml et al., 2004; Lee & Cho, 2006). PID directly phosphorylates PIN1, suggesting that phosphorylation of PIN1 by PID is required for normal PIN1 localization (Michniewicz et al., 2007). The maize *barren inflorescence2 (bif2)* gene is co-orthologous to *PID* and *bif2* mutants also have pin-like inflorescences that lack lateral structures (McSteen et al., 2007). BIF2 is required for normal localization of ZmPIN1a and BIF2 physically interacts with and phosphorylates ZmPIN1a in vitro, suggesting that BIF2 also regulates auxin transport via PIN1 regulation in maize (Skirpan et al., 2009). Functional characterization of polar auxin transport in maize comes from experiments utilizing the auxin transport inhibitor, *N*-1-naphthylphthalamic acid (NPA). Maize inflorescences treated with NPA resemble auxin transport or signalling mutants and lack or have greatly reduced numbers of lateral structures, including branches and spikelets (Gallavotti, Yang, et al., 2008; Wu & McSteen, 2007).

The formation of local auxin maxima and initiation of lateral primordia requires auxin synthesis in addition to polar auxin transport. Two maize mutants, *vanishing tassel2 (vt2)* and *sparse infloresence1 (spi1)*, have mutations in enzymes proposed to function at distinct steps of the tryptophan-dependent auxin synthesis pathway (Gallavotti, Barazrsh, et al., 2008; Phillips et al., 2011). Both *vt2* and *spi1* mutants make unbranched tassels and initiate only a few spikelets, confirming the critical role of auxin synthesis in lateral primordia initiation (Gallavotti, Barazesh, et al., 2008; Phillips et al., 2011). *vt2* encodes a co-ortholog of TRYPTOPHAN AMINOTRANSFERASE OF *ARABIDOPSIS*1 (TAA1) (Phillips et al., 2011), which converts tryptophan to indole-3-pyruvic acid (IPA) (Stepanova et al., 2008; Tao et al., 2008; Yamada, Greenham, Prigge, Jensen, & Estelle, 2009). *spi1* encodes a *YUCCA*-like flavin monooxygenase (Gallavotti, Barazesh, et al., 2008), however the exact role of YUCCA-Like enzymes in auxin biosynthesis is currently unclear (Nonhebel et al., 2011; Tivendale et al., 2010). Genetic analysis, however, suggests that *spi1* and *vt2* function in the same auxin biosynthesis pathway (Phillips et al., 2011). Interestingly, while both *spi1* and *vt2* single mutants show dramatic phenotypes consistent with reduced auxin, single mutants in *Arabidopsis TAA1*-like or *YUCCA* genes exhibit no or very mild phenotypes (Cheng, Dai, & Zhao, 2006; Stepanova et al., 2008). Thus, maize is an excellent system to probe the function of specific genes in auxin biosynthesis.

The auxin biosynthesis pathway also intersects with the auxin transport pathway. In *spi1* mutants, *ZmPin1a* is not localized to incipient lateral primordia (Gallavotti, Yang, et al., 2008), consistent with the model that auxin affects its own transport (Petrasek & Friml, 2009; Tanaka, Dhonukshe, Brewer, & Friml, 2006). Double mutants between genes in the auxin synthesis and transport pathways, including *vt2; bif2* and *spi1; bif2* double mutants show synergistic genetic interactions demonstrating the overlapping roles of auxin synthesis and transport in development (Gallavotti, Barazesh, et al., 2008; Phillips et al., 2011).

Auxin elicits biological responses at least in part by up-regulating transcription of early auxin responsive genes in the presence of auxin. Based on work mainly in *Arabidopsis*, a model for auxin-mediated signalling has emerged (Chapman & Estelle, 2009). In the absence of auxin, auxin response factors (ARFs) form heterodimers with AUXIN-INDOLE3 ACETIC ACID (Aux-IAA) proteins and bind to the regulatory DNA regions of early auxin responsive genes. The ARF/Aux-IAA complex acts as a bridge to recruit the co-repressor TOPLESS (TPL), which is thought to inhibit transcription by regulating chromatin structure (Krogan, Hogan, & Long, 2012; Long, Ohno, Smith, & Meyerowitz, 2006; Szemenyei, Hannon, & Long, 2008; Wang, Kim, & Somers, 2013). In the presence of auxin, auxin binds to a co-receptor complex, which includes the F-box protein, TRANSPORT RESPONSE1/AUXIN SIGNALING F-BOX (TIR1/AFB) and AUX-IAA proteins (Dharmasiri, Dharmasiri, & Estelle, 2005; Kepinski & Leyser, 2005). Binding of auxin to its receptor triggers the degradation of Aux-IAA proteins, which in turn allows ARFs to promote transcription of the early auxin responsive genes (Gray, Kepinski, Rouse, Leyser, & Estelle, 2001; Tan et al., 2007).

Much less is known about the downstream components of the auxin signalling pathway in maize. The maize genome contains multiple copies of TIR1-like receptors, Aux-IAA, and ARF proteins; however these genes have not been functionally characterized and their role in development may be difficult to uncover due to genetic redundancy (Wang, Deng, Bian, Lv, & Xie, 2010; Wang et al., 2012; Xing et al., 2011). A mutation in the *TPL* homologue encoded by the *ramosa1 enhancer locus 2 (rel2)* gene has a role in the regulation of inflorescence branching and enhances branching phenotypes of *ramosa1 (ra1), ra2* or *ra3* single mutants (see below) (Gallavotti et al., 2010). Whether or not *rel2* plays a direct role in auxin signalling is unknown.

The *barren stalk1 (ba1)* mutant also fails to initiate lateral meristems in the inflorescence and phenotypically resembles other auxin synthesis and signalling mutants (Ritter, Padilla, & Schmidt, 2002). *ba1* encodes a non-canonical bHLH protein that at least in vitro is also phosphorylated by the BIF2 kinase (Gallavotti et al., 2004; Skirpan, Wu, & McSteen, 2008), suggesting that BA1 may also function in the auxin pathway. *ba1* RNA is not expressed when auxin transport is disrupted by NPA treatment, leading to the hypothesis that BA1 functions downstream of auxin transport (Wu & McSteen, 2007). The exact role of BA1 in auxin transport or signalling is still unknown.

9.4.2 Identity and Determinacy of the SPM

Meristem identity and determinacy are interrelated concepts that can be used to describe the activity of a given meristem. Meristem identity is determined by the number and type of primordia produced. For example, FMs initiate floral organ primordia, but do not initiate meristem primordia. In contrast, meristem determinacy refers to the ongoing activity of a meristem and usually is reflected by the number of primordia initiated by a meristem. For example, IM and BM are indeterminate meristems because they continuously initiate SPM, whereas SPM, SM and FM are determinate meristems because they initiate a limited number of meristem or organ primordia and stem cells are consumed in the production of primordia. Many genes that regulate the identity and determinacy of specific meristems in the developing maize inflorescence have been identified, largely through mutant analysis. These genes are being assembled into genetic networks to describe the molecular mechanisms that pattern maize inflorescences.

SPM determinacy is regulated by the *ramosa (ra)* genes, defined by three classical mutants, *ra1, ra2* and *ra3* (Figure 9.4). *ra1, ra2,* and *ra3* single mutants have similar phenotypes, characterized by increased branching in both the ear and tassel, which is caused by loss of SPM determinacy. In strong alleles of the *ra* mutants, SPM are transformed into BM or an intermediate meristem type between BM and SPM. All three *ra* genes have been cloned. *ra1* and *ra2* both encode transcriptional regulators; *ra1* encodes a plant-specific C_2H_2 zinc-finger TF and *ra2* encodes a lateral organ boundary-domain containing TF (Vollbrecht, Springer, Goh, Buckler, & Martienssen, 2005; Bortiri et al., 2006). *ra3* encodes a trehalose-6-phosphate phosphatase, which has been proposed to modify a sugar signal or as a transcriptional regulator (Satoh-Nagasawa, Nagasawa, Malcomber, Sakai, & Jackson, 2006). Genetic and molecular analysis suggest a model in which *ra2* and *ra3* function in parallel pathways, upstream of *ra1*. *ra2* and *ra3* single mutants both have decreased levels of *ra1* and *ra1* levels are further reduced in the *ra2; ra3* double mutant, indicating that *ra2* and *ra3* function in parallel

(Bortiri et al., 2006; Satoh-Nagasawa et al., 2006). Interestingly, none of the *ra* genes are expressed in the SPM per se; rather they are expressed in a cup-shaped domain directly subtending the SPM. *ra2* is the most broadly expressed of the *ra* genes and *ra2* expression marks initiating BM, SPM and SM (but not FM) even before there is any morphological indication of meristem primordia initiation (Bortiri et al., 2006). *ra1* is expressed after *ra2* at the base of initiating SPM (Vollbrecht et al., 2005). *ra3* is expressed at the base of SPM and SM, and also in a stripe between the UFM and LFM (Satoh-Nagasawa et al., 2006). Consistent with the restricted expression pattern of *ra1*, *ra1* mutants only affect SPM determinacy, while *ra2* and *ra3* also have defects in SM determinacy. While *ra2* is conserved in other grasses, *ra1* and *ra3* are only found in the Andropogenae, a tribe characterized by the presence of paired rather than single spikelets (Vollbrecht et al., 2005; Kellogg, 2007). It has been proposed that *ra1* and *ra3* function in a regulatory module in the Andropogenae to impose determinacy on the SPM.

Recent transcriptional profiling and ChIP-sequencing experiments have begun to elucidate the gene networks governed by the *ra* genes (Eveland et al., 2014). Genes that are differentially modulated in *ra1, ra2* and *ra3* single mutants largely overlap, suggesting that RA TFs converge on a similar developmental pathway to regulate SPM determinacy and inflorescence branching. This data is also consistent with the model that RA2 and RA3 regulate *ra1* expression. RA1 directly targets genes involved in GA biosynthesis and catabolism, indicating that RA1 fine-tunes GA levels to balance cell division and differentiation within the SPM. Many RA1 target genes contain binding sites for other TF families, including bZIP and MADS-box TFs, and RA1 and KN1 directly regulate nearly 200 common target genes (Bolduc et al., 2012; Eveland et al., 2014). Thus RA1 likely functions in concert with other TF families to regulate gene expression and control SPM determinacy.

A powerful strategy to uncover additional components of genetic pathways is through the use of enhancer and suppressor screens. The *ramosa* pathway member, *RAMOSA ENHANCER LOCUS2,* was identified as an enhancer of *ra1*, and *rel2* mutants increase the SPM determinacy defects of weak *ra1* alleles. *rel2* was positionally cloned and encodes a gene with similarity to *TOPLESS (TPL)* in *Arabidopsis* (Gallavotti et al., 2010). *TPL* is required for apical/basal polarity in the *Arabidopsis* embryo and as discussed previously functions as a co-repressor with Aux-IAA proteins in auxin signalling (Long et al., 2006; Szemenyei et al., 2008). The REL2 protein physically interacts with RA1 and RA1 has been proposed to recruit

REL2 to specific target loci (Gallavotti et al., 2010). It is unknown if REL2 directly interacts with the auxin pathway, although the auxin reporter *DR5* is upregulated in *ra* mutants (Eveland et al., 2014) and the *ra* and auxin pathways clearly must interact at some level to regulate inflorescence branching.

Most loss-of-function mutants that alter meristem determinacy increase indeterminacy in the affected meristem, indicating that the normal gene function is to impose determinacy on an indeterminate meristem fate. In contrast, the dominant antimorph (dominant negative) mutant, *Suppressor of sessile spikelet1 (Sos1),* exhibits increased determinacy, suggesting that the normal function of the *sos1* gene is to promote meristem indeterminacy. In normal plants, the SPM initiate two SM that will form the spikelet pair. In *Sos1* mutants, however, SPM initiate only a single SM and no sessile spikelet is formed (Wu, Skirpan, & McSteen, 2009). *Sos1* genetically interacts with the *ra* mutants, indicating that *sos1* functions in a common pathway with the *ra* genes to regulate SPM determinacy (Wu et al., 2009). Interestingly, *Sos1* exhibits different interactions with each of the *ra* mutants, indicating that *sos1* and the *ra* genes do not function in a simple linear pathway. *Sos1* remains uncloned and the molecular underpinnings of the *Sos1* defects are unknown. It is also unclear if or how *Sos1* interacts with the auxin pathway to promote initiation of the sessile spikelet.

9.4.3 Boundary Formation in the Phytomer Controls Meristem Determinacy

In plants, axillary meristems are formed in the axils of leaves or bracts. Together, the axillary meristem, bract and stem form the phytomer, the repeating unit of plant growth. Differences in how cells are allocated to components within the phytomer are determinates of morphological differences between species and within species at different developmental stages. For example in maize, bracts grow out during vegetative growth to form prominent leaves while outgrowth of axillary meristems is suppressed. In contrast, bract outgrowth is often suppressed in inflorescences and axillary meristems such as the BM and SPM grow out to form the long and short branches of the tassel. The balance between bract outgrowth and suppression impacts the morphology of mature inflorescences both by directly controlling whether or not a visible bract is present, but also indirectly by impacting meristem determinacy and phyllotaxy.

In maize, at least four genes, defined by the *tasselsheath (tsh)* mutants, are required to suppress bract outgrowth in the inflorescence. In addition to bract suppression defects, *tsh* mutants exhibit decreased branching, indicating a role for the *tsh* mutants in lateral meristem initiation and determinacy (Whipple et al., 2010). Bract outgrowth also alters phyllotaxy as derepressed bracts in *tsh* mutants force SPM apart and change the phyllotaxy from ordered rows of SPM to a spiral phyllotaxy (Chuck, Whipple, Jackson, & Hake, 2010). Two *tsh* genes have been cloned: *tsh1* encodes a GATA zinc-finger protein that is homologous to *HANABA TARANU* of *Arabidopsis* (Whipple et al., 2010) and *tsh4* encodes an SPB-box TF (Chuck et al., 2010). SPB-box TFs, including *tsh4,* are targeted by the closely related microRNAs (miRNAs) miR156 and miR529, and this regulation is critical to set up the boundaries within the phytomer. The dominant *Corngrass1 (Cg1)* mutant is caused by overexpression of miR156 and has common phenotypes with *tsh4,* including derepressed bract growth and spiral phyllotaxy in the inflorescence (Chuck, Cigan, Saeteurn, & Hake, 2007). TSH4 expression is expanded in *ra* mutants and the *tsh4* mutant is epistatic to *ra* single mutants, indicating that the *ra* genes promote meristem determinacy at least in part by restricting expression of TSH4 (Chuck et al., 2010). A model has been proposed in which there is a competition for cells within lateral primordia to form the three components of the phytomer; the distribution of cells to these three components ultimately impacts the final morphology of the phytomer. For example, if more cells are allocated to the bract, as in *tsh* mutants, fewer cells are allocated to the axillary meristem and stem, which results in decreased branching. Thus, all components of the phytomer are integrated and alteration of one domain of the phytomer can indirectly alter growth and development of the other domains.

9.4.4 Identity and Determinacy of the SM

The spikelet is the fundamental unit of grass inflorescences. In maize, spikelets initiate two FMs that produce the floral organs. Tassel spikelets contain two florets at maturity, however ear spikelets contain a single floret due to abortion of the lower floret. Genes that are required for SM identity (*branched silkeless1 (bd1)*) and determinacy (*indeterminate spikelet1 (ids1), sister of indeterminate spikelet1 (sid1)* and *tasselseed4 (ts4)*) have also been identified (Chuck, Meeley, & Hake, 1998; Chuck, Muszynski, Kellogg, Hake, & Schmidt, 2002; Chuck, Meeley, Irish, Sakai, & Hake, 2007) (Figure 9.4).

The *bd1* gene is required for SM identity (Chuck et al., 2002). In *bd1* mutant ears, SM do not produce FM, but rather are converted to indeterminate long branches and in *bd1* mutant tassels, SM produce additional SM. *bd1* encodes an ethylene-responsive element-binding factor (ERF) TF and is expressed at the base of the glume and SM junction in both the ear and tassel, where it presumably represses indeterminate branch growth (Chuck et al., 2002). Like the *ra* genes, *bd1* is not expressed in the meristem in which it functions, but rather is expressed in the cells subtending the SM. The transcriptional targets of BD1 are still unknown.

SM determinacy is controlled by a miRNA-regulated network that includes the miRNA *ts4*/miR172e and the AP2-like genes, *ids1* and *sid1*. The *tasselseed4 (ts4)* mutant contains a mutation in miR172e, a unique member of the miR172 family and has indeterminate SM and FM as well as sex determination defects. *ts4*/mir172e targets two *AP2*-like genes, *indeterminate spikelet1 (ids1)* and *sister of ids1 (sid1)* (Chuck, Meeley, et al., 2007; Chuck, Meeley, & Hake, 2008). The importance of *ts4*/miR172e regulation of *ids1* is underscored by the dominant, gain-of-function *Ts6* mutant, which has phenotypes similar to *ts4,* including indeterminate SM and FM and feminized tassels. *Ts6* contains a mutation in the miRNA-binding site of *ids1,* which results in increased IDS1 activity and indeterminate SM and FM (Chuck, Meeley, et al., 2007). Interestingly, loss-of-function mutants in *ids1* also have indeterminate SM (Chuck et al., 1998) and resemble at least some aspects of *ts4* and *Ts6* mutants, which are caused by increased IDS1 activity. This seemingly paradoxical result—that gain-of-function and loss-of-function mutations in *ids1* both cause SM indeterminacy—was explained by the discovery of the *ids1* paralogue, *sister of ids1 (sid1).* In *ids1* single mutants, *sid1* levels increase, suggesting that the net IDS1/SID1 activity is increased in *ids1* single mutants. Importantly, *ids1; sid1* double mutants completely suppress the *ts4*/miR172e phenotype, indicating that upregulation of these two target mRNAs are responsible for *ts4* mutant defects. Furthermore, *ids1; sid1* double mutants fail to initiate FM, and spikelets in the double mutants terminate in a floret-like structure that contains bract-like organs, but no floral organs (Chuck et al., 2008). Therefore, the normal role of *ids1/sid1* is to promote FM initiation and increased IDS1/SID1 activity (*ids1, ts4* or *Ts6* single mutants) results in additional FM, whereas lack of IDS1/SID1 activity (*ids1; sid1* double mutants) results in no FM.

Other mutants that affect SM determinacy have been identified, including *indeterminate floral apex1 (ifa1)* and *reversed germ orientation1 (rgo1)* (Laudencia-Chingcuanco & Hake, 2002; Kaplinsky & Freeling, 2003). *rgo1* has phenotypes similar to *ids1* in which SM initiate extra FM, indicating a role for *rgo1* in promoting SM determinacy. *rgo1* and *ids1* exhibit nonallelic non-complementation, indicating that these genes likely function in a common pathway to promote SM determinacy (Kaplinsky & Freeling, 2003). *ifa1* is required for determinacy in all meristem types in the inflorescence, including SPM, SM, and FM and functions redundantly with other inflorescence regulators to promote determinate meristem fates. For example, in *ids1; ifa1* double mutants SM are replaced with BM and in *ifa1; zag1* double mutants, FM are replaced with BM (Laudencia-Chingcuanco & Hake, 2002). Both *ifa1* and *rgo1* are uncloned, and a more complete understanding of their roles in promoting meristem determinacy awaits determining their molecular identity.

9.5 GENETIC CONTROL OF FLORAL DEVELOPMENT

The grasses produce highly derived flowers called florets. Like eudicot flowers, florets contain stamens and carpels, however florets lack petals and sepals. Instead, florets have lemma, palea and lodicules, which cannot be assigned analogous structures in eudicots based on morphology alone. The last three decades have resulted in a wealth of knowledge about the genes that regulate floral development in eudicot species, such as *Arabidopsis* and Antirrhinum. In comparison, relatively little is known about floral development in monocots, although recent work in grass species including maize and rice has started to elucidate the gene regulatory networks that control floral development in monocots and the grasses in particular.

In maize, the SM initiates an UFM and LFM that produce the two florets in the spikelet (Figure 9.4). The florets are for the most part morphologically indistinguishable, although development of the LFM is delayed relative to the UFM and the lower floret is slightly smaller than the upper floret (Hsu & Peterson, 1991). In the ear, the LFM aborts, albeit relatively late in development, to produce a spikelet containing a single floret. Sex determination occurs through stamen arrest in the ear and carpel abortion in the tassel.

MADS-box TFs are key regulators of flowering time, floral meristem fate and floral organ identity in all angiosperms studied (Ciaffi, Paolacci, Tanzarella, & Porceddu, 2011; Heijmans, Morel, & Vandenbussche, 2012; Krizek & Fletcher, 2005; Ng & Yanofsky, 2001). MADS-box TFs function in a combinatorial fashion by forming dimers and higher order complexes to regulate the transcription of target genes (Egea-Cortines, Saedler, & Sommer, 1999; Fan, Hu, Tudor, & Ma, 1997; Honma & Goto, 2001). In *Arabidopsis*, floral organs are arranged in four concentric whorls and the ABC-class MADS-box genes function in a combinatorial fashion to specify organ identity in each whorl: class A genes specify whorl 1 (sepals), class A and B genes specify whorl 2 (petals), class B and C genes specify whorl 3 (stamens) and class C genes alone specify whorl 4 (carpels); class C genes have an additional function in promoting meristem determinacy (Coen & Meyerowitz, 1991). The class E genes function redundantly with the A, B, and C class genes to specify floral organ identity in all four whorls, and class D genes have been proposed to specify a fifth whorl, or ovules (Theissen, 2001; Krizek & Fletcher, 2005). The quartet model proposes that two MADS-box TF dimers form a tetramer and this tetramer is the functional MADS-box transcriptional regulatory unit (Theissen, 2001). A few examples of interactions between MADS-box TFs and non-MADS domain proteins have been reported in plants, including interactions with receptor-like kinases, other classes of TFs, RNA-binding proteins and chromatin remodelling proteins (see Immink, Kaufmann, & Angenent, 2010 and references therein). In addition, at least one MADS-box TF is present in a high molecular weight protein complex in excess of a MADS-box tetramer (Helliwell, Wood, Robertson, James Peacock, & Dennis, 2006), suggesting that other proteins also function in MADS-box mediated transcriptional regulation.

The ABC model of floral organ identity is a simple and elegant model that explains how floral organ identity is specified in *Arabidopsis* remarkably well. It is clear however, that the MADS-box transcriptional network is more complex than indicated by the ABC model both in *Arabidopsis* and other species. Understanding this transcriptional regulatory network in maize as well as other species requires an in-depth understanding of the composition and transcriptional targets of different MADS-box protein complexes, as well as the function of target genes.

There has been considerable interest in how MADS-box genes regulate floral development in monocots and the grasses in particular and which, if any, aspects of the ABC model apply to species outside *Arabidopsis* and Antirrhinum. Indeed, MADS-box transcriptional regulation is also critical in maize, and three maize mutants with defects in floral development encode MADS-box TFs. Analysis of these mutants, in combination with expression and molecular analysis of additional MADS-box TFs indicates that some aspects of the ABC model are conserved in maize, however the MADS-box transcriptional regulatory network is undoubtedly complex and key differences exist between maize and *Arabidopsis* (Thompson & Hake, 2009; Heijmans et al., 2012).

In contrast to *Arabidopsis*, only a handful of maize mutants that affect floral development have been characterized and cloned. Two floral mutants, *silky1 (si1)* and *zea agamous1 (zag1)* encode MADS-box genes and analysis of these mutants indicates the B and C class functions are largely conserved in maize. *si1* encodes a gene co-orthologous to the class B gene, *apetala3 (ap3)* (Ambrose et al., 2000). In *Arabidopsis*, B class function is defined by two genes, *apetala3 (ap3)* and *pistillata (pi)* (Jack, Brockman, & Meyerowitz, 1992; Goto & Meyerowitz, 1994). AP3 and PI bind DNA as obligate heterodimers and together with the class A, C and E genes specify whorl two and whorl three floral organs (petals and stamens, respectively). *si1* mutants exhibit similar homeotic conversions in which lodicules are transformed into lemma/palea-like organs and stamens are transformed into carpels, indicating that *si1* has similar functions in *Arabidopsis* and maize (Ambrose et al., 2000). Furthermore, SI1 and one PI homologue, ZMM16 have similar biochemical activities to AP3 and PI, indicating that the same molecular module specifies petals and lodicules in eudicots and maize respectively (Whipple et al., 2004). Conservation of B class function in specifying petals and lodicules also indicates that petals and lodicules are homologous structures, despite their distinct morphologies.

In *Arabidopsis*, class C function is defined by a single gene, *AGAMOUS (AG),* which is responsible for both class C functions: specification of organ identity and promotion of meristem determinacy (Yanofsky et al., 1990). Maize contains three *ag* homologues, *zag1* and the recently duplicated genes, *zmm2* and *zmm23* (Munster et al., 2002). *zag1* mutants have indeterminate floral meristems and produce an excess of carpels in the ear, but do not have floral organ identity defects (Mena et al., 1996). Thus it appears that class C function has been subfunctionalized in the grasses, with *zag1* functioning in meristem determinacy and the other two *ag* homologues, *zmm2* and *zmm23* functioning primarily in organ identity. Indeed, class C function has largely been subfunctionalized in rice in which *OsMADS3* plays a more critical role in floral organ identity and *OsMADS58* plays a more critical role in meristem determinacy (Yamaguchi et al., 2006). To date, no mutants in *zmm2* or *zmm23* have been reported and as a result functional analysis of *zmm2* and *zmm23* in floral development has not been possible.

In contrast to class B and C genes, the function of which is at least partially conserved in maize, the function of class A genes in other plants is more complex. In *Arabidopsis*, class A function is defined by the MADS-box gene, *AP1* and the non-MADS-box TF *AP2* (Mandel, Gustafson-Brown, Savidge, & Yanofsky, 1992; Jofuku, den Boer, Van Montagu, & Okamuro, 1994). In *ap1* mutants, sepals and petals are transformed to carpelloid and stamenoid-like structures, indicating that *ap1* is required to specify whorl 1 and 2 organs (Mandel et al., 1992). In addition, *ap1* is required to promote floral meristem fate, and functions partially redundantly with the closely related genes, *cauliflower (cal)* and *fruitful (ful)* (Bowman, Alvarez, Weigel, Meyerowitz, & Smyth, 1993; Kempin, Savidge, & Yanofsky, et al., 1995; Ferrandiz, Gu, Martienssen, & Yanofsky, 2000). Classical class A function, however, has not been defined in a species outside the Brassicaceae and the loss of sepal identity in *ap1* mutants has been interpreted as loss of floral meristem identity, suggesting that the ancestral role for *AP1*-like genes is to promote floral meristem identity (Litt, 2007). No mutants in an *AP1*-like gene have been reported in maize, so the function of *AP1*-like genes is unclear. The *AP1* homologue, *zap1a* is expressed in florets, but excluded from stamens and carpels, suggesting that *zap1a* might have a function in the development or specification of the outer whorls (Mena, Mandel, Lerner, Yanofsky, & Schmidt, 1995).

Other MADS-box TFs have key roles in maize floral development. The maize *bearded-ear* mutant has defects in multiple aspects of floral development and encodes a MADS-box TF belonging to the *AGL6* clade. In *Arabidopsis*, *AGL6* does not have a clear role in floral development, but *AGL6*-like genes are critical for floral development in both maize and rice (Ohmori et al., 2009; Thompson et al., 2009). Interestingly, *bde* mutants differentially affect development in the UFM and LFM, suggesting that BDE has distinct roles in the UFM and LFM. In the UFM, *bde* is required for FM determinacy and normal floral organ development, while in the LFM, *bde* is required for FM identity (Thompson et al., 2009). One other mutant, *restorer of fertility2 (rf2)*, also has different phenotypes in the upper and lower florets (Liu, Cui, Horner, Weiner, & Schnable, 2001) and at least two MADS-box genes, *zmm8* and *zmm14* are only detectable in the UFM by in situ hybridisation (Cacharrón, Saedler, & Theissen, 1999). Thus, the UFM and LFM utilize distinct gene regulatory networks to control floral development.

MADS-box genes are likely required for other aspects of maize floral development. Recent work in rice has elucidated the role of many MADS-box genes, which function in multiple aspects of floral development including FM determinacy, organ identity and development (for review, see Hirano, Tanaka, & Toriba, 2014). In maize, many MADS-box genes are expressed during floral development (Heuer et al., 2001; Zhao et al., 2011), however functional characterization of many MADS-box genes in maize is particularly difficult due to the high degree of gene duplication and genetic redundancy. A goal for the future is to elucidate the function of MADS-box genes in maize using reverse genetics approaches.

Mis-expression of MADS-box genes can also lead to dramatic effects in the inflorescence. The dominant *Tunicate1 (Tu1)* mutant, also known as pod corn, has striking inflorescence phenotypes including elongated outer glumes, sex determination defects in the tassel and increased branching in the ear (Eyster, 1921; Langdale, Irish, & Nelson, 1994). *Tu1* phenotypes are the result of ectopic expression of the MADS-box gene, *zmm19* (Han, Jackson, & Martienssen, 2012; Wingen et al., 2012). In normal maize plants, *zmm19* expression is restricted to husk and leaf tissue, but in *Tu1* plants *zmm19* is ectopically expressed in developing inflorescences in a domain similar to *ra1* and *ra3* (Han et al., 2012). Intriguingly, several developmental regulators including *ra1, ra2, ra3, ts1, ts2* and *ts4* interact genetically with *Tu1* (Langdale et al., 1994) and all contain putative MADS-box binding elements (CArG-box) in their 5′ flanking sequence, raising the intriguing possibility that ZMM19 directly represses these genes in *Tu1* mutants to produce the dramatic pleiotropic *Tu1* phenotypes (Han et al., 2012).

Non-MADS-box genes are also critical for maize floral development. The *LEAFY (LFY)* homologues *zfl1* and *zfl2* are critical for multiple aspects of floral development including FM identity, FM determinacy, and floral organ identity (Bomblies & Doebley, 2006). In *Arabidopsis*, *LFY* is required for floral meristem identity at least in part by promoting expression of the ABC MADS-box genes (Huala & Sussex, 1992; Weigel, Alvarez, Smyth, Yanofsky, & Meyerowitz, 1992). In *zfl1; zfl2* double mutants, FM branch and form additional meristem-like structures indicating that FM identity is lost (Bomblies & Doebley, 2006). *zfl1; zfl2* double mutants have additional defects including decreased branch number in the tassel and changes in phyllotaxy, suggesting that *zfl1/zfl2* might also have novel roles in the grasses (Bomblies & Doebley, 2006).

As yet uncloned mutants will likely shed light on additional genes involved in floral development. As mentioned previously, the uncloned gene, *ifa1,* is required for FM determinacy in addition to the SPM and SM determinacy (Laudencia-Chingcuanco & Hake, 2002). *ifa1* maps near the MADS-box gene, *zmm8,* but recent mapping data indicates that *zmm8* is outside the *ifa1* interval and not responsible for the *ifa1* phenotype (B. Thompson, unpublished). The dominant mutant, *Polytypic1 (Pt1)* is also uncloned and has indeterminate FM as well as sex determination defects (Postlethewait & Nelson, 1964). A more complete understanding of the floral regulatory network will require isolation and characterization of additional floral mutants and as well as reverse genetic approaches to uncover the roles of known floral regulators, such as the MADS-box genes.

9.6 GENETIC REGULATION OF SEX DETERMINATION

Maize is a monoecious plant with male and female flowers separated on distinct inflorescences. The tassel produces staminate (male) flowers, while the ear produces pistillate (female) flowers. Floral meristems in both the tassel and ear initiate perfect flowers that contain both stamens and carpels and sex determination occurs through carpel abortion in the tassel and stamen arrest in the ear. Sex determination is under genetic control and mutants that affect sex determination in both the tassel and the ear have been identified. Several of these mutants have now been cloned, implicating multiple plant hormones in the sex determination process, including GA, jasmonic acid (JA), and BRs (Figure 9.5). Other mutants have no clear link to hormones and it is unclear if these mutants impinge on hormonal regulation of sex determination or if they function in a hormone-independent pathway. Determining the molecular basis of uncloned sex determination mutants may provide crucial links between the hormonal and genetic pathways that govern sex determination.

GA plays a pivotal role in maize sex determination. Mutants in the GA biosynthetic pathway including *anther ear1 (an1)* and the *dwarf (d)* mutants, *d1, d2, d3* and *d5*, all make fully developed anthers in the ear (Emerson & Emerson, 1922; Bensen et al., 1995) and these defects can be suppressed by the addition of exogenous GA (Phinney, 1956; Phinney & West, 1960). The dominant mutants, *D8* and *D9* contain mutations in *DELLA* homologues,

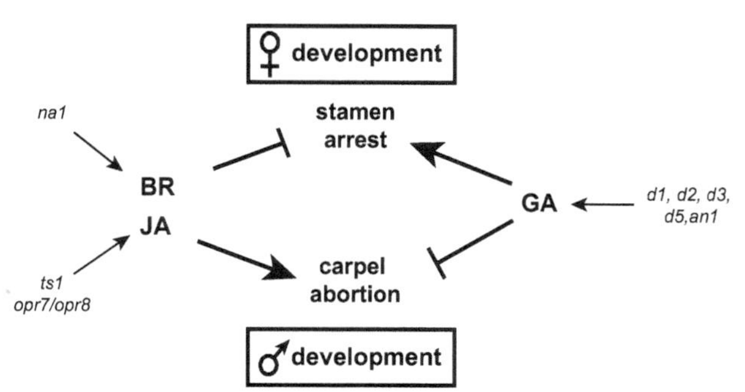

Figure 9.5 Schematic of hormonal regulation of maize sex determination. Brassinosteroids (BR) and jasmonic acid (JA) promote male development in the tassel by promoting carpel abortion and preventing stamen arrest. In contrast, Gibberellin (GA) promotes female development by promoting stamen arrest in the ear and preventing carpel abortion in the tassel. Genes required for BR, JA, and GA biosynthesis are indicated.

which function in GA signalling, and also make fully developed anthers in the ear (Harberd & Freeling, 1989; Peng et al., 1999). Therefore, both GA synthesis and signalling is required for stamen arrest in the ear. All mutants defective in GA synthesis or response have other phenotypes consistent with lack of GA such as a short stature, and broad short leaves. Application of exogenous GA has the opposite effect on sex determination and promotes carpel development in the tassel (Nickerson, 1959). Thus, GA is critical to regulate both stamen arrest in the ear and carpel abortion in the tassel.

Carpel abortion in the tassel requires the activity of the *tasselseed (ts)* genes, defined by six *ts* mutants. Of these six allelic groups, three *ts* mutants are dominant (*Ts3, Ts5, Ts6*) and three are recessive (*ts1, ts2* and *ts4*). *Ts3, ts4, Ts5* and *Ts6* make perfect flowers, whereas *ts1* and *ts2* exhibit more severe feminization and do not make stamens at all in the tassel. The *ts* mutants can be grouped into two categories based on their effect on inflorescence branching (Irish, Langdale, & Nelson, 1994). *ts1, ts2, Ts3* and *ts5* affect sex determination, but do not affect inflorescence branching. In contrast, *ts4* and *Ts6* are required for SPM, SM and FM determinacy and produce highly branched inflorescences in addition to pistillate tassels florets.

Cloning of *ts1* and *ts2* genes indicates that JA is required for carpel abortion in the tassel. *ts1* encodes a plastid-targeted 13-lipoxegenase that functions in the first step of JA synthesis in the conversion of linoleic acid to JA (Acosta et al., 2009). JA levels are reduced ~10-fold in *ts1* mutants compared to heterozygous siblings and application of JA can restore carpel abortion in *ts1* mutants, confirming the role of *ts1* in JA synthesis. *ts2* encodes a short chain alcohol dehydrogenase, the substrate of which is unknown (DeLong, Calderon-Urrea, & Dellaporta, 1993). *ts2* mutants can also be rescued by application of exogenous JA, suggesting that *ts2* may also function in the JA biosynthesis pathway, perhaps catalysing a β-oxidation step in the pathway (Acosta et al., 2009). *ts2* mRNA is reduced in *ts1* mutants compared to normal plants, suggesting that *ts1* functions upstream of *ts2* transcription (Calderon-Urrea & Dellaporta, 1999), although it is possible there is a positive feedback loop whereby JA induces expression of enzymes required for its synthesis. The expression patterns of *ts1* and *ts2* appear to be distinct, however. *ts1* is expressed at the base of spikelets in a stripe of the central inflorescence to the base of the spikelet pedicel, and *ts2* expressed at the base of the developing gynoecium (DeLong et al., 1993; Acosta et al., 2009). The non-overlapping expression patterns of *ts1* and *ts2* appear at odds with a model in which both TS1 and TS2 function in the same JA biosynthesis pathway, although it is possible that one or more JA intermediates is mobile.

Further experimentation is required to understand if and how *ts2* functions in JA synthesis and the site of JA production.

The importance of JA in carpel abortion in the tassel is supported by analysis of the two oxophytodienoate genes, *opr7* and *opr8,* which also function in the JA biosynthetic pathway. *opr7; opr8* double mutants have dramatically reduced JA levels and exhibit a severe tasselseed phenotype in which pistils fail to abort in the tassel and development of stamens is arrested (Yan et al., 2012). Furthermore, sex determination defects in *opr7; opr8* double mutants can be suppressed by application of exogenous JA. Together, analysis of the *ts1* and the *opr7; opr8* double mutant clearly demonstrate the critical role of JA in sex determination. The site of JA production and the molecular mechanism by which JA promotes sex determination remain unclear and require further experimentation.

In normal ears, carpel abortion is prevented by the *silkless1 (sk1)* gene (Jones, 1925). In *sk1* mutants, carpels abort in both the tassel and ear following a *tasselseed*-mediated cell death program (Calderon-Urrea & Dellaporta, 1999). Furthermore, *ts2* is epistatic to *sk1,* indicating that *sk1* prevents carpel abortion in the ear by regulating *ts2* activity (Calderon-Urrea & Dellaporta, 1999). The molecular identity of *sk1* is still unknown and determining the precise molecular role of *sk1* in sex determination awaits cloning of the *sk1* mutant.

In addition to JA, the BRs plant hormones are required for carpel abortion in the tassel. The classical *nana1 plant1(na1)* mutant contains a mutation in the *DE-ETIOLATED2 (DET2)* homologue, a well-characterized gene in the BR biosynthetic pathway (Hartwig et al., 2011). *na1* mutants exhibit a tasselseed phenotype in which pistil abortion fails and stamens fail to fully develop in the tassel. *na1* mutants have greatly reduced levels of BR and sex determination defects can be rescued by the application of exogenous BR, consistent with *na1* functioning in BR biosynthesis. In addition to sex determination defects, *na1* mutants are dwarves and have decreased cell elongation, reminiscent of BR mutants in *Arabidopsis* (Chory, Nagpal, & Peto, 1991; Bishop & Koncz, 2002; Nemhauser & Chory, 2004). *na1* is expressed in developing stamens and in the outer cell layers of the carpel primordia prior to pistil abortion, suggesting that BR does not directly promote pistil abortion. Rather, the normal role of BR may be to promote stamen development and the lack of pistil abortion in the tassel may be a secondary effect of disrupted stamen development (Hartwig et al., 2011). Precisely how the GA, JA, and BR pathways intersect to orchestrate pistil abortion in the tassel and stamen arrest in the ear is still unknown and an area for future research.

The two other cloned tasselseed mutants, *ts4* and *Ts6,* have no clear links to hormone signalling. *ts4* and *Ts6* are also unique among the *ts* mutants in that they are required to restrict branching in the inflorescence in addition to their roles in sex determination (Irish et al., 1994). As discussed in section 9.4 above, *ts4* is a loss-of-function mutation in miR172e and *Ts6* contains a mutation in the *ts4*/mir172e binding site of the AP2-like gene, *ids1* (Chuck, Meeley, et al., 2007). How upregulation of *ids1* and its paralogue, *sid1* results in failure of carpel abortion is still unclear. In *Arabidopsis*, GA is required to promote FM identity (Okamuro, Szeto, Lotys-Prass, & Jofuku, 1997) and thus GA might be a key link between the sex determination and meristem identity pathways.

9.7 CONCLUDING REMARKS

The last decade has seen a proliferation in the identification of genes that regulate patterning and development of the maize inflorescences. Mutant analysis and cloning have proven extraordinarily productive endeavours and in combination with emerging molecular and genomic tools, these analyses are beginning to elucidate the regulatory pathways and networks that govern inflorescence development. A common theme of recent studies is the intimate link between transcriptional and hormone regulatory networks in the control of initiation and activity of meristems. A major challenge for the future is to more fully understand how these key networks are integrated to ultimately determine inflorescence architecture. Some links have been made between key transcriptional regulators and hormonal signalling pathways. For example, *rel2* mutants enhance *ra1* mutants and *rel2* encodes a co-repressor proposed to function in auxin signalling, potentially linking auxin signalling and the RA1 TF. Furthermore, both the KN1 and RA1 TFs modulate hormone levels within the meristem at least in part by directly regulating genes that function in GA synthesis and metabolism (Bolduc et al., 2012; Eveland et al., 2014). The targets of other key TFs that function in inflorescence development including IDS/SID1, BD1, and MADS-box TFs are largely unknown and very little is known about the downstream targets of hormone signalling. Nevertheless, these efforts have amassed an impressive framework that describes the molecular regulation of maize inflorescence development. While gaps in the regulatory pathways and networks remain, these gaps will undoubtedly continue to be filled by mutant analysis and cloning coupled with molecular, biochemical, and genomic approaches. These endeavours will produce a more complete understanding of fundamental aspects of plant biology as well as identify new targets for genetic manipulation to potentially increase yield.

REFERENCES

Acosta, I. F., Laparra, H., Romero, S. P., Schmelz, E., Hamberg, M., Mottinger, J. P., et al. (2009). Tasselseed1 is a lipoxygenase affecting jasmonic acid signaling in sex determination of maize. *Science, 323*, 262–265.

Ambrose, B. A., Lerner, D. R., Ciceri, P., Padilla, C. M., Yanofsky, M. F., & Schmidt, R. J. (2000). Molecular and genetic analyses of the silky1 gene reveal conservation in floral organ specification between eudicots and monocots. *Molecular Cell, 5*, 569–579.

Benjamins, R., Quint, A., Weijers, D., Hooykaas, P., & Offringa, R. (2001). The PINOID protein kinase regulates organ development in *Arabidopsis* by enhancing polar auxin transport. *Development, 128*, 4057–4067.

Bennett, S. R. M., Alvarez, J., Bossinger, G., & Smyth, D. R. (1995). Morphogenesis in pinoid mutants of *Arabidopsis thaliana. Plant Journal, 8*, 505–520.

Bensen, R. J., Johal, G. S., Crane, V. C., Tossberg, J. T., Schnable, P. S., Meeley, R. B., et al. (1995). Cloning and characterization of the maize An1 gene. *Plant Cell, 7*, 75–84.

Betsuyaku, S., Takahashi, F., Kinoshita, A., Miwa, H., Shinozaki, K., Fukuda, H., et al. (2011). Mitogen-activated protein kinase regulated by the CLAVATA receptors contributes to shoot apical meristem homeostasis. *Plant and Cell Physiology, 52*, 14–29.

Bharathan, G., Janssen, B. J., Kellogg, E. A., & Sinha, N. (1997). Did homeodomain proteins duplicate before the origin of angiosperms, fungi, and metazoa? *Proceedings of the National Academy of Sciences of the United States of America, 94*, 13749–13753.

Bharathan, G., Janssen, B. J., Kellogg, E. A., & Sinha, N. (1999). Phylogenetic relationships and evolution of the KNOTTED class of plant homeodomain proteins. *Molecular Biology and Evolution, 16*, 553–563.

Bishop, G. J., & Koncz, C. (2002). Brassinosteroids and plant steroid hormone signaling. *Plant Cell, 14*(Suppl.), S97–S110.

Bolduc, N., & Hake, S. (2009). The maize transcription factor KNOTTED1 directly regulates the gibberellin catabolism gene ga2ox1. *Plant Cell, 21*, 1647–1658.

Bolduc, N., Tyers, R. G., Freeling, M., & Hake, S. (2014). Unequal redundancy in maize knotted1 homeobox genes. *Plant Physiology, 164*, 229–238.

Bolduc, N., Yilmaz, A., Mejia-Guerra, M. K., Morohashi, K., O'Connor, D., Grotewold, E., et al. (2012). Unraveling the KNOTTED1 regulatory network in maize meristems. *Genes & Development, 26*, 1685–1690.

Bomblies, K., & Doebley, J. F. (2006). Pleiotropic effects of the duplicate maize FLORICAULA/LEAFY genes zfl1 and zfl2 on traits under selection during maize domestication. *Genetics, 172*, 519–531.

Bommert, P., Nagasawa, N. S., & Jackson, D. (2013). Quantitative variation in maize kernel row number is controlled by the FASCIATED EAR2 locus. *Nature Genetics, 45*, 334–337.

Bommert, P., Je, B. I., Goldshmidt, A., & Jackson, D. (2013). The maize Galpha gene COMPACT PLANT2 functions in CLAVATA signalling to control shoot meristem size. *Nature, 502*, 555–558.

Bommert, P., Lunde, C., Nardmann, J., Vollbrecht, E., Running, M., Jackson, D., et al. (2005). Thick tassel dwarf1 encodes a putative maize ortholog of the *Arabidopsis* CLAVATA1 leucine-rich repeat receptor-like kinase. *Development, 132*, 1235–1245.

Bortiri, E., Chuck, G., Vollbrecht, E., Rocheford, T., Martienssen, R., & Hake, S. (2006). Ramosa2 encodes a LATERAL ORGAN BOUNDARY domain protein that determines the fate of stem cells in branch meristems of maize. *Plant Cell, 18*, 574–585.

Bowman, J. L., Alvarez, J., Weigel, D., Meyerowitz, E. M., & Smyth, D. R. (1993). Control of flower development in *Arabidopsis thaliana* by apetala1 and interacting genes. *Development, 119*, 721–743.

Cacharrón, J., Saedler, H., & Theissen, G. (1999). Expression of MADS box genes ZMM8 and ZMM14 during inflorescence development of *Zea mays* discriminates between the upper and the lower floret of each spikelet. *Development and Genes Evolution, 209*, 411–420.

Calderon-Urrea, A., & Dellaporta, S. L. (1999). Cell death and cell protection genes determine the fate of pistils in maize. *Development, 126*, 435–441.

Chapman, E. J., & Estelle, M. (2009). Mechanism of auxin-regulated gene expression in plants. *Annual Review of Genetics, 43*, 265–285.

Cheng, P. C., Greyson, R. I., & Walden, D. B. (1983). Organ initiation and the development of unisexual flowers in the tassel and ear of *Zea mays*. *American Journal of Botany, 70*, 450–462.

Cheng, Y., Dai, X., & Zhao, Y. (2006). Auxin biosynthesis by the YUCCA flavin monooxygenases controls the formation of floral organs and vascular tissues in *Arabidopsis*. *Genes & Development, 20*, 1790–1799.

Chory, J., Nagpal, P., & Peto, C. A. (1991). Phenotypic and genetic analysis of det2, a new mutant that affects light-regulated seedling development in *Arabidopsis*. *Plant Cell, 3*, 445–459.

Chuck, G., Meeley, R. B., & Hake, S. (1998). The control of maize spikelet meristem fate by the APETALA2-like gene indeterminate spikelet1. *Genes & Development, 12*, 1145–1154.

Chuck, G., Meeley, R., & Hake, S. (2008). Floral meristem initiation and meristem cell fate are regulated by the maize AP2 genes ids1 and sid1. *Development, 135*, 3013–3019.

Chuck, G., Cigan, A. M., Saeteurn, K., & Hake, S. (2007). The heterochronic maize mutant corngrass1 results from overexpression of a tandem microRNA. *Nature Genetics, 39*, 544–549.

Chuck, G., Whipple, C., Jackson, D., & Hake, S. (2010). The maize SBP-box transcription factor encoded by tasselsheath4 regulates bract development and the establishment of meristem boundaries. *Development, 137*, 1243–1250.

Chuck, G., Muszynski, M., Kellogg, E., Hake, S., & Schmidt, R. J. (2002). The control of spikelet meristem identity by the branched silkless1 gene in maize. *Science, 298*, 1238–1241.

Chuck, G., Meeley, R., Irish, E., Sakai, H., & Hake, S. (2007). The maize tasselseed4 microRNA controls sex determination and meristem cell fate by targeting tasselseed6/ indeterminate spikelet1. *Nature Genetics, 39*, 1517–1521.

Ciaffi, M., Paolacci, A. R., Tanzarella, O. A., & Porceddu, E. (2011). Molecular aspects of flower development in grasses. *Sexual Plant Reproduction, 24*, 247–282.

Clark, R. M., Wagler, T. N., Quijada, P., & Doebley, J. (2006). A distant upstream enhancer at the maize domestication gene tb1 has pleiotropic effects on plant and inflorescent architecture. *Nature Genetics, 38*, 594–597.

Clark, S. E., Williams, R. W., & Meyerowitz, E. M. (1997). The CLAVATA1 gene encodes a putative receptor kinase that controls shoot and floral meristem size in *Arabidopsis*. *Cell, 89*, 575–585.

Coen, E. S., & Meyerowitz, E. M. (1991). The war of the whorls: genetic interactions controlling flower development. *Nature, 353*, 31–37.

DeLong, A., Calderon-Urrea, A., & Dellaporta, S. L. (1993). Sex determination gene TASSELSEED2 of maize encodes a short-chain alcohol dehydrogenase required for stage-specific floral organ abortion. *Cell, 74*, 757–768.

Dharmasiri, N., Dharmasiri, S., & Estelle, M. (2005). The F-box protein TIR1 is an auxin receptor. *Nature, 435*, 441–445.

Egea-Cortines, M., Saedler, H., & Sommer, H. (1999). Ternary complex formation between the MADS-box proteins SQUAMOSA, DEFICIENS and GLOBOSA is involved in the control of floral architecture in *Antirrhinum majus*. *Embo Journal*, *18*, 5370–5379.
Emerson, R. A., & Emerson, S. H. (1922). Genetic interrelations of two andromonoecious types of maize, dwarf and anther ear. *Genetics*, *7*, 203–236.
Eveland, A. L., Goldshmidt, A., Pautler, M., Morohashi, K., Liseron-Monfils, C., Lewis, M.W., et al. (2014). Regulatory modules controlling maize inflorescence architecture. *Genome Research*, *24*(3), 431–443.
Eyster, W. H. (1921). The linkage relations between the factors for tunicate ear and starchy-sugary endosperm in maize. *Genetics*, *6*, 209–240.
Fan, H.Y., Hu, Y., Tudor, M., & Ma, H. (1997). Specific interactions between the K domains of AG and AGLs, members of the MADS domain family of DNA binding proteins. *The Plant Journal*, *12*, 999–1010.
Ferrandiz, C., Gu, Q., Martienssen, R., & Yanofsky, M. F. (2000). Redundant regulation of meristem identity and plant architecture by FRUITFULL, APETALA1 and CAULIFLOWER. *Development*, *127*, 725–734.
Fletcher, J. C., Brand, U., Running, M. P., Simon, R., & Meyerowitz, E. M. (1999). Signaling of cell fate decisions by CLAVATA3 in *Arabidopsis* shoot meristems. *Science*, *283*, 1911–1914.
Forestan, C., Farinati, S., & Varotto, S. (2012). The maize PIN gene family of auxin transporters. *Frontiers in Plant Science*, *3*, 16.
Friml, J., Yang, X., Michniewicz, M., Weijers, D., Quint, A., Tietz, O., et al. (2004). A PINOID-dependent binary switch in apical-basal PIN polar targeting directs auxin efflux. *Science*, *306*, 862–865.
Gallavotti, A., Barazesh, S., Malcomber, S., Hall, D., Jackson, D., Schmidt, R. J., et al. (2008). sparse inflorescence1 encodes a monocot-specific YUCCA-like gene required for vegetative and reproductive development in maize. *Proceedings of the National Academy Sciences of the United States of America*, *105*, 15196–15201.
Gallavotti, A., Long, J. A., Stanfield, S., Yang, X., Jackson, D., Vollbrecht, E., et al. (2010). The control of axillary meristem fate in the maize ramosa pathway. *Development*, *137*, 2849–2856.
Gallavotti, A., Yang, Y., Schmidt, R. J., & Jackson, D. (2008). The Relationship between auxin transport and maize branching. *Plant Physiology*, *147*, 1913–1923.
Gallavotti, A., Zhao, Q., Kyozuka, J., Meeley, R. B., Ritter, M. K., Doebley, J. F., et al. (2004). The role of barren stalk1 in the architecture of maize. *Nature*, *432*, 630–635.
Galweiler, L., Guan, C., Muller, A., Wisman, E., Mendgen, K., Yephremov, A., et al. (1998). Regulation of polar auxin transport by AtPIN1 in *Arabidopsis* vascular tissue. *Science*, *282*, 2226–2230.
Goto, K., & Meyerowitz, E. M. (1994). Function and regulation of the *Arabidopsis* floral homeotic gene PISTILLATA. *Genes & Development*, *8*, 1548–1560.
Gray, W. M., Kepinski, S., Rouse, D., Leyser, O., & Estelle, M. (2001). Auxin regulates SCF(TIR1)-dependent degradation of AUX/IAA proteins. *Nature*, *414*, 271–276.
Han, J. J., Jackson, D., & Martienssen, R. (2012). Pod corn is caused by rearrangement at the tunicate1 locus. *Plant Cell*, *24*, 2733–2744.
Harberd, N. P., & Freeling, M. (1989). Genetics of dominant gibberellin-insensitive dwarfism in maize. *Genetics*, *121*, 827–838.
Hartwig, T., Chuck, G. S., Fujioka, S., Klempien, A., Weizbauer, R., Potluri, D. P., et al. (2011). Brassinosteroid control of sex determination in maize. *Proceedings of the National Academy of Sciences of the United States of America*, *108*, 19814–19819.
Hay, A., & Tsiantis, M. (2010). KNOX genes: versatile regulators of plant development and diversity. *Development*, *137*, 3153–3165.
Heijmans, K., Morel, P., & Vandenbussche, M. (2012). MADS-box genes and floral development: the dark side. *Journal of Experimental Botany*, *63*, 5397–5404.

Helliwell, C. A., Wood, C. C., Robertson, M., James Peacock, W., & Dennis, E. S. (2006). The *Arabidopsis* FLC protein interacts directly in vivo with SOC1 and FT chromatin and is part of a high-molecular-weight protein complex. *The Plant Journal, 46*, 183–192.
Heuer, S., Hansen, S., Bantin, J., Brettschneider, R., Kranz, E., Lorz, H., et al. (2001). The maize MADS box gene ZmMADS3 affects node number and spikelet development and is co-expressed with ZmMADS1 during flower development, in egg cells, and early embryogenesis. *Plant Physiology, 127*, 33–45.
Hirano, H.Y., Tanaka, W., & Toriba, T. (2014). Grass flower development. *Methods in Molecular Biology, 1110*, 57–84.
Honma, T., & Goto, K. (2001). Complexes of MADS-box proteins are sufficient to convert leaves into floral organs. *Nature, 409*, 525–529.
Hsu, S. U., & Peterson, P. A. (1991). The upper and lower florets of spikelets in maize. *Journal of Genetics and Breeding, 45*, 215–222.
Huala, E., & Sussex, I. M. (1992). LEAFY interacts with floral homeotic genes to regulate *Arabidopsis* floral development. *Plant Cell, 4*, 901–913.
Immink, R. G. H., Kaufmann, K., & Angenent, G. C. (2010). The 'ABC' of MADS domain protein behaviour and interactions. *Seminars in Cell & Developmental Biology, 21*, 87–93.
Irish, E. E., Langdale, J. A., & Nelson, T. M. (1994). Interactions between tassel seed genes and other sex-determining genes in maize. *Developmental Genetics, 15*, 155–171.
Jack, T., Brockman, L. L., & Meyerowitz, E. M. (1992). The homeotic gene APETALA3 of *Arabidopsis thaliana* encodes a MADS box and is expressed in petals and stamens. *Cell, 68*, 683–697.
Jackson, D., Veit, B., & Hake, S. (1994). Expression of maize *KNOTTED1* related homeobox genes in the shoot apical meristem predicts patterns of morphogenesis in the vegetative shoot. *Development, 120*, 405–413.
Jofuku, K. D., den Boer, B. G., Van Montagu, M., & Okamuro, J. K. (1994). Control of *Arabidopsis* flower and seed development by the homeotic gene APETALA2. *Plant Cell, 6*, 1211–1225.
Jones, D. F. (1925). Heritable characters of maize XXIII-silkless. *Journal of Heredity, 16*, 339–341.
Kaplinsky, N. J., & Freeling, M. (2003). Combinatorial control of meristem identity in maize inflorescences. *Development, 130*, 1149–1158.
Kellogg, E. (2000). Molecular and morphological evolution in the *Andropogoneae*. In S. W. L. Jacobs, & J. Everett (Eds.), *Grasses: Systematics and evolution* (pp. 149–158). Collingwood, Vic: Commonwealth Scientific and Industrial Research Organization (CSIRO).
Kellogg, E. A. (2007). Floral displays: genetic control of grass inflorescences. *Current Opinion in Plant Biology, 10*, 26–31.
Kempin, S. A., Savidge, B., & Yanofsky, M. F. (1995). Molecular-basis of the cauliflower phenotype in *Arabidopsis*. *Science, 267*, 522–525.
Kepinski, S., & Leyser, O. (2005). The *Arabidopsis* F-box protein TIR1 is an auxin receptor. *Nature, 435*, 446–451.
Kerstetter, R. A., Laudencia-Chingcuanco, D., Smith, L. G., & Hake, S. (1997). Loss-of-function mutations in the maize homeobox gene, knotted1, are defective in shoot meristem maintenance. *Development, 124*, 3045–3054.
Kinoshita, A., Betsuyaku, S., Osakabe, Y., Mizuno, S., Nagawa, S., Stahl, Y., et al. (2010). RPK2 is an essential receptor-like kinase that transmits the CLV3 signal in *Arabidopsis*. *Development, 137*, 3911–3920.
Krizek, B. A., & Fletcher, J. C. (2005). Molecular mechanisms of flower development: an armchair guide. *Nature Reviews Genetics, 6*, 688–698.
Krogan, N. T., Hogan, K., & Long, J. A. (2012). APETALA2 negatively regulates multiple floral organ identity genes in *Arabidopsis* by recruiting the co-repressor TOPLESS and the histone deacetylase HDA19. *Development, 139*, 4180–4190.

Langdale, J. A., Irish, E. E., & Nelson, T. M. (1994). Action of the tunicate locus on maize floral development. *Developmental Genetics, 15*, 176–187.

Laudencia-Chingcuanco, D., & Hake, S. (2002). The indeterminate floral apex1 gene regulates meristem determinacy and identity in the maize inflorescence. *Development, 129*, 2629–2638.

Lee, S. H., & Cho, H. T. (2006). PINOID positively regulates auxin efflux in *Arabidopsis* root hair cells and tobacco cells. *Plant Cell, 18*, 1604–1616.

Litt, A. (2007). An evaluation of A-function: evidence from the APETALA1 and APETALA2 gene lineages. *International Journal of Plant Sciences, 168*, 73–91.

Liu, F., Cui, X., Horner, H. T., Weiner, H., & Schnable, P. S. (2001). Mitochondrial aldehyde dehydrogenase activity is required for male fertility in maize. *Plant Cell, 13*, 1063–1078.

Long, J. A., Moan, E. I., Medford, J. I., & Barton, M. K. (1996). A member of the KNOTTED class of homeodomain proteins encoded by the STM gene of *Arabidopsis*. *Nature, 379*, 66–69.

Long, J. A., Ohno, C., Smith, Z. R., & Meyerowitz, E. M. (2006). TOPLESS regulates apical embryonic fate in *Arabidopsis*. *Science, 312*, 1520–1523.

Mandel, M. A., Gustafson-Brown, C., Savidge, B., & Yanofsky, M. F. (1992). Molecular characterization of the *Arabidopsis* floral homeotic gene APETALA1. *Nature, 360*, 273–277.

Mayer, K. F., Schoof, H., Haecker, A., Lenhard, M., Jurgens, G., & Laux, T. (1998). Role of WUSCHEL in regulating stem cell fate in the *Arabidopsis* shoot meristem. *Cell, 95*, 805–815.

McConnell, J. R., & Barton, M. K. (1998). Leaf polarity and meristem formation in *Arabidopsis*. *Development, 125*, 2935–2942.

McSteen, P., Malcomber, S., Skirpan, A., Lunde, C., Wu, X., Kellogg, E., et al. (2007). barren inflorescence2 Encodes a co-ortholog of the PINOID serine/threonine kinase and is required for organogenesis during inflorescence and vegetative development in maize. *Plant Physiology, 144*, 1000–1011.

Mena, M., Mandel, M. A., Lerner, D. R., Yanofsky, M. F., & Schmidt, R. J. (1995). A characterization of the MADS-box gene family in maize. *The Plant Journal, 8*, 845–854.

Mena, M., Ambrose, B. A., Meeley, R. B., Briggs, S. P., Yanofsky, M. F., & Schmidt, R. J. (1996). Diversification of C-function activity in maize flower development. *Science, 274*, 1537–1540.

Meyer, R. S., & Purugganan, M. D. (2013). Evolution of crop species: genetics of domestication and diversification. *Nature Reviews Genetics, 14*, 840–852.

Michniewicz, M., Zago, M. K., Abas, L., Weijers, D., Schweighofer, A., Meskiene, I., et al. (2007). Antagonistic regulation of PIN phosphorylation by PP2A and PINOID directs auxin flux. *Cell, 130*, 1044–1056.

Millar, A. A., & Gubler, F. (2005). The *Arabidopsis* GAMYB-like genes, MYB33 and MYB65, are microRNA-regulated genes that redundantly facilitate anther development. *Plant Cell, 17*, 705–721.

Miwa, H., Betsuyaku, S., Iwamoto, K., Kinoshita, A., Fukuda, H., & Sawa, S. (2008). The receptor-like kinase SOL2 mediates CLE signaling in *Arabidopsis*. *Plant & Cell Physiology, 49*, 1752–1757.

Muller, R., Bleckmann, A., & Simon, R. (2008). The receptor kinase CORYNE of *Arabidopsis* transmits the stem cell-limiting signal CLAVATA3 independently of CLAVATA1. *Plant Cell, 20*, 934–946.

Munster, T., Deleu, W., Wingen, L. U., Ouzunova, M., Cacharron, J., Faigl, W., et al. (2002). Maize MADS-box genes galore. *Maydica, 47*, 287–301.

Nemhauser, J. L., & Chory, J. (2004). Bring it on: new insights into the mechanism of brassinosteroid action. *Journal of Experimental Botany, 55*, 265–270.

Ng, M., & Yanofsky, M. F. (2001). Function and evolution of the plant MADS-box gene family. *Nature Reviews Genetics, 2*, 186–195.

Nickerson, N. H. (1959). Sustained treatment with gibberellin acid of five different kinds of maize. *Annals of the Missouri Botanical Garden, 47,* 19–37.

Nonhebel, H., Yuan, Y., Al-Amier, H., Pieck, M., Akor, E., Ahamed, A., et al. (2011). Redirection of tryptophan metabolism in tobacco by ectopic expression of an *Arabidopsis* indolic glucosinolate biosynthetic gene. *Phytochemistry, 72,* 37–48.

Ohmori, S., Kimizu, M., Sugita, M., Miyao, A., Hirochika, H., Uchida, E., et al. (2009). MOSAIC FLORAL ORGANS1, an AGL6-like MADS box gene, regulates floral organ identity and meristem fate in rice. *Plant Cell, 21,* 3008–3025.

Okada, K., Ueda, J., Komaki, M. K., Bell, C. J., & Shimura, Y. (1991). Requirement of the auxin polar transport system in early stages of *Arabidopsis* floral bud formation. *Plant Cell, 3,* 677–684.

Okamuro, J. K., Szeto, W., Lotys-Prass, C., & Jofuku, K. D. (1997). Photo and hormonal control of meristem identity in the *Arabidopsis* flower mutants apetala2 and apetala1. *Plant Cell, 9,* 37–47.

Peng, J., Richards, D. E., Hartley, N. M., Murphy, G. P., Devos, K. M., Flintham, J. E., et al. (1999). 'Green revolution' genes encode mutant gibberellin response modulators. *Nature, 400,* 256–261.

Petrasek, J., & Friml, J. (2009). Auxin transport routes in plant development. *Development, 136,* 2675–2688.

Phillips, K. A., Skirpan, A. L., Liu, X., Christensen, A., Slewinski, T. L., Hudson, C., et al. (2011). Vanishing tassel2 encodes a grass-specific tryptophan aminotransferase required for vegetative and reproductive development in maize. *Plant Cell, 23,* 550–566.

Phinney, B. O. (1956). Growth response of single-gene dwarf mutants in maize to gibberellic acid. *Proceedings of the National Academy of Sciences of the United States of America, 42,* 185–189.

Phinney, B. O., & West, C. A. (1960). Gibberellins as native plant growth regulators. *Annual Review of Plant Physiology, 11,* 411–436.

Postlethewait, S. N., & Nelson, O. E. (1964). Characterization of development in maize through use of mutants. I. Polytypic (Pt) + Ramosa-1 (Ra1) Mutants. *American Journal of Botany, 51,* 238.

Reiser, L., Sanchez-Baracaldo, P., & Hake, S. (2000). Knots in the family tree: evolutionary relationships and functions of knox homeobox genes. *Plant Molecular Biology, 42,* 151–166.

Ritter, M. K., Padilla, C. M., & Schmidt, R. J. (2002). The maize mutant barren stalk1 is defective in axillary meristem development. *American Journal of Botany, 89,* 203–210.

Satoh-Nagasawa, N., Nagasawa, N., Malcomber, S., Sakai, H., & Jackson, D. (2006). A trehalose metabolic enzyme controls inflorescence architecture in maize. *Nature, 441,* 227–230.

Schoof, H., Lenhard, M., Haecker, A., Mayer, K. F., Jurgens, G., & Laux, T. (2000). The stem cell population of *Arabidopsis* shoot meristems is maintained by a regulatory loop between the CLAVATA and WUSCHEL genes. *Cell, 100,* 635–644.

Skirpan, A., Wu, X., & McSteen, P. (2008). Genetic and physical interaction suggest that BARREN STALK 1 is a target of BARREN INFLORESCENCE2 in maize inflorescence development. *The Plant Journal, 55,* 787–797.

Skirpan, A., Culler, A. H., Gallavotti, A., Jackson, D., Cohen, J. D., & McSteen, P. (2009). BARREN INFLORESCENCE2 interaction with ZmPIN1a suggests a role in auxin transport during maize inflorescence development. *Plant & Cell Physiology, 50,* 652–657.

Stepanova, A. N., Robertson-Hoyt, J., Yun, J., Benavente, L. M., Xie, D.Y., Dolezal, K., et al. (2008). TAA1-mediated auxin biosynthesis is essential for hormone crosstalk and plant development. *Cell, 133,* 177–191.

Szemenyei, H., Hannon, M., & Long, J. A. (2008). TOPLESS mediates auxin-dependent transcriptional repression during *Arabidopsis* embryogenesis. *Science, 319,* 1384–1386.

Taguchi-Shiobara, F., Yuan, Z., Hake, S., & Jackson, D. (2001). The fasciated ear2 gene encodes a leucine-rich repeat receptor-like protein that regulates shoot meristem proliferation in maize. *Genes & Development*, *15*, 2755–2766.

Tan, X., Calderon-Villalobos, L. I., Sharon, M., Zheng, C., Robinson, C. V., Estelle, M., et al. (2007). Mechanism of auxin perception by the TIR1 ubiquitin ligase. *Nature*, *446*, 640–645.

Tanaka, H., Dhonukshe, P., Brewer, P. B., & Friml, J. (2006). Spatiotemporal asymmetric auxin distribution: a means to coordinate plant development. *Cellular and Molecular Life Sciences: CMLS*, *63*, 2738–2754.

Tao, Y., Ferrer, J. L., Ljung, K., Pojer, F., Hong, F., Long, J. A., et al. (2008). Rapid synthesis of auxin via a new tryptophan-dependent pathway is required for shade avoidance in plants. *Cell*, *133*, 164–176.

Theissen, G. (2001). Development of floral organ identity: stories from the MADS house. *Current Opinion in Plant Biology*, *4*, 75–85.

Thompson, B. E., & Hake, S. (2009). Translational biology: from *Arabidopsis* flowers to grass inflorescence architecture. *Plant Physiology*, *149*, 38–45.

Thompson, B. E., Bartling, L., Whipple, C., Hall, D. H., Sakai, H., Schmidt, R., et al. (2009). Bearded-ear encodes a MADS box transcription factor critical for maize floral development. *Plant Cell*, *21*, 2578–2590.

Tivendale, N. D., Davies, N. W., Molesworth, P. P., Davidson, S. E., Smith, J. A., Lowe, E. K., et al. (2010). Reassessing the role of N-hydroxytryptamine in auxin biosynthesis. *Plant Physiology*, *154*, 1957–1965.

Tsuda, K., Ito, Y., Sato, Y., & Kurata, N. (2011). Positive autoregulation of a KNOX gene is essential for shoot apical meristem maintenance in rice. *Plant Cell*, *23*, 4368–4381.

Ulmasov, T., Murfett, J., Hagen, G., & Guilfoyle, T. J. (1997). Aux/IAA proteins repress expression of reporter genes containing natural and highly active synthetic auxin response elements. *Plant Cell*, *9*, 1963–1971.

Vollbrecht, E., Reiser, L., & Hake, S. (2000). Shoot meristem size is dependent on inbred background and presence of the maize homeobox gene, *knotted1*. *Development*, *127*, 3161–3172.

Vollbrecht, E., Springer, P. S., Goh, L., Buckler, E. S. t., & Martienssen, R. (2005). Architecture of floral branch systems in maize and related grasses. *Nature*, *436*, 1119–1126.

Wang, H., Nussbaum-Wagler, T., Li, B., Zhao, Q., Vigouroux, Y., Faller, M., et al. (2005). The origin of the naked grains of maize. *Nature*, *436*, 714–719.

Wang, L., Kim, J., & Somers, D. E. (2013). Transcriptional corepressor TOPLESS complexes with pseudoresponse regulator proteins and histone deacetylases to regulate circadian transcription. *Proceedings of the National Academy of Sciences of the United States of America*, *110*, 761–766.

Wang, Y., Deng, D., Bian, Y., Lv, Y., & Xie, Q. (2010). Genome-wide analysis of primary auxin-responsive Aux/IAA gene family in maize (*Zea mays* L.). *Molecular Biology Reports*, *37*, 3991–4001.

Wang, Y., Deng, D., Shi, Y., Miao, N., Bian, Y., & Yin, Z. (2012). Diversification, phylogeny and evolution of auxin response factor (ARF) family: insights gained from analyzing maize ARF genes. *Molecular Biology Reports*, *39*, 2401–2415.

Weigel, D., Alvarez, J., Smyth, D. R., Yanofsky, M. F., & Meyerowitz, E. M. (1992). LEAFY controls floral meristem identity in *Arabidopsis*. *Cell*, *69*, 843–859.

Whipple, C. J., Ciceri, P., Padilla, C. M., Ambrose, B. A., Bandong, S. L., & Schmidt, R. J. (2004). Conservation of B-class floral homeotic gene function between maize and *Arabidopsis*. *Development*, *131*, 6083–6091.

Whipple, C. J., Hall, D. H., DeBlasio, S., Taguchi-Shiobara, F., Schmidt, R. J., & Jackson, D. P. (2010). A conserved mechanism of bract suppression in the grass family. *Plant Cell*, *22*, 565–578.

Wills, D. M., Whipple, C. J., Takuno, S., Kursel, L. E., Shannon, L. M., Ross-Ibarra, J., et al. (2013). From many, one: genetic control of prolificacy during maize domestication. *PLoS Genet*, *9*, e1003604.

Wingen, L. U., Munster, T., Faigl, W., Deleu, W., Sommer, H., Saedler, H., et al. (2012). Molecular genetic basis of pod corn (Tunicate maize). *Proceedings of the National Academy of Sciences of the United States of America*, *109*, 7115–7120.

Wu, M. F., Tian, Q., & Reed, J. W. (2006). *Arabidopsis* microRNA167 controls patterns of ARF6 and ARF8 expression, and regulates both female and male reproduction. *Development*, *133*, 4211–4218.

Wu, X., & McSteen, P. (2007). The role of auxin transport during inflorescence development in maize (*Zea mays*, Poaceae). *American Journal of Botany*, *94*, 1745–1755.

Wu, X., Skirpan, A., & McSteen, P. (2009). Suppressor of sessile spikelets1 functions in the ramosa pathway controlling meristem determinacy in maize. *Plant Physiology*, *149*, 205–219.

Xing, H., Pudake, R. N., Guo, G., Xing, G., Hu, Z., Zhang, Y., et al. (2011). Genome-wide identification and expression profiling of auxin response factor (ARF) gene family in maize. *BMC Genomics*, *12*, 178.

Yamada, M., Greenham, K., Prigge, M. J., Jensen, P. J., & Estelle, M. (2009). The TRANSPORT INHIBITOR RESPONSE2 gene is required for auxin synthesis and diverse aspects of plant development. *Plant Physiology*, *151*, 168–179.

Yamaguchi, T., Lee, D.Y., Miyao, A., Hirochika, H., An, G., & Hirano, H.Y. (2006). Functional diversification of the two C-class MADS box genes OSMADS3 and OSMADS58 in *Oryza sativa*. *Plant Cell*, *18*, 15–28.

Yan, Y., Christensen, S., Isakeit, T., Engelberth, J., Meeley, R., Hayward, A., et al. (2012). Disruption of OPR7 and OPR8 reveals the versatile functions of jasmonic acid in maize development and defense. *Plant Cell*, *24*, 1420–1436.

Yanofsky, M. F., Ma, H., Bowman, J. L., Drews, G. N., Feldmann, K. A., & Meyerowitz, E. M. (1990). The protein encoded by the *Arabidopsis* homeotic gene agamous resembles transcription factors. *Nature*, *346*, 35–39.

Zhao, Y., Li, X. Y., Chen, W. J., Peng, X. J., Cheng, X. A., Zhu, S. W., et al. (2011). Whole-genome survey and characterization of MADS-box gene family in maize and sorghum. *Plant Cell, Tissue and Organ Culture*, *105*, 159–173.

CHAPTER TEN

Molecular Control of Inflorescence Development in Asteraceae

Suvi K. Broholm, Teemu H. Teeri and Paula Elomaa[1]
Department of Agricultural Sciences, University of Helsinki, Helsinki, Finland
[1]Corresponding author: e-mail address: paula.elomaa@helsinki.fi

Contents

Abstract

Flowers and their arrangement in inflorescences have a key role for the reproductive success of plants as well as for crop yield. An example of the huge variation in floral and inflorescence structures is seen in the sunflower family, Asteraceae that bears complex inflorescences (capitula) that are comprised of structurally and functionally different types of flowers. In contrast to the classical models such as *Arabidopsis*, *Antirrhinum* or *Petunia* with single flower forms in their inflorescences, very little is known of the mechanisms regulating the capitulum organisation in Asteraceae. Molecular studies have mostly focused on *Gerbera hybrida* (Mutisieae), *Helianthus annuus* (Heliantheae) and *Senecio* sp. (Senecioneae) as models. Functional studies indicate that the duplicated *CYCLOIDEA (CYC)/TEOSINTE BRANCHED1(TB1)*-like genes encoding TCP domain transcription factors have evolved novel functions in Asteraceae to regulate flower type

Advances in Botanical Research, Volume 72
ISSN 0065-2296
http://dx.doi.org/10.1016/B978-0-12-417162-6.00010-9

identity but they also perform late functions regulating flower organ growth. MADS-box genes basically follow the 'standard' ABCE model in regulating organ identities in Asteraceae, although the presence of an A function cannot be affirmed. Moreover, examples of sub- or neo-functionalisation of duplicated MADS-box gene family members are apparent. In summary, a few key components that regulate Asteraceae flower and inflorescence development have already been discovered. However, the detailed molecular networks involved in patterning the complex Asteraceae inflorescence, flower primordia initiation and coordination of their differential fate and growth are still largely unresolved.

10.1 INTRODUCTION

Flowers and their arrangement within inflorescences have a central role for the reproductive fitness of plants as well as for human sustenance. The underlying molecular mechanisms connected with flower and inflorescence development in angiosperms have mostly been studied using the classical eudicot models, *Arabidopsis thaliana*, *Antirrhinum majus* and *Petunia hybrida*, while *Oryza sativa* and *Zea mays* represent the most studied models in monocots. Molecular studies in diverse plant lineages have suggested that evolution of regulatory mechanisms occurred through gene and whole genome duplication events as well as convergent co-option of the key developmental control genes and are behind the enormous morphological variation apparent in the floral and inflorescence forms in nature (reviewed by Preston, Hileman, & Cubas, 2011; Rosin & Kramer, 2009). Evolutionary developmental biology aims at explaining the diversity in animal and plant body plans. By comparing the molecular basis of development in various species, genetic similarities or differences can be linked with the observed morphological variation. Changes in the function of key developmental genes are important for generating morphological novelty. Such changes may result from altered spatial or temporal expression domains due to changes in *cis*-regulatory regions of the genes and/or by changes in *trans*-acting regulatory proteins. Furthermore, modifications in the protein coding sequences may affect protein partner or target gene specificity. Comparative studies using species representing various plant lineages are a prerequisite for comprehensive understanding of the development of diverse floral forms during evolution of plants.

In this chapter, we summarise the molecular work conducted to enlighten the development of the complex structure of Asteraceae inflorescences. Asteraceae represents a highly derived plant family that comprises approximately 10% of all angiosperm species and includes many economically important crop plants (sunflower, lettuce, safflower, artichoke, endive) as well as ornamentals (gerbera, chrysanthemum, *Osteospermum*, *Tagetes*, marigold). It is a relatively young family that shows rapid tribal radiation connected with whole genome duplication events during the last 40 My (Barker et al., 2008; Kim, Choi, & Jansen, 2005). Analysis of over 1 million expressed sequence tags (ESTs) from 18 species indicated a single paleopolyploidisation event at the base of Asteraceae as well as independent whole genome duplication events in the tribes Heliantheae and Mutisieae that may have promoted diversification of gene duplicates through sub- and/or neofunctionalisation (Barker et al., 2008). The apparent reason for the evolutionary success of this family relates to its showy inflorescence, the capitulum that provides an attractive pollination target. The capitulum in Asteraceae is a pseudanthium, 'false flower' that resembles a single, solitary flower, but is comprised of different types of flowers with specialised roles. This feature distinguishes Asteraceae from the conventional model species that bear only single flower forms in their inflorescences. Molecular studies have mostly focused on gerbera (*Gerbera hybrida*), groundsel (*Senecio vulgaris*) and sunflower (*Helianthus annuus*), and to some extent on chrysanthemum (*Chrysanthemum* sp.), as models to understand how floral polymorphy is determined and what specifies flower type identity and directs their differential growth. Gene discovery has greatly been facilitated by the rapid progress in genome sequencing and transcriptome analyses in these species (e.g. http://compgenomics.ucdavis.edu/index.php; Barker et al., 2008; Hegarty et al., 2005; Hodgins et al., 2014; Huang et al., 2012; Kane et al., 2011; Kuang, Li, Peng, Sun, & Wang, 2013; Laitinen et al., 2005; Wang, Jiang, et al., 2013; Wang, Ma, et al., 2013). Although genetic transformation protocols have been developed for many Asteraceous species, they still seem to be distinctly inefficient when compared to the classical models such as *Arabidopsis* in terms of high-throughput functional studies. In this chapter, we summarise the studies focusing on the functional roles of TCP and MADS domain transcription factors in regulating Asteraceae inflorescence development as well as discuss the nature of the unknown signals that primarily define the early stages of inflorescence development and the differential fate of the flower types.

10.2 STRUCTURE AND EVOLUTION OF THE CAPITULUM

Asteraceae is among the largest families of plants with more than 23,000 species that occupy all the continents except Antarctica (Jeffrey, 2007). The most distinctive feature of Asteraceae is their inflorescence structure, the highly compressed inflorescence branch system called the capitulum or flower head, in which all the flowers are attached to a receptacle that is surrounded by involucral bracts. The capitulum forms a pseudanthium, a structure that resembles a single large flower as shown for gerbera and sunflower in Figure 10.1. Other characters that are diagnostic to the family include anthers united in a tube and inferior ovary position (Harris, 1995).

Figure 10.1 Flower heads in *Gerbera hybrida* (top) and *Helianthus annuus* (below) are compressed structures comprised of distinct flower types. The marginal ray flowers (RF) are large and bilaterally symmetrical. The gerbera ray flowers are female while those in sunflower are neutral in sex. The ray flowers have showy ventral ligule (vl) formed by three fused petals while the two dorsal petals are rudimentary (dp). Gerbera has intermediate trans flowers (TF) similar to ray flowers but just smaller in size. In both species, the central disc flowers (DF) are hermaphrodite and bear carpels and pollen-producing stamens. In gerbera the disc flowers are gradually changing from bilateral symmetry towards radial symmetry the more central their position in the capitulum is, while in sunflower all the disc flowers are fully radially symmetrical (tubular). (See the colour plate.)

Modification of the outer floral whorl into pappus bristles that take part in seed dispersal is also a highly common character within the family.

The Asteraceae flower heads can be homogamous or heterogamous. In homogamous heads all the flowers are similar, whereas in heterogamous heads, flowers in the margins of the head are morphologically and functionally distinct from the centrally located flowers (Figure 10.1). The marginal flowers of the heterogamous heads can be sterile or female and they have showy petals with zygomorphic (bilateral) symmetry. The central disc flowers are hermaphrodite with functional stamens and carpels while their petals are modest and most often actinomorphic (radially symmetrical). Several studies have shown that the presence of the showy marginal flowers in heterogamous heads is adaptive and increases the attractiveness of the inflorescence to pollinators and affects the rate of outcrossing, thereby increasing genetic diversity and fitness (Andersson, 2008; Marshall & Abbott, 1984; Stuessy, Spooner, & Evans, 1986; Sun & Ganders, 1990).

Both homogamous and heterogamous heads are present in the most basal Asteraceae subfamilies, the Barnadesioideae and Mutisieae (Gillies, Cubas, Coen, & Abbott, 2002). The ancestral Asteraceae corolla (the petal whorl) shape is considered to be tubular (actinomorphic), from which the zygomorphic bilabiate (three outer and two inner petals) and ray (strap-shaped fused ligule with maximum three apical teeth) corolla have evolved (Stuessy & Urtubey, 2006). The bilabiate corolla shape is considered ancestral to the ray corolla since the subfamilies Barnadesioideae and Mutisieae bear bilabiate corollas and marginal, 'true' ray flowers are found only after the split of the rest of the family from these basalmost subfamilies. The inner petals of the bilabiate corolla have been considered relicts from a less condensed inflorescence structure and the ray corolla shape as an adaptation to the condensed capitulum structure (Gillies et al., 2002). However, often in literature and also in this chapter, the term 'ray flower' is used to refer to marginal flowers independently of their morphology.

The Asteraceae capitulum provides a beautiful example of spiral phyllotaxis in which the number of spirals follows the Fibonacci series (1, 2, 3, 5, 8, 13, each number being the sum of the preceding two). The floral primordia develop in the capitulum so that the newest primordium approaches 137.5° (the golden angle) from the preceding primordium, leading to left and right turning spiral rows (parastichies) of floral primordia (Figure 10.2). The initiation and development of individual flowers in the Asteraceae heads is often considered to be uniformly acropetal, i.e. proceeding from the margins towards the apex. In a broad morphological analysis of early

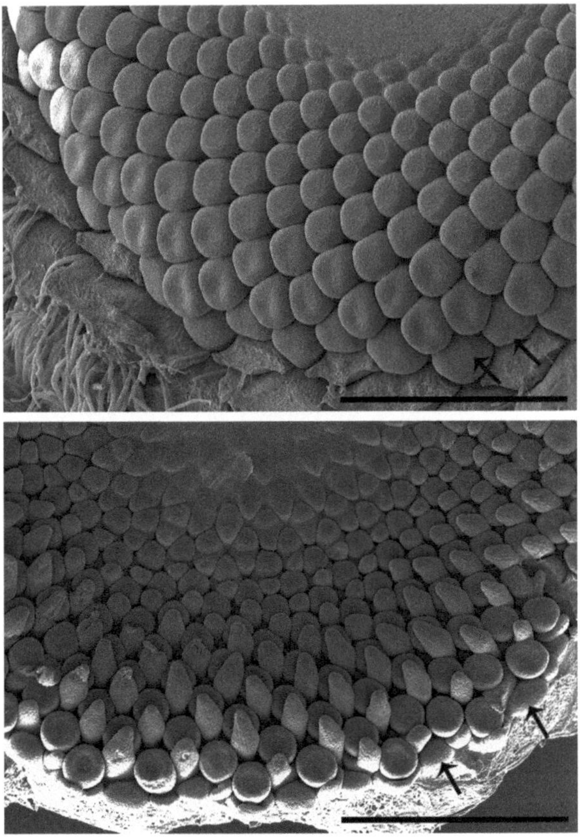

Figure 10.2 Early development of heterogamous heads in *Gerbera hybrida* (top) and *Helianthus annuus* (below). The capitula show spiral phyllotaxis with left and right turning rows of floral primordia. In both species, there is a developmental delay in marginal ray flowers compared to the adjacent trans or disc flowers, respectively. Examples of developing ray flower primordia are marked with arrows in the figures. Scale bars 500 μm.

capitulum development throughout the Asteraceae family, Harris (1995) has shown that the initiation and/or development of floral primordia consistently differs between homogamous and heterogamous capitula. In homogamous capitula, the floral primordia always initiate in an acropetal order. In contrast, in species with heterogamous capitula, the pattern of primordia initiation differs between the flower types. The disc flower primordia are the first ones to initiate and their initiation proceeds in an acropetal manner. The initiation and subsequent development of the marginal ray flower primordia is delayed in comparison to that of the disc flower primordia and proceeds basipetally towards the margins of the inflorescence meristem

(IM) (Bello, Álvarez, Torices, & Fuertes-Aguilar, 2013; Harris, 1991; 1995; Thomas, Rudall, Ellis, Savolainen, & Glover, 2009). A milder version of this pattern, in which the primordia initiation is acropetal but the development of the marginal ray flowers is delayed in comparison to the disc flowers, is also common for heterogamous capitula (Harris, 1995, Figure 10.2).

The correlation between heterogamous heads and the loss of uniformly acropetal primordia initiation and development, as found by the morphological studies of 39 species in 12 tribes of Asteraceae during inflorescence and flower ontogeny by Harris (1995), is intriguing in terms of the evolution of the capitulum structure. Harris' study included one heterogamous species (*Mutisia coccinea*) from the basal subfamily Mutisieae, but unfortunately no species from the basalmost subfamily Barnadesioideae. Our morphological analyses on gerbera (also a member of the subfamily Mutisieae) are in line with Harris' results on *M. coccinea* in that the delay in development of the marginal flowers in comparison to the disc flowers is less pronounced than in heterogamous species of the more derived subfamilies, such as in sunflower (Figure 10.2). Harris interprets this to correlate with the primitive nature of the marginal flowers and the less prominent differences between the flower types in the basal subfamily Mutisieae. To strengthen this hypothesis, it is essential to study the ontogeny of floral initiation and development in heterogamous species of Barnadesioideae.

Recent botanical studies have raised new hypotheses related to IM organisation in Asteraceae. Claßen-Bockhoff & Bull-Hereñu (2013) present an ontogeny-based concept describing the various inflorescence types. They propose a concept of 'floral unit meristem' (FUM). In their model, two types of IMs can be identified: IMs that produce flower meristems in an acropetal manner (such as the *Arabidopsis* raceme) and FUMs that give rise to dense, often flower-like inflorescence structures (such as the Asteraceae flower heads). Based on morphological data, they conclude that FUM resembles more floral meristems (FMs) than IMs. Both FMs and FUMs are determinate (with histologically visible mantle core organisation), development occurs in centripetal direction, internodes are inhibited and both meristems are enlarged, dome-shaped or flat and initially naked (FM without flower organs or FUM without flower primordia, respectively). They propose that FUM evolved from FMs. Molecular studies from a developmental perspective may help to clarify this view.

Another interesting view on Asteraceae inflorescence structure is based on inflorescence morphologies of the families that are basal to Asteraceae, including Menyanthaceae, Goodeniaceae and Calycereceae that together form a phylogenetically well-supported clade (the MGCA-clade) within Asterales (Lundberg & Bremer, 2003; Tank & Donoghue, 2010). Pozner, Zanotti, and Johnson (2012) presented a morphological characterisation of young developing inflorescences of several Calyceraceae species (a sister family of Asteraceae; Lundberg & Bremer, 2003) and uncovered that the basic inflorescence structure of Calyceraceae is a very condensed thyrsoid that combines both racemose and cymose branching patterns. The peripheral flowers are organised into a cymose pattern whereas the central flowers have a racemose pattern. Calyceraceae inflorescences show no further differentiation between peripheral and central flowers, in contrast to the heterogamous inflorescences in Asteraceae. Pozner et al. (2012) propose that the Asteraceae inflorescences could have remnants of the cymose and racemose developmental controls that are present in Calyceraceae, and that this could be related to the delayed and basipetal development of ray flowers in the periphery and the acropetal development of disc flowers in the heterogamous Asteraceae inflorescences. Furthermore, they postulate that differences between the peripheral and central parts of the IM may be the evolutionary basis for the consequent development of distinct flower types in Asteraceae. Molecular evidence for these suggestions remains to be elucidated.

10.3 CYCLOIDEA/TEOSINTE BRANCHED1-LIKE TCP DOMAIN TRANSCRIPTION FACTORS IN REGULATION OF CAPITULUM DEVELOPMENT IN ASTERACEAE

The major morphological trait that distinguishes individual flower types in Asteraceae is their symmetry. Typically, the marginal ray flowers are strongly bilaterally symmetrical while the centremost disc flowers are radially symmetrical. The change in symmetry may be very sharp as observed in sunflower with functionally neutral, strongly bilateral ray flowers and hermaphrodite and radially symmetrical, tubular disc flowers (Figure 10.1). In gerbera, the change in the symmetry occurs gradually along the capitulum radius and many cultivars occupy intermediate, bilaterally symmetrical trans flowers between the marginal and central flowers. Furthermore, in many cases most of the disc flowers are bilaterally symmetrical and in fact only few of the centremost ones are (almost) radially symmetrical (Figure 10.1). Coen et al. (1995) pointed out the resemblance of the Asteraceae capitula with *centroradialis*-like (*cen*) mutant in *A. majus*. In *cen*, a fully radially symmetrical flower is terminating the growth of the inflorescence axis while the flowers in the axillary positions remain asymmetric. The radially symmetrical terminal flower in *cen* is similar to axillary flowers in a *cycloidea* mutant suggesting that *cyc* is not active in the apical meristem (Coen et al., 1995). The similarity to Asteraceae capitula with symmetrical, central flowers and bilaterally symmetrical peripheral flowers around the inflorescence apex raised the hypothesis that CYCLOIDEA (CYC)-like TCP domain transcription factors may control flower type identity in Asteraceae (Coen et al., 1995; reviewed in Gillies et al., 2002).

Bilateral symmetry (monosymmetry) of flowers has evolved multiple times in diverse angiosperm lineages from radially symmetrical ancestors (Endress, 1999; 2001) and has been shown to be regulated by the *CYC*-like genes (reviewed by Busch & Zachgo, 2009; Martín-Trillo & Cubas, 2010; Preston & Hileman, 2009). Asteraceae, together with orchids, legumes, Dipsacales and Lamiales are the most successful clades with predominantly monosymmetric flowers (Endress, 2001). The classical studies in *Antirrhinum,* where the first to show that the *CYC* gene and its paralog *DICHOTOMA* (*DICH*) are the key genes that act partially redundantly to establish bilateral symmetry of flowers (Luo et al., 1999; Luo, Carpenter, Vincent, Copsey, & Coen, 1996). *CYC* and *DICH* are expressed in the dorsal domain of the flowers where they modulate development of the dorsal petals as well as arrest of the dorsal most (odd) stamen development (Luo et al., 1999, 1996). Interestingly, *CYC* reduces the growth of the dorsal domain of the FM at early developmental stages while later on, it increases the growth rate of the petals (Luo et al., 1996). *DICH* expression instead, is localised to the dorsal half of the dorsal petals to establish internal asymmetry (Luo et al., 1999). Both *CYC* and *DICH* encode TCP domain regulatory proteins that are ancient plant-specific proteins, which contain a conserved basic–helix–loop–helix motif (TCP domain) involved in DNA binding and interaction with other proteins (Cubas, Lauter, Doebley, & Coen, 1999; Kosugi & Ohashi, 1997). TCP proteins have been named after the first identified members of the gene family; *TEOSINTE BRANCHED1* (*TB1*) of maize (Doebley, Stec, & Hubbard, 1997), *CYC* and *PROLIFERATING CELL FACTOR (PCF1/2*) of rice (Kosugi & Ohashi, 1997). Phylogenetically PCF clade proteins form one subfamily (class I) while the CYC/TB1 clade and CINCINNATA (CIN) clade proteins form the second subfamily (class II) (Aguilar-Martinez, Poza-Carrion, & Cubas, 2007). CYC/TB1 clade genes are specific to angiosperms whereas PCF and CIN genes are considered more ancient as they are present also for example in mosses (*Physcomitrella*) and lycophytes (*Selaginella*) (Navaud, Dabos, Carnus, Tremousaygue, & Herve, 2007). Especially the members of the class II TCP transcription factor genes have been shown to control plant growth and development by modulating cell proliferation patterns, acting either as repressors or activators of downstream target genes. However, also other functional roles for TCP factors are emerging (reviewed in Martín-Trillo & Cubas, 2010).

In core eudicots, the CYC/TB1 clade, also called as the 'ECE' clade, has experienced two major duplication events and can be divided into three subclades: CYC1, CYC2 and CYC3 (Howarth & Donoghue, 2006). The *Arabidopsis* genome contains altogether 24 TCP genes, but only three CYC/TB1 clade genes representing each of the three subclades are present (Martín-Trillo & Cubas, 2010). *TCP18/BRC1/TBL1* and *TCP12/BRC2* represent the CYC1 and CYC3 subclades, respectively (Howarth & Donoghue, 2006). Like *TB1* in maize, the *Arabidopsis BRC1* is controlling shoot branching, especially affecting the outgrowth capacity of the buds (Aguilar-Martinez et al., 2007; Finlayson, 2007; Finlayson, Krishnareddy, Kebrom, & Casal, 2010), while *BRC2* functions in coordinating growth among branches (Finlayson et al., 2010). The *Arabidopsis TCP1* belongs to the CYC2 subclade and is most similar to the *Antirrhinum CYC* and *DICH* (Howarth & Donoghue, 2006). Although the *Arabidopsis TCP1* is expressed asymmetrically in the adaxial part of flower meristems, this pattern is not maintained during flower development (Cubas, Coen, & Zapater, 2001), and the *tcp1* mutant does not show any obvious changes in phenotype (Cubas, 2004). This has been thought to account for the development of radially symmetrical flowers in *Arabidopsis* but also, it suggests that incipient asymmetry may have existed in radially symmetrical ancestral flowers and used for generation of asymmetry multiple times during evolution (Cubas et al., 2001). *TCP1* has, however been shown to affect the longitudinal elongation of the petioles, rosette leaves and inflorescence stems in *Arabidopsis* (Koyama, Sato, & Ohme-Takagi, 2010).

Interestingly, both in sunflower and in gerbera (Chapman, Leebens-Mack, & Burke, 2008; Tähtiharju et al., 2012), the *CYC/TB1*-like gene subfamily shows extensive expansion and 10 gene family members have been discovered in these species, dating back to a single paleopolyploidisation event near the origin of the Asteraceae family, and to independent genome duplications in the tribes Heliantheae (sunflower tribe) and Mutisieae (gerbera tribe) (Barker et al., 2008). Therefore, it is reasonable to suspect that these duplicated genes have diverged and experienced sub- and/or neo-functionalisation contributing to the unique characteristics of the Asteraceae inflorescence. Several lines of evidence indicate the involvement of *CYC*-like genes in regulating flower type identity in Asteraceae and will be summarised next.

10.3.1 Gerbera

The first functional studies to verify the role of *CYC*-like genes in regulation of flower type identity were conducted in gerbera (Broholm et al., 2008). Ectopic expression of *GhCYC2*, a member of the CYC2 subclade of *CYC/TB1*-like genes in transgenic gerbera converted disc flowers into ray-like. The petals in disc flowers were longer and showed a ligular structure, i.e. enhanced bilateral symmetry. Furthermore, stamen development was disrupted and the modified flowers did not release any pollen (Broholm et al., 2008). Both traits are typical of the marginal ray flowers in the wild type. In addition, in ray flowers ectopic expression of *GhCYC2* resulted in reduced petal length indicating that *GhCYC2* was able to either promote or suppress growth depending on the site of its expression. The transgenic lines strongly resembled the 'crested' or 'double' gerbera cultivars with enlarged, male sterile trans and disc flowers (Figure 10.3). The crested trait has been shown to be controlled by a single locus corresponding to two different dominant alleles; Cr^d being dominant to *Cr* and both of them dominant to the recessive *cr* allele (Kloos, George, & Sorge, 2004). However, the crested cultivars did not show any differences in *GhCYC2* expression. Furthermore, suppression of *GhCYC2* expression in transgenic gerbera did not affect ray flower development, suggesting that *GhCYC2* alone was not sufficient to regulate flower type identity that most likely involves other redundantly acting genes.

Broholm et al. (2008) showed that the expression of *GhCYC2* followed the radial organisation of the gerbera capitulum and was prominently expressed in emerging ray flower primordia while no expression was detected in the centremost disc flower primordia. Furthermore, in contrast to many *CYC*-like genes in diverse plant lineages with bilaterally symmetrical flowers, *GhCYC2* expression was absent from the dorsal rudimentary petals and showed ventralised pattern concentrating to the large ligule that is formed when three of the petals are fused together. In transgenic lines with suppressed *GhCYC2* expression occasional splitting of trans flower ligules into five to eight separate petals were observed (Broholm et al., 2008). The correlation of the *GhCYC2* expression pattern and the transgenic phenotypes with organ fusion indicated further functional differentiation of *GhCYC2* in gerbera.

Tähtiharju et al. (2012) extended the analysis of the *CYC/TB1* gene family by showing that, as observed in sunflower (Chapman et al., 2008), also gerbera comprises 10 family members, falling into the three subclades

Figure 10.3 Transgenic line that ectopically expresses the gerbera CYC-like gene GhCYC2 (top) highly resembles a gerbera cultivar showing the 'crested' or 'double' phenotype with enlarged, male sterile trans and disc flowers (below). (See the colour plate.)

characterised in core eudicots (Howarth & Donoghue, 2006). Phylogenetic analyses however indicated that the gerbera and sunflower genes result from species-specific gene duplications and losses (Tähtiharju et al., 2012). The results further showed that the CYC2 clade shows the highest number of paralogs both in gerbera and sunflower. Interestingly, all the six CYC2 clade genes in gerbera and the five genes in sunflower showed high level of conservation in their expression patterns and were differentially regulated in the developing flower types. All of them were upregulated in ray flower primordia during early stages of development, suggesting that they may have functions in ray flower differentiation. In the case of gerbera, most of the CYC2 clade genes were also expressed in the intermediate trans flower primordia with the exception of *GhCYC3* whose expression was localised into ray flower primordia only. Pairwise protein–protein interaction studies in yeast indicated the existence of a complex network of interactions involving CYC/TB1-like proteins. Tähtiharju et al. (2012) observed both homodimerisation and heterodimerisation of CYC-like proteins occurring within and between the three subclades. In general, compared to the sunflower proteins, gerbera proteins showed higher capacity for homodimerisation while heterodimer formation, especially for the CYC2 clade proteins, was more similar in both species (Tähtiharju et al., 2012). In addition to apparent early functions during flower type differentiation, the expression patterns of CYC2-like genes indicate that most of them may also have late functions during flower organ development especially in ovaries, carpels and stamens.

10.3.2 Helianthus

Sunflower has heterogamous inflorescences consisting of bilaterally symmetrical, sterile ray flowers and radially symmetrical, hermaphrodite disc flowers. The *chrysanthemoides* (*chry*) or *double* (*dbl*) mutant of sunflower resembles the *crested* mutant described in gerbera. The *chry* and *dbl* mutants bear flowers that are all of ray identity although the zygomorphy is diminishing in the more internal whorls where the flowers still show petal fusion extending up to 80% of the corolla length (Berti, Fambrini, Turi, Bertini, & Pugliesi, 2005; Fambrini, Bertini, & Pugliesi, 2003). In addition to changes in the symmetry of petals, the modified flowers in the mutant show unbranched stigmas, lack ovules and bear small anthers that produce only few pollen grains. In the *tubular ray flower* (*turf*) or *tubular-rayed* (*tub*) mutant instead, changes are restricted to ray flowers that are tubular and almost radially symmetrical showing reduced size of the corolla (Berti et al., 2005; Chapman et al., 2012). The male and female reproductive organs develop as in wild type in these modified marginal flowers while the petal number is altered; instead of three-lobed corollas normally found in ray flowers most corollas are five-lobed as typical of disc flowers. Genetic studies indicate that the *chry* trait is controlled by a single semidominant major gene and an unknown number of modifier genes (Fambrini et al., 2003), while *turf* is controlled by a single recessive gene (Berti et al., 2005). Analysis of the sunflower mutants thus indicates the involvement of both *TURF* and *CHRY* genes in establishing the corolla symmetry in ray and disc flowers but also their involvement in carpel and stamen development.

Molecular analysis of the *turf* mutant by Fambrini, Salvini, & Pugliesi (2011) identified the CYC2 clade gene *HaCYC2c* as a key regulator of ray flower identity. They found a large insertion (5787 bp) in the basic motif of the TCP domain that was a part of a CACTA-like transposon and led to a premature stop codon. The inheritance of the *turf* allele correlated with the segregating phenotypes. Chapman et al. (2012) gave further molecular evidence of the involvement of *HaCYC2c* in regulating flower type identity. Analysis of the *dbl* and *tub* mutants revealed that both phenotypes were caused by independent mutations in *HaCYC2c*. Also the expression analysis by Tähtiharju et al. (2012) suggested that the *HaCYC2c* gene is the potential candidate (together with *HaCYC2d*) regulating ray flower development in sunflower. *HaCYC2c* is very specifically expressed in ray flower primordia during early stages of flower development while its expression is lacking from disc flower primordia at corresponding developmental stages. Genetic mapping by Chapman et al. (2012) indicated that both traits mapped to the same genomic region where three CYC2-clade genes (*HaCYC2b*, *HaCYC2c* and *HaCYC2e*) were previously mapped (Chapman et al., 2008). Sequence analysis of the *HaCYC2c* genes from *dbl* and *tub* mutant lines revealed an insertion upstream of the start codon in *dbl* and an additional transposon insertion in the coding region of the *HaCYC2c-tub* allele. As a consequence, both mutations altered the expression of the *HaCYC2c* gene resulting in ectopic activation of *HaCYC2c* transcription in all flower types of the *dbl* mutant and reduced expression in the *tub* mutant, respectively. Three other *dbl* lines showed the same mutated form of *HaCYC2c* gene indicating that it may have arisen only once and has been later incorporated into multiple sunflower cultivars. However, an independent *tub* mutant line showed a unique retrotransposon insertion in an intron in the 3′ UTR of the gene resulting in complete loss of expression of *HaCYC2c*. Interestingly both *dbl* and *tub* alleles contain the same insertion in their promoter regions indicating that the *tub* allele is derived from the *dbl* allele recently during evolution. Altogether, the data indicate that *HaCYC2c* is the key gene that functions in specifying ray vs disc identity in sunflower (Chapman et al., 2012). Sunflower has altogether five CYC2 clade genes, which have arisen through four independent duplication events (Chapman et al., 2008). All of them are upregulated in emerging ray flower primordia (Tähtiharju et al., 2012) but the functions of other *HaCYC2* genes are still not known. The expression analyses conducted during later developmental stages of various floral tissues, however, suggest functional diversification for duplicated sunflower genes at the level of single flowers (Chapman et al., 2008). Chapman et al. (2008) also give evidence that positive selection has promoted divergence of *HaCYC2a, b* and *c* and characterised four amino acids in the TCP and R domain as being under selection.

Phylogenetic analysis between gerbera and sunflower *CYC*-like genes indicates that the orthologue for the *HaCYC2c* could be the gerbera *GhCYC5*. Both these proteins show similar properties and are, in fact, not able to form homo- or heterodimers with CYC-like proteins in yeast two-hybrid assays (Tähtiharju et al., 2012). Recent analyses in gerbera however indicate functional redundancy between the gene family members and suggest that in addition to *GhCYC2, GhCYC3* and *GhCYC4* are involved in regulating ray flower identity rather than *GhCYC5* (Juntheikki-Palovaara et al., unpublished data). Thereby, it seems that orthologous genes in these two species do not necessarily share the same function.

10.3.3 Senecio

In *S. vulgaris* flower head development shows natural variation that has been linked to interspecific hybridisation events. The diploid species *Senecio squalidus* (the Oxford ragwort) that was introduced to British Isles at the beginning of eighteenth century (Harris, 2002) has been shown to occasionally hybridise with the native, allotetraploid *S. vulgaris* (groundsel) (Marshall & Abbott, 1980). The hybrid variant *S. vulgaris* var. *hibernicus* Syme produces radiate instead of discoid flower heads found in the native *S. vulgaris* var. *vulgaris*, indicating that the radiate condition may have arisen through introgression from *S. squalidus* (Abbott, 1992; Abbott, James, Milne, & Gillies, 2003). The radiate form of *S. vulgaris* can also be produced artificially by crossing *S. vulgaris* var. *vulgaris* and *S. squalidus* (Ingram, Weir, & Abbott, 1980). Genetic studies have shown that this variation is under control of a single genetic locus called *RAY*, so that the radiate condition is dominant to the discoid (Ingram & Taylor, 1982; Trow, 1912).

Kim et al. (2008) isolated *CYC*-like genes from *S. vulgaris* as candidates for the *RAY* locus. Two duplicated genes, *RAY1* and *RAY2,* were specifically expressed in the outer flower primordia, however, in a similar pattern in radiate and non-radiate genotypes. Interestingly, both genes were more highly expressed in the non-radiate genotypes. The genes were shown to be tightly linked and they co-segregated with flower-head phenotypes thus confirming the association with the *RAY* locus. Sequencing of the *RAY1* and *RAY2* from *S. squalidus* further suggested that the radiate form of *S. vulgaris* may have arisen through hybridisation between the species. Constitutive expression of an internal fragment of *RAY1* in the radiate *S. vulgaris* background repressed ray flower development to variable extent while constitutive expression of *RAY2* promoted ventral identity and resulted in development of tubular ray flowers (Kim et al., 2008). Phylogenetic analyses indicate that *RAY2* is the orthologue of gerbera *GhCYC2* (Chapman et al., 2012; Kim et al., 2008; Tähtiharju et al., 2012), however, in contrast to *GhCYC2* in transgenic gerbera, ectopic expression of *RAY2* did not alter disc flower morphology in transgenic *Senecio*.

10.3.4 Functional Diversification of CYC-like Genes in Asteraceae

The functional studies discussed in previous paragraphs that exploited transgenic plants and/or mutants in gerbera, sunflower and *Senecio* show recruitment of CYC2 clade homologues to regulate flower type identity in Asteraceae. The expansion and apparent functional conservation of the CYC2 clade genes is associated with the evolution of the complex inflorescence architecture in Asteraceae. Both in gerbera and sunflower, the CYC2 clade contains the largest number of gene duplicates, all of which are upregulated in developing ray flower primordia, indicating a heterotopic change in the expression domain at the inflorescence level. This is also true for the duplicated *RAY1* and *RAY2* genes in *Senecio*. Whether this specificity is caused by modifications in the non-coding *cis*-regulatory elements or in *trans*-acting factors is currently not known. In addition, at the level of single flowers, the gerbera *GhCYC2* was shown to be expressed in the large ventral ligule and the expression was specifically lacking from the dorsal domain of the flower indicating establishment of a novel function in petal fusion and growth of the ligule (Broholm et al., 2008). Thus far, outside Asteraceae, all CYC2 clade genes have been shown to control either dorsal or dorsal and lateral domains of the flowers (Busch & Zachgo, 2007; Citerne, Pennington, & Cronk, 2006; Feng et al., 2006; Wang et al., 2008). Therefore, it is likely that the spatial change in the expression domain of the CYC2 clade genes in Asteraceae has resulted in novel growth regulation at the level of single flowers by co-option of these genes to pattern ventral morphology of the flowers.

Recent studies in transgenic gerbera indicate redundancy among selected gene family members in regulating flower type identity (Juntheikki-Palovaara et al., unpublished data), but further functional studies are needed to verify whether additional functional divergence has occurred among the gene duplicates. Expression analysis by Tähtiharju et al. (2012) showed that in sunflower, a CYC3 clade gene, *HaCYC3a* and a CYC1 clade gene, *HaCYC1a* were also expressed in developing flower primordia and were upregulated in ray flower or disc flower primordia, respectively. Besides the control of shoot branching, the functions for CYC1 and CYC3 clade genes are poorly understood, and it is possible that also these genes may affect inflorescence or flower development. The CYC1 and CYC3 clade genes show relatively high expression in vegetative tissues of gerbera and sunflower but, together with the CYC2 clade genes, also during late stages of reproductive organ development (Tähtiharju et al., 2012). The capacity to form protein–protein complexes among the different CYC clades further indicates that the proteins may have shared functions during late stages of floral differentiation, e.g. in ovaries (Tähtiharju et al., 2012).

10.4 FLOWER ORGAN DETERMINATION: THE ABCE-MODEL IN ASTERACEAE

The regulatory processes related to flower organ development are well understood. According to the ABCE model of flower development, the four basic organ types are specified by overlapping A, B, C and E functions programmed by transcription factors. The code is combinatorial and mutations in genes encoding the transcription factors are homeotic, i.e. in the mutant phenotype "something has been changed into the likeness of something else" (Bateson, 1894).

The early model of flower organ determination included only the ABC genes (Coen & Meyerowitz, 1991) with the well-known scheme where A-class genes specified sepal, AB petal, BC stamen and C carpel determination, respectively. Although A, B and C functions are necessary, they are not sufficient to determine floral organ identity. The missing factors were found in *Arabidopsis* to be encoded by four redundantly acting genes, *SEPALLATA 1-4 (SEP1-4)* (Ditta, Pinyopich, Robles, Pelaz, & Yanofsky, 2004; Pelaz, Ditta, Baumann, Wisman, & Yanofsky, 2000). A quadruple mutant developed leaves where floral organs would normally form, and ectopic expression of A, B and C genes together with a SEP factor converted *Arabidopsis* rosette leaves into floral organs (Honma & Goto, 2001). The SEP function was named E since, meanwhile, the genetic factors that determine ovule identity had been identified and assigned to the D class (Angenent et al., 1995).

The ABCE model of floral organ identity determination is shared by all flowering plants, although different plant lineages show specific features (Bowman, 1997; Soltis, Chanderbali, Kim, Buzgo, & Soltis, 2007). Very importantly, at least the A function seems to be limited to *Arabidopsis* and its close relatives. Whether an A function can be defined in other plants or whether a general model works better without an A function, as in the original *Antirrhinum* model that today would be written as 0BC

(Schwarz-Sommer, Huijser, Nacken, Saedler, & Sommer, 1990), is not settled. Apart from one of the A function encoding genes (*APETALA2 (AP2)*, Jofuku, den Boer, Van Montagu, & Okamuro, 1994), all ABCE function genes belong to the type II (MIKC^C) MADS-box gene family, represented by ~40 members in plant genomes analysed so far (Díaz-Riquelme, Lijavetzky, Martínez-Zapater, & Carmona, 2009; de Folter et al., 2005; Leseberg et al., 2008). That means, only a fraction of MADS-box genes are involved in floral organ development (11 in *Arabidopsis*), while others have distinct functions related to different aspects of the flowering process and a whole lot are unknown for their function.

The ABCE MADS domain proteins form dimers and are believed to bind their consensus binding site in DNA (the CArG box) as a pair of dimers, i.e. as a protein quartet binding two CArG boxes simultaneously and causing concomitant DNA bending (Melzer, Verelst, & Theißen, 2009; Theiβen & Saedler, 2001). An E function MADS domain protein is involved in all quartets and sometimes called the 'glue' of the higher order complexes (Immink et al., 2009). The E proteins also typically contain a transcriptional activation domain. The classical ABC code for floral organ determination therefore becomes, in *Arabidopsis*, AA–EE (or AE–AE) for sepals, AE–BB for petals, BB–CE for stamens, CC–EE (or CE–CE) for carpels and finally CE–DE for ovules (Immink, Kaufmann, & Angenent, 2010). While the A function may be lacking in most flowering plants (sepals being encoded by EE–EE), the BB dimer is always a heterodimer of two different B function MADS domain proteins.

In following sections, functional studies conducted in Asteraceae ABCE MADS-box genes will be summarised (see also Table 10.1). We will first discuss the most conserved MADS-box genes encoding the B and C functions, we will then discuss the E function and finally present evidences for the function of A-class genes, that appears to be less conserved in Asteraceae.

10.4.1 The B-Class MADS-Box Genes

The B-class MADS-box genes are defined as homeotic regulators of petal and stamen organ identities, through combinatorial interaction with other MADS-box genes. The first B-class genes identified were the *A. majus DEFICIENS* (*DEF*) and *GLOBOSA* (*GLO*) (Sommer et al., 1990; Tröbner et al., 1992), and the functional orthologues in *Arabidopsis*, *AP3* and *PISTILLATA* (*PI*) (Goto & Meyerowitz, 1994; Jack, Brockman, & Meyerowitz, 1992). Heterodimerisation of these two B-class proteins was shown to be crucial for their capability to bind DNA (Goto & Meyerowitz, 1994;

Table 10.1 Selected MADS-Box Genes Involved in Regulating Flower Organ Identities or Inflorescence Development in Asteraceae. G; ***Gerbera hybrida***, CDM or C; Chrysanthemum (***Dendranthema grandiflorum***), ClM or Cl; ***Chrysanthemum lavandulifolium***, HAM or Ha; ***Helianthus annuus***

Gene	Function	Reference
TM3-Like		
GSOC1	Late petal development; overexpression leads to partial loss of floral identity	Ruokolainen et al., 2011
GSOC2-3	Phylogenetically close to *AtSOC1* and *PhUNS*, functions not verified	Ruokolainen et al., 2011
ClSOC1-1, 1-2	Expression in leaves and shoot apices, overexpression promotes flowering in *Arabidopsis*	Fu, Qi, Yang, Dai, & Dai, 2013
AP1 Clade		
GSQUA1, 3	Functions not verified	Ruokolainen, Ng, Broholm, et al., 2010; Yu et al., 1999
CDM111	Partially complements *Arabidopsis ap1-1*, overexpression leads to early flowering in transgenic chrysanthemum	Shchennikova et al., 2004; Shulga et al., 2011
HAM75, 92	Overexpression leads to early flowering in transgenic chrysanthemum	Shulga et al., 2008; Shulga et al., 2011
FUL Clade		
GSQUA2	Overexpression leads to accelerated flowering, dwarfism and vegetative abnormalities in transgenic gerbera	Ruokolainen, Ng, Broholm, et al., 2010
GSQUA4,5,6	Functions not verified	Ruokolainen, Ng, Broholm, et al., 2010
CDM8	Expression in both vegetative and floral tissues	Shchennikova et al., 2004
ClM8	Promotes flowering in *Arabidopsis*	Wang, Jiang, et al., 2013
CDM41	Expression in vegetative and floral tissues	Shchennikova et al., 2004
PI Clade		
GGLO1	B-function, petal and stamen identity	Yu et al., 1999; Broholm et al., 2010
CDM86	PI-like, function not verified	Shchennikova et al., 2004

Continued

Table 10.1 Selected MADS-Box Genes Involved in Regulating Flower Organ Identities or Inflorescence Development in Asteraceae. G; ***Gerbera hybrida***, CDM or C; Chrysanthemum (***Dendranthema grandiflorum***), ClM or Cl; ***Chrysanthemum lavandulifolium***, HAM or Ha; ***Helianthus annuus***—cont'd

Gene	Function	Reference
HaPI, HAM31	B-function, functions not verified	Dezar et al., 2003; Shulga et al., 2008
euAP3 Clade		
GDEF2	B-function, petal and stamen identity	Yu et al., 1999; Broholm et al., 2010
GDEF3	Paralogue of GhDEF2; function not verified	Broholm et al., 2010
CDM19, 115	AP3-like, function not verified	Shchennikova et al., 2004
HaAP3, HAM2, HAM63	AP3-like, functions not verified	Dezar et al., 2003; Shulga et al., 2008
Tm6 Clade		
GDEF1	Stamen development	Broholm et al., 2010
HAM91	TM6-like, function not verified	Shulga et al., 2008
AG Clade		
GAGA1-2	C-function, stamen and carpel identity	Yu et al., 1999
CDM37	AG-like, function not verified	Shchennikova et al., 2004
CAG1	97% Similarity with CDM37, stamen and pistil identity	Aida et al., 2008
HaAG, HAM45, HAM59	AG-like, functions not verified	Dezar et al., 2003; Shulga et al., 2008
SEP Clade		
GRCD1	Stamen identity	Kotilainen et al., 2000
GRCD2	Carpel identity, floral and inflorescence meristem determinacy	Uimari et al., 2004
GRCD3	Function not verified	Ruokolainen, Ng, Albert, et al., 2010
GRCD4, 5	General E-function, function not verified	Ruokolainen, Ng, Albert, et al., 2010
CDM44	SEP3-like, function not verified	Shchennikova et al., 2004
HAM137	SEP3-like, function not verified	Shulga et al., 2008

Schwarz-Sommer et al., 1992; Tröbner et al., 1992). Moreover, the heterodimers bind to their own promoters and establish a positive autoregulatory loop that is suggested to restrict the function of B-class genes to their own expression domains (Riechmann, Krizek, & Meyerowitz, 1996; Schwarz-Sommer et al., 1992).

Most core eudicot species have B-class genes that belong to three different lineages: the *PI*, eu*AP3* and *TM6* lineages, although both *Arabidopsis* and *Antirrhinum* have lost the *TM6*-like genes (Kramer, Dorit, & Irish, 1998). In gerbera, there are four B-class genes: *GGLO1* in the *PI* lineage, *GDEF2* and *GDEF3* in the eu*AP3* and *GDEF1* in the *TM6* lineage (Broholm et al., 2010). Sunflower shows orthologues for all the gerbera B genes, but has also an additional *PI*-like gene (Dezar, Tioni, Gonzalez, & Chan, 2003; Shulga, Shchennikova, Angenent, & Skryabin, 2008, Table 10.1) that has either been lost from gerbera or is derived specifically in the sunflower tribe due to whole genome duplication events at the base of this tribe (Barker et al., 2008). Functional analyses for the Asteraceae B-class genes have been performed only in gerbera, whereas some protein interaction studies have been done with the three B-class genes isolated in chrysanthemum (*Dendranthema grandiflorum*). In gerbera, GGLO1 was shown to be able to form heterodimers with all the other B-class proteins in yeast (Broholm et al., 2010), similarly to their orthologues in chrysanthemum (Shchennikova, Shulga, Immink, Skryabin, & Angenent, 2004). However, the capacity to form higher order protein complexes has diverged between these two Asteraceae species. In chrysanthemum, only one of the two B-protein dimer combinations is able to form ternary complexes with a C-class protein (Shchennikova et al., 2004), whereas in gerbera all the three B-protein dimers form ternary complexes with the C-class proteins (Broholm et al., 2010; Ruokolainen, Ng, Albert, Elomaa, & Teeri, 2010).

Functional studies were conducted for all the gerbera B-class genes except *GDEF3* by analysing transgenic gerbera lines. Both suppressed and ectopic expression of *GGLO1* leads to the homeotic conversions expected for a true B function gene, i.e. formation of pappus bristles in whorl two and carpeloid structures in whorl three in *GGLO1*-suppression lines, and conversion of whorl one into petaloid structures and whorl four to stamen-like in the *GGLO1*-overexpression lines (Broholm et al., 2010; Yu et al., 1999). In contrast, the ectopic expression of neither *GDEF2* nor *GDEF1* led to homeotic conversions, which may be due to the lack of *GGLO1* expression in the whorls one and four in these lines. The expression of *GGLO1* is strictly restricted to whorls two and three while *GDEF1, 2* and *3* are expressed in all four floral whorls and are also induced by ectopic *GGLO1* expression. Moreover, the suppression of *GDEF2*, and especially *GDEF1* expression led to milder homeotic changes than those seen in the *GGLO1* suppression lines. The *GDEF2* suppression lines showed a novel phenotype with ovary wall-like cells in the abaxial petal epidermis. This can be interpreted as an indication of a sepaloid origin for the ovary wall, and to support an appendicular origin of the inferior ovary walls (Gustafsson & Albert, 1999) that are characteristic to Asteraceae species. The mild stamen phenotype caused by suppressed *GDEF1* expression suggests that *GDEF1* takes part in regulating stamen development, as its expression pattern also indicates, but that its function is partially redundant with the other gerbera B-class genes (Broholm et al., 2010).

10.4.2 The C-Class MADS-Box Genes

Gerbera expresses two apparently redundant C-class MADS-box genes, *GAGA1* and *GAGA2*. The coding sequences of these genes share 85% identity whereas their amino acid sequences share 90% identity. The genes are paralogues (not allelic), since selfed offspring of gerbera invariably carries both forms. *GAGA1* and *GAGA2* are typical C-class MADS-box genes. They are expressed in the centre of flower primordia which gives rise to stamens and carpels (Yu et al., 1999) and the GAGA1 and GAGA2 proteins interact with gerbera E-class MADS domain proteins in pairwise yeast 2-hybrid assays. In a yeast 3-hybrid assay, GAGA1 and GAGA2 interacted also with all the gerbera B dimers GGLO1–GDEF2, GGLO1–GDEF1 and GGLO1–GDEF3 (Broholm et al., 2010; Ruokolainen, Ng, Albert, et al., 2010).

Ectopic expression of *GAGA1* or *GAGA2* in gerbera causes a dramatic loss of ray flower petals, which initiates development but result in withered dry structures, a fate shared by stamens in ray flowers. In disc flowers the petals are converted to colourless organs, forming a fused structure that covers whorl three and four organs. These organs resemble stamens developed in disc flowers; however, they never form pollen. Furthermore, transgenic gerbera lines carrying the antisense constructs of either *GAGA1* or *GAGA2* show phenotypes typical of loss of the C function. Stamens are converted to petals and in the ray flowers grow to full-length in the strongest downregulation lines without forming withered dry structures. In addition to organ identity determination, the C function has a role in controlling determinacy of the FM, whereby mutations in C-class MADS-box genes lead to indeterminate flowers with reiterative arrangements of whorl 1- and 2-like organs (sepals and petals). This also happens in gerbera when *GAGA1/ GAGA2* genes are downregulated. Pappus hairs and petal structures can be found in the centre of the flower, although these structures do not expand in size (Yu et al., 1999). *AGAMOUS*-like MADS-box genes have also been isolated from sunflower (Dezar et al., 2003; Shulga et al., 2008) and chrysanthemum (Aida, Komano, Saito, Nakase, & Murai, 2008). In chrysanthemum, transgenic downregulation of the *CAG* gene led to conversion of both stamen and pistil into corolla-like tissue (Aida et al., 2008).

10.4.3 The E-Class MADS-Box Genes

The first MADS-box gene identified in gerbera was *GERBERA REGULATOR OF CAPITULUM DEVELOPMENT1 (GRCD1)*, isolated from a petal cDNA library (Kotilainen et al., 2000). Phylogenetic reconstructions grouped it with the *Arabidopsis* genes *SEP1, 2* and *3*. The gerbera *SEP*-like genes were named as *GRCD*s. *GRCD1* has a very specific role as a necessary C function activator that is restricted to whorl three of the marginal flowers. In transgenic gerbera lines where *GRCD1* was downregulated by antisense technology, the staminodia in the ray flowers convert to petals and are not aborted. They show very high resemblance to whorl three organs of ray flowers, in which *GAGA1/2* are partially downregulated. However, unlike *GAGA1/2* downregulation, the phenotype of *GRCD1* downregulation is restricted to whorl 3 of the ray flowers. Carpels develop normally, and so do disc flower stamens that remain fertile. It is not known if disc flower stamen development requires less GRCD1 protein than ray flower stamen development or if GRCD1 is a negatively acting regulator responsible for abortion of ray flower stamens and outcompeted in disc flowers. *GRCD1* is indeed expressed more strongly in the marginal than in the central flower primordia (Laitinen, Broholm, Albert, Teeri, & Elomaa, 2006). Phylogenetic analyses (Zahn et al., 2005) show that also chrysanthemum and sunflower have a *GRCD1*-like gene. The three Asteraceae genes form a separate clade, sister to *SEP3*-like genes of all flowering plants. The *GRCD1* lineage encodes an unusually large number of unique amino acid changes in the MADS and K domains of the corresponding protein molecules, indicating that they may have acquired novel functions during evolution (Zahn et al., 2005). This predicts that *GRCD1*-like genes may be involved in male sterility of marginal flower also in chrysanthemum and sunflower.

GRCD2 is a related gerbera *SEP*-like gene whose activity is restricted to whorl 4, when considering its ability to specify organ identity. Transgenic gerbera lines where *GRCD2* expression is downregulated do not affect organs in whorl 1, 2 and 3, but carpels are converted either to leaf-like organs in mildly or to petal-like organs in strongly downregulated lines, respectively (Uimari et al., 2004). Although *GRCD2* acts as a whorl 4 specific C-function activator, loss of floral determinacy, typical of C function mutants, does not occur in *GRCD2* downregulated plants. Instead, a dramatic conversion of the ovary takes place and the structure becomes filled with floral organs and floral primordia. This can be interpreted as floral reversion and the flower meristem that normally is consumed after carpel determination reverts to an IM. Interestingly, although the *Arabidopsis* triple mutant *sep1/2/3* shows a similar phenotype (reversion), it had been interpreted as a loss of determinacy (Pelaz et al., 2000). The inferior position of the gerbera ovary made it easy to distinguish these separate phenomena. *GRCD2* showed a third function in the downregulation lines. The number of flowers in gerbera inflorescence is set during early development of the capitulum. Loss of *GRCD2* activity led to persistence of an undifferentiated region in the centre of the developing inflorescence, which continued to produce extra disc flowers until the inevitable senescence of the whole capitulum (Uimari et al., 2004). This was interpreted as conversion of a determinate to an indeterminate capitulum, causing some discordance with the traditional botanical terminology, where a determinate inflorescence develops the terminal flower first. Regardless of this, *GRCD2* is important for the plant to be able to consume the IM in an orderly fashion and to produce a reasonably fixed number of flowers (which however fluctuates a lot between seasons).

Gerbera has altogether five *SEP*-like genes. A protein–protein interaction analysis in yeast suggested that similarly to *Arabidopsis*, also gerbera has redundant genes encoding the E function. In fact, GRCD1 and 2 behaved very differently from typical E function proteins since they interacted nearly exclusively with the GAGA proteins, did not contain transcriptional activation sequences and did not homodimerise. Instead, GRCD4 and GRCD5 showed all these features and could interact with several gerbera MADS proteins, forming hubs within the interaction map (Ruokolainen, Ng, Albert, et al., 2010), similarly to SEP1 and SEP3 proteins in the *Arabidopsis* interaction map (de Folter et al., 2005). We predict that the concomitant downregulation of *GRCD4* and *GRCD5* will have strong effects. The fifth gerbera SEP-like protein GRCD3, most similar to AGL6 of *Arabidopsis*, is a weak interactor with a limited number of interacting partners (Ruokolainen, Ng, Albert, et al., 2010). *GRCD3* shows similarity to the *Petunia PhAGL6* (Rijpkema, Zethof, Gerats, & Vandenbussche, 2009) and is strongly expressed in ovules, carpels and petals, and weakly in stamens and pappus bristles (Ruokolainen, Ng, Albert, et al., 2010). However, its function in transgenic gerbera has not been verified.

10.4.4 The A-Class MADS-Box Genes

Although it has become well established that the homeotic A function is limited to *Arabidopsis* and its close relatives (Litt, 2007), flowering plants typically have one or several MADS-box genes falling into the same clade with the *Arabidopsis* A function MADS-box gene *AP1*. The first A-class gene isolated was *SQUAMOSA* from *Antirrhinum*, which similarly to *AP1* of *Arabidopsis* is expressed in whorls 1 and 2 of *Antirrhinum* flowers (Huijser et al., 1992); yet it does not contribute to defining sepal and petal identity. The key of the homeotic A function (needed for sepal and petal development) may simply be restriction of the C function to the inner whorls. In *Arabidopsis* this is accomplished by both *AP1* and *AP2*. It has not been established whether plants other than *Arabidopsis* share a mechanism for this task. It is however clear that MADS-box genes belonging to the *AP1* clade are usually not involved.

The *Arabidopsis AP1* and the *Antirrhinum SQUAMOSA* genes share of the capacity to determine flower meristem identity. Mutations in these genes cause impaired transition from IMs to flower meristems (Bowman, Alvarez, Weigel, Meyerowitz, & Smyth, 1993; Huijser et al., 1992) and ectopic expression of *AP1* in *Arabidopsis* causes early flowering (Mandel & Yanofsky, 1995). Gerbera has at least six *AP1*-like genes, *GSQUA1-6* (Ruokolainen, Ng, Broholm, et al., 2010). Their expression does not suggest a homeotic A function. The expression patterns vary, but many are expressed in the inner whorls and none in the perianth whorls only. Functional analyses in transgenic gerbera showed a role for only one of these, *GSQUA2*, whose ectopic expression led to early flowering and vegetative aberrations (dwarfism, elongated internodes, poor root development) (Ruokolainen, Ng, Broholm, et al., 2010). It is remarkable that while many MADS-box genes cause early flowering in *Arabidopsis* when overexpressed, this has never happened in gerbera except with *GSQUA2*. Also in transgenic chrysanthemum, ectopic expression of a chrysanthemum *AP1*-like gene, *CDM111*, as well as heterologous sunflower genes, *HAM75* and *HAM92*, led to early flowering (Shulga, Mitiouchkina, Schennikova, Skryabin, & Dolgov, 2011). *CDM111*, *HAM75* and *HAM92* are *euAP1*-like genes while the gerbera *GSQUA2* is a *paleoAP1/FUL*-like (Ruokolainen, Ng, Broholm, et al., 2010).

While the functions of the other *GSQUA* genes remain obscure, it can be concluded that *GSQUA2* perhaps shares a primordial function of *AP1*-like genes in controlling flowering. In gerbera, it remains unresolved what factors restrict the C-function outside of the perianth. When expressed in the outer whorls under the viral 35S promoter, C-genes readily convert gerbera petals to structures that have stamen-like features, indicating that ectopic expression of C-genes can overcome the restrictive mechanism.

10.5 CONNECTIONS BETWEEN MADS AND TCP REGULATORS

The studies conducted in Asteraceae so far indicate that the CYC2 clade genes have early functions during flower type specification, but also late functions affecting the growth patterns of petals and stamens as well as carpels and ovary, at least in sunflower. It is also clear that several MADS-box genes in diverse plant species are controlling the late stages of organ growth (reviewed in Dornelas, Patreze, Angenent, & Immink, 2010). In *Antirrhinum* flowers *CYC* expression is initiated in whorls one to three but is maintained only in whorls two and three at later developmental stages (Clark & Coen, 2002). Expression analyses in the *Antirrhinum* B-class mutant (*def*) indicated that *CYC* expression was normal at early developmental stages; however it declined later on in whorl 2 indicating that B activity is required to maintain *CYC* expression in whorl 2 but not in whorl 3. It was postulated that the C gene *PLENA* may contribute to maintenance of *CYC* in whorl 3 (Clark & Coen, 2002). Altogether with the functional data obtained from Asteraceae it seems that in single flowers *CYC*-like genes do not affect flower organ identity determination but rather function downstream of MADS-box genes by either promoting or repressing the growth of the organs. Several recent studies have identified a large number of genes associated with cell proliferation and cell growth among the target genes of *Arabidopsis* MADS-box proteins (reviewed in Dornelas et al., 2010). Kaufmann et al. (2009) showed that 15 transcription factor families were significantly overrepresented among SEP3 targets. Interestingly, among them there were 11 TCP domain transcription factor genes. Wellmer, Alves-Ferreira, Dubois, Riechmann, and Meyerowitz (2006) identified seven TCP genes belonging to class II, among the targets of AP1. Dornelas et al. (2010) also suggest that MADS domain transcription factors may act in concert with TCP factors in larger complexes.

cDNA microarray analyses comparing gene expression in gerbera ray and disc flower primordia revealed that similarly to *CYC2* clade genes also several MADS-box transcription factor genes show differential expression between the flower types (Laitinen et al., 2006). The microarray analyses showed that the *SEP* clade MADS-box genes *GRCD1*, *GRCD5* and *GRCD3* as well as *GhSOC1* were upregulated during ray flower development compared to the corresponding developmental stage of disc flowers. The expression patterns of these genes were also highly correlated not only during flower primordia development but also in various floral organs (Laitinen et al., 2006). Downregulation of *GRCD1* converted the sterile staminoids in ray flowers into petals indicating a specific homeotic function in whorl 3 (Kotilainen et al., 2000). It is not known whether *CYC2* clade gene expression is altered in these lines but it is equally possible that *GRCD1* is a target gene for (a) CYC2 clade regulator(s). Overexpression of *GhSOC1* instead affected the growth of the petals; the ray flower petals were shorter and stamen development in disc flowers was disrupted (Ruokolainen, Ng, Albert, Elomaa, & Teeri, 2011). It is tempting to speculate that *GhSOC1* may function upstream in the regulatory cascade imposed to *CYC2* clade genes. Gene duplication events during CYC and MADS-box gene evolution have been proposed to be tightly correlated with one another as well as with major changes of floral organisation in angiosperms (Howarth & Donoghue, 2006). However, the detailed regulatory connections between these key floral regulators during early flower development and later in controlling the size and shape of individual floral organs are yet to be discovered.

10.6 MORPHOGENETIC CONTROL OF INFLORESCENCE DEVELOPMENT

Although some species of Asteraceae bear homogamous capitula with monomorphic flowers, the predominant form is heterogamous capitula with morphologically distinct flowers that impose interesting developmental questions. How differential development is controlled and, especially, how is the differential fate of the flowers determined? The mechanisms of developmental control have recently been slightly elucidated; as the family of TCP transcription factors were shown to be involved in flower type specific development (see Section 10.3). Signals that determine the developmental progression are still essentially obscure.

Models of chemical control of morphogenesis date back to the beginnings of the previous century. Alan Turing showed through mathematical analysis that patterns can be spontaneously formed by postulating interactions of two diffusible factors (two-component reaction–diffusion systems,

Turing, 1952). Lewis Wolpert introduced the French Flag model (Wolpert, 1969), where the key is that cells react differentially to different concentrations of a diffusible morphogen. Hans Meinhardt added to the concept by introducing stabilisation of the gradient through involvement of autoactivated response factors (Gierer & Meinhardt, 1972). These concepts are central to biological pattern formation, although the actual networks in development are much more complicated – e.g. the French Flag model, despite of its successful application in many cases and ability to explain pattern regeneration, is not scale invariant (Jaeger, Irons, & Monk, 2008).

Models of morphogenic control of development do not address the nature of the morphogen (or morphogenic trigger, see Benková, Ivanchenko, Friml, Shishkova, & Dubrovsky, 2009), but several examples show that these factors can be small molecular weight compounds (phytohormones), proteins (e.g. transcription factors) or RNA molecules (small RNAs). Further, passive diffusion is not the only way to distribute these factors throughout the tissues: active transport of auxin by the PIN proteins forms an elegant and fundamental example (Vieten, Sauer, Brewer, & Friml, 2007). Large molecules, on the other hand, are thought to move from cell to cell through plasmodesmata, whose presence, aperture and gating can be controlled. In addition to chemical gradients, physical gradients (stress and strain) have received attention recently (Boudaoud, 2010). Such gradients would explain how a relatively rapid signal can be transmitted (a relevant concern for large structures such as developing flower heads in Asteraceae), but are more difficult to approach. Conversion of physical signals to chemical signals (e.g. gene expression) is thought to take place through action of mechanosensors, plasma membrane located mechanosensitive Ca^{2+} channels (Pickard, 2007) or through regulation of the orientation of microtubules in the shoot apex (Hamant et al., 2008).

How do these concepts relate to our understanding of development of the flower head of Asteraceae? Initiation of leaf primordia on an apical meristem is dependent on efflux-dependent auxin gradients. Inhibiting formation of auxin gradients, either chemically or genetically in *pin* mutants, prevents formation of primordia (Reinhardt, Mandel, & Kuhlemeier, 2000). The auxin gradient hypothesis readily explains spiral phyllotaxis, where subsequent primordia appear at angles of about 137° (Smith et al., 2006). This hypothesis can explain the overall positioning of the flowers in the capitulum of Asteraceae, including the spiral parastichies that occur in series of Fibonacci numbers (Mitchison, 1977). Still, it is difficult to understand how the model would apply to a large capitulum surface, where the margin is packed with emerging flower primordia. Mathematical analysis shows that the principle leads to best packing of flower primordia on the circular surface of the capitulum (Douady & Couder, 1993). It is possible that evolution has optimised best packing by some other means, which then leads to a superficial resemblance of spiral leaf phyllotaxis with the feature of the generative angle being close to the golden angle (137.5°) and the number of opposite winding parastichies being two consecutive Fibonacci numbers. Hauntingly, the few black-spotted petals on inflorescences of *Gorteria* occur always at angles of 137.5° (Thomas et al., 2009), as if these flowers were 'successive' in their initiation.

In spite of their explanatory power in specifying primordia positioning, efflux-dependent auxin gradients do not explain how the identity of different flower types (and involucral bracts) is defined. During usual plant propagation, a set of ray flowers sometimes emerges among the disc flowers. In sunflower, this can be related to boron deficiency that causes cracks in the IM (Blamey, 1976), or in gerbera to insect damage of developing inflorescences (our own observations, see Figure 10.4). In a classical series of wounding experiments on developing sunflower inflorescences, John Palmer and his co-workers showed that wound margins take the role of the inflorescence margin and induce formation of flower primordia in the successive and natural order of bracts-rays-discs (Hernandez & Palmer, 1988; Hernandez & Palmer, 1990; Palmer & Marc, 1982). They concluded that the capitulum margin is the organising centre for inflorescence development, but that the margin as such is not the key. Rather, the fact that the margin interrupts the plasmodesmatal connection between cells (the symplasmic

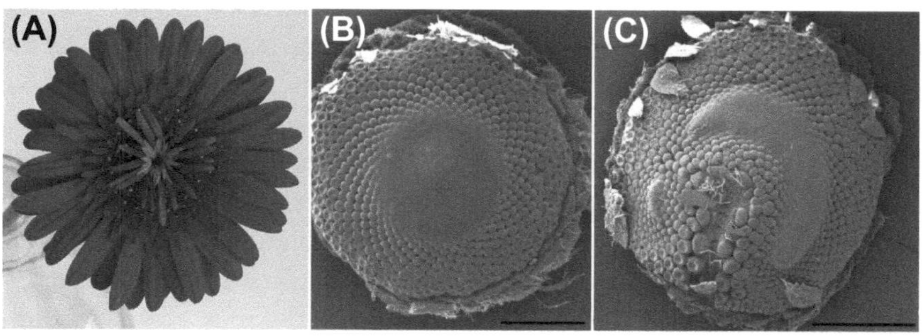

Figure 10.4 (A) A gerbera cultivar displaying ectopic involucral bracts and ray flowers in the centre of the inflorescence. (B) Scanning electron microscopy (SEM) image of a wild type undamaged capitulum of gerbera showing normal initiation of flower primordia. (C) SEM image of a wounded wild type capitulum shows that wounding creates new margins that re-initiate bracts (coloured in green) and induce flower primordia in an abnormal pattern. Scale bars 1 mm. (See the colour plate.)

domain), re-establishes the organising centre at the wound site (Hernandez & Palmer, 1990). Two important conclusions can be made from these experiments. First, the organising centre (capitulum or wound margin) is able to send a propagating signal that controls both the initiation of floral primordia and their identity. Second, although the IM is responsive to these signals during a particular developmental window, there is no constraint in where the marginal organs (bracts or ray flowers) can develop. This is demonstrated in the insect-wounded gerbera IM in Figure 10.4.

10.7 CONCLUDING REMARKS

Molecular studies in Asteraceae demonstrate how common developmental regulators have been recruited to control complex traits characteristic for the capitula development. Both in the case of MADS-box and TCP domain transcription factor genes, functional evolution involving spatial and/or temporal specialisation of the duplicated genes as well as the capacity to form different protein complexes are the major forces behind the development of phenotypic novelties. In Asteraceae *CYC*-like genes show major roles in regulating early stages of flower type differentiation while MADS-box genes show both conserved and diverged roles in organ identity determination. Moreover, both gene family members are apparently contributing to the late stages of organ growth, defining the final size and shape of the organs. What is still largely unresolved is how the complex Asteraceae inflorescence is organised and how the initiation of flower primordia and their differential fate and growth is coordinated. The highly compressed inflorescence in Asteraceae has been considered as a condensed raceme, however, practically nothing is known of its organisation at the molecular level. What are the detailed mechanisms connecting CYC and MADS protein functions? What specifies the ray flower abundant expression of CYC2 clade genes during the early IM development? Regulatory evolution through gene duplication and convergent co-option of conserved genetic modules forms a central theme in the evolution of morphological novelties in plants. Thereby, we postulate that the genes that regulate inflorescence patterning in *Arabidopsis* or petunia are the most obvious candidates that have been co-opted to function in Asteraceae.

ACKNOWLEDGEMENTS

We thankfully acknowledge all present and previous members of the Gerbera Lab who have contributed to this work. Dr Anneke Rijpkema is especially thanked for providing figures for this chapter. Prof. Victor A. Albert is thanked for our long-term collaboration on the evolutionary aspects of our work. The plant breeding company Terra Nigra B.V., the Netherlands, is acknowledged for providing plant material. The Gerbera Lab has been funded by the Academy of Finland (grants 115849 and 1139092 for P. E. and grant 1139513 and the Programme for Centres of Excellence in Research for T. T.).

REFERENCES

Abbott, R. J. (1992). Plant invasions, interspecific hybridization and the evolution of new plant taxa. *Trends in Ecology & Evolution*, 7, 401–405.

Abbott, R. J., James, J. K., Milne, R. I., & Gillies, A. C. M. (2003). Plant introductions, hybridization and gene flow. *Philosophical Transactions of the Royal Society of London. Series B: Biological Sciences*, *358*, 1123–1132.

Aguilar-Martinez, J. A., Poza-Carrion, C., & Cubas, P. (2007). *Arabidopsis* BRANCHED1 acts as an integrator of branching signals within axillary buds. *The Plant Cell*, *19*, 458–472.

Aida, R., Komano, M., Saito, M., Nakase, K., & Murai, K. (2008). Chrysanthemum flower shape modification by suppression of chrysanthemum *AGAMOUS* gene. *Plant Biotechnology*, *25*, 55–59.

Andersson, S. (2008). Pollinator and nonpollinator selection on ray morphology in *Leucanthemum vulgare* (oxeye daisy, Asteraceae). *American Journal of Botany*, *95*, 1072–1078.

Angenent, G. C., Franken, J., Busscher, M., van Dijken, A., van Went, J. L., Dons, H. J., et al. (1995). A novel class of MADS box genes is involved in ovule development in petunia. *The Plant Cell*, 7, 1569–1582.

Barker, M. S., Kane, N. C., Matvienko, M., Kozik, A., Michelmore, R. W., Knapp, S. J., et al. (2008). Multiple paleopolyploidizations during the evolution of the compositae reveal parallel patterns of duplicate gene retention after millions of years. *Molecular Biology and Evolution*, *25*, 2445–2455.

Bateson, W. (1894). *Materials for the study of variation treated with especial regard to discontinuity in the origin of species*. London: Macmillan.

Bello, M. A., Álvarez, I., Torices, R., & Fuertes-Aguilar, J. (2013). Floral development and evolution of capitulum structure in *Anacyclus* (Anthemideae, Asteraceae). *Annals of Botany*, *112*, 1597–1612.

Benková, E., Ivanchenko, M. G., Friml, J., Shishkova, S., & Dubrovsky, J. G. (2009). A morphogenetic trigger: is there an emerging concept in plant developmental biology? *Trends in Plant Science*, *14*, 189–193.

Berti, F., Fambrini, M., Turi, M., Bertini, D., & Pugliesi, C. (2005). Mutations of corolla symmetry affect carpel and stamen development in *Helianthus annuus*. *Canadian Journal of Botany*, *83*, 1065–1072.

Blamey, F. P. C. (1976). Boron nutrition of sunflowers (*Helianthus annuus* L.) on an avalon medium sandy loam. *Agrochemophysica*, *8*, 5–10.

Boudaoud, A. (2010). An introduction to the mechanics of morphogenesis for plant biologists. *Trends in Plant Science*, *15*, 353–360.

Bowman, J. L. (1997). Evolutionary conservation of angiosperm flower development at the molecular and genetic levels. *Journal of Biosciences*, *22*, 515–527.

Bowman, J. L., Alvarez, J., Weigel, D., Meyerowitz, E. M., & Smyth, D. R. (1993). Control of flower development in *Arabidopsis thaliana* by *APETALA1* and interacting genes. *Development*, *119*, 721–743.

Broholm, S. K., Pöllänen, E., Ruokolainen, S., Tähtiharju, S., Kotilainen, M., Albert, V. A., et al. (2010). Functional characterization of B class MADS-box transcription factors in *Gerbera hybrida*. *Journal of Experimental Botany*, *61*, 75–85.

Broholm, S. K., Tähtiharju, S., Laitinen, R. A., Albert, V. A., Teeri, T. H., & Elomaa, P. (2008). A TCP domain transcription factor controls flower type specification along the radial axis of the gerbera (Asteraceae) inflorescence. *Proceedings of the National Academy of Sciences of the United States of America*, *105*, 9117–9122.

Busch, A., & Zachgo, S. (2007). Control of corolla monosymmetry in the Brassicaceae *Iberis amara*. *Proceedings of the National Academy of Sciences of the United States of America, 104*, 16714–16719.

Busch, A., & Zachgo, S. (2009). Flower symmetry evolution: towards understanding the abominable mystery of angiosperm radiation. *BioEssays, 31*, 1181–1190.

Chapman, M. A., Leebens-Mack, J. H., & Burke, J. M. (2008). Positive selection and expression divergence following gene duplication in the sunflower *CYCLOIDEA* gene family. *Molecular Biology and Evolution, 25*, 1260–1273.

Chapman, M. A., Tang, S., Draeger, D., Nambeesan, S., Shaffer, H., Barb, J. G., et al. (2012). Genetic analysis of floral symmetry in Van Gogh's sunflowers reveals independent recruitment of *CYCLOIDEA* genes in the Asteraceae. *PLoS Genetics, 8*, e1002628.

Citerne, H. L., Pennington, R. T., & Cronk, Q. C. (2006). An apparent reversal in floral symmetry in the legume *Cadia* is a homeotic transformation. *Proceedings of the National Academy of Sciences of the United States of America, 103*, 12017–12020.

Clark, J. I., & Coen, E. S. (2002). The *cycloidea* gene can respond to a common dorsoventral prepattern in *Antirrhinum*. *The Plant Journal, 30*, 639–648.

Claßen-Bockhoff, R., & Bull-Hereñu, K. (2013). Towards an ontogenetic understanding of inflorescence diversity. *Annals of Botany, 112*, 1523–1542.

Coen, E. S., & Meyerowitz, E. M. (1991). The war of the whorls: genetic interactions controlling flower development. *Nature, 353*, 31–37.

Coen, E. S., Nugent, J. M., Luo, D., Bradley, D., Cubas, P., Chadwick, M., et al. (1995). Evolution of floral symmetry. *Philosophical Transactions of the Royal Society of London. Series B: Biological Sciences, 350*, 35–38.

Cubas, P. (2004). Floral zygomorphy, the recurring evolution of a successful trait. *Bioessays, 26*, 1175–1184.

Cubas, P., Coen, E., & Zapater, J. M. (2001). Ancient asymmetries in the evolution of flowers. *Current Biology, 11*, 1050–1052.

Cubas, P., Lauter, N., Doebley, J., & Coen, E. (1999). The TCP domain: a motif found in proteins regulating plant growth and development. *The Plant Journal, 18*, 215–222.

Dezar, C. A., Tioni, M. F., Gonzalez, D. H., & Chan, R. L. (2003). Identification of three MADS-box genes expressed in sunflower capitulum. *Journal of Experimental Botany, 54*, 1637–1639.

Díaz-Riquelme, J., Lijavetzky, D., Martínez-Zapater, J. M., & Carmona, M. J. (2009). Genome-wide analysis of MIKC^{C}-type MADS box genes in grapevine. *Plant Physiology, 149*, 354–369.

Ditta, G., Pinyopich, A., Robles, P., Pelaz, S., & Yanofsky, M. F. (2004). The *SEP4* gene of *Arabidopsis thaliana* functions in floral organ and meristem identity. *Current Biology, 14*, 1935–1940.

Doebley, J., Stec, A., & Hubbard, L. (1997). The evolution of apical dominance in maize. *Nature, 386*, 485–488.

Dornelas, M. C., Patreze, C. M., Angenent, G. C., & Immink, R. G. H. (2010). MADS: the missing link between identity and growth? *Trends in Plant Science, 16*, 89–97.

Douady, S., & Couder, Y. (1993). Phyllotaxis as a self-organized growth process. In J. M. Garcia-Ruiz, E. Louis, P. Meakin, & L. M. Sander (Eds.), *Growth patterns in physical sciences and biology* (pp. 341–352). US: Springer.

Endress, P. K. (1999). Symmetry in flowers: diversity and evolution. *International Journal of Plant Sciences, 160*(Suppl. 6), S3–S23.

Endress, P. K. (2001). Evolution of floral symmetry. *Current Opinion in Plant Biology, 4*, 86–91.

Fambrini, M., Bertini, D., & Pugliesi, C. (2003). The genetic basis of a mutation that alters the floral symmetry in sunflower. *Annals of Applied Biology, 143*, 341–347.

Fambrini, M., Salvini, M., & Pugliesi, C. (2011). A transposon-mediate inactivation of a *CYCLOIDEA*-like gene originates polysymmetric and androgynous ray flowers in *Helianthus annuus*. *Genetica, 139*, 1521–1529.

Feng, X., Zhao, Z., Tian, Z., Xu, S., Luo, Y., Cai, Z., et al. (2006). Control of petal shape and floral zygomorphy in *Lotus japonicus*. *Proceedings of the National Academy of Sciences of the United States of America, 103*, 4970–4975.

Finlayson, S. A. (2007). *Arabidopsis* TEOSINTE BRANCHED1-like 1 regulates axillary bud outgrowth and is homologous to monocot TEOSINTE BRANCHED1. *Plant & Cell Physiology, 48*, 667–677.

Finlayson, S. A., Krishnareddy, S. R., Kebrom, T. H., & Casal, J. J. (2010). Phytochrome regulation of branching in *Arabidopsis*. *Plant Physiology, 152*, 1914–1927.

de Folter, S., Immink, R. G., Kieffer, M., Pařenicová, L., Henz, S. R., Weigel, D., et al. (2005). Comprehensive interaction map of the *Arabidopsis* MADS box transcription factors. *The Plant Cell, 17*, 1424–1433.

Fu, J., Qi, S., Yang, L., Dai, Y., & Dai, S. (2013). Characterization of Chrysanthemum *ClSOC1-1* and *ClSOC1-2*, homologous genes of *SOC1*. *Plant Molecular Biology Reporter*. http://dx.doi.org/10.1007/s11105-013-0679-8.

Gierer, A., & Meinhardt, H. (1972). A theory of biological pattern formation. *Kybernetik, 12*, 30–39.

Gillies, A. C. M., Cubas, P., Coen, E., & Abbott, R. J. (2002). Making rays in the Asteraceae: genetics and evolution of radiate versus discoid flower heads. In Q. C. B. Cronk, R. M. Bateman, & J. A. Hawkins (Eds.), *Developmental genetics and plant evolution* (pp. 233–246). London: Taylor and Francis.

Goto, K., & Meyerowitz, E. M. (1994). Function and regulation of the *Arabidopsis* floral homeotic gene *PISTILLATA*. *Genes and Development, 8*, 1548–1560.

Gustafsson, M. H. G., & Albert, V. A. (1999). Inferior ovaries and angiosperm diversification. In P. M. Hollingsworth, R. M. Bateman, & R. J. Gornall (Eds.), *Molecular systematics and plant evolution, the systematics association special volume series* (pp. 403–431). New York, USA: Taylor and Francis.

Hamant, O., Heisler, M. G., Jonsson, H., Krupinski, P., Uyttewaal, M., Bokov, P., et al. (2008). Developmental patterning by mechanical signals in *Arabidopsis*. *Science, 322*, 1650–1655.

Harris, E. M. (1991). Floral initiation and early development in *Erigeron philadelphicus* (Asteraceae). *American Journal of Botany, 78*, 108–121.

Harris, E. M. (1995). Inflorescence and floral ontogeny in Asteraceae: a synthesis of historical and current concepts. *Botanical Review, 61*, 3–278.

Harris, S. A. (2002). Introduction of Oxford ragwort *Senecio squalidus* L. (Asteraceae), to the United Kingdom. *Watsonia, 24*, 31–43.

Hegarty, M. J., Jones, J. M., Wilson, I. D., Barker, G. L., Coghill, J. A., Sanchez-Baracaldo, P., et al. (2005). Development of anonymous cDNA microarrays to study changes to the *Senecio* floral transcriptome during hybrid speciation. *Molecular Ecology, 14*, 2493–2510.

Hernandez, L. F., & Palmer, J. H. (1988). Regeneration of the sunflower capitulum after cylindrical wounding of the receptacle. *American Journal of Botany, 75*, 1253–1261.

Hernandez, L. F., & Palmer, J. H. (1990). Colchicine-induced displacement of floral organ regeneration sites in the wounded sunflower capitulum. *Microscopía Electrónica Y Biología Celular, 14*, 159–164.

Hodgins, K. A., Lai, Z., Oliveira, L. O., Still, D. W., Scascitelli, M., Barker, M. S., et al. (2014). Genomics of Compositae crops: reference transcriptome assemblies and evidence of hybridization with wild relatives. *Molecular Ecology Resources, 14*, 166–177.

Honma, T., & Goto, K. (2001). Complexes of MADS-box proteins are sufficient to convert leaves into floral organs. *Nature, 409*, 525–529.

Howarth, D. G., & Donoghue, M. J. (2006). Phylogenetic analysis of the "ECE" (CYC/TB1) clade reveals duplications predating the core eudicots. *Proceedings of the National Academy of Sciences of the United States of America*, *103*, 9101–9106.
Huang, H., Wang, Y., Wang, S., Wu, Z., Yang, K., Niu, Y., et al. (2012). Transcriptome-wide survey and expression analysis of stress-responsive *NAC* genes in *Chrysanthemum lavandulifolium*. *Plant Science*, *193–194*, 18–27.
Huijser, P., Klein, J., Lönnig, W. E., Meijer, H., Saedler, H., & Sommer, H. (1992). Bracteomania, an inflorescence anomaly, is caused by the loss of function of the MADS-box gene *Squamosa* in *Antirrhinum majus*. *The EMBO Journal*, *11*, 1239–1249.
Immink, R. G., Kaufmann, K., & Angenent, G. C. (2010). The 'ABC'of MADS domain protein behaviour and interactions. In *Seminars in cell & developmental biology* (Vol. 21, pp. 87–93). Academic Press.
Immink, R. G. H., Tonaco, I. A., de Folter, S., Shchennikova, A., van Dijk, A. D. J., Busscher-Lange, J., et al. (2009). SEPALLATA3: the 'glue' for MADS box transcription factor complex formation. *Genome Biology*, *10*, R24.
Ingram, R., & Taylor, L. (1982). The genetic control of a non-radiate condition in *Senecio squalidus* L. and some observations on the role of ray florets in the Compositae. *New Phytologist*, *91*, 749–756.
Ingram, R., Weir, J., & Abbott, R. J. (1980). New evidence concerning the origin of inland radiate groundsel, *S. vulgaris* L. var. *hibernicus* Syme. *New Phytologist*, *84*, 543–546.
Jack, T., Brockman, L. L., & Meyerowitz, E. M. (1992). The homeotic gene APETALA3 of *Arabidopsis thaliana* encodes a MADS box and is expressed in petals and stamens. *Cell*, *68*, 683–697.
Jaeger, J., Irons, D., & Monk, N. (2008). Regulative feedback in pattern formation: towards a general relativistic theory of positional information. *Development*, *135*, 3175–3183.
Jeffrey, C. (2007). Compositae: Introduction with key to tribes. In J. J. W. Kadereit, & C. Jeffrey (Eds.), *Flowering plants: Eudicots; asterales* (Vol. 8, pp. 61–87). Berlin: Springer Verlag.
Jofuku, K. D., den Boer, B. G., Van Montagu, M., & Okamuro, J. K. (1994). Control of *Arabidopsis* flower and seed development by the homeotic gene *APETALA2*. *The Plant Cell*, *6*, 1211–1215.
Kane, N. C., Gill, N., King, M. G., Bowers, J. E., Berges, H., Gouzy, J., et al. (2011). Progress towards a reference genome for sunflower. *Botany*, *89*, 429–437.
Kaufmann, K., Muiño, J. M., Jauregui, R., Airoldi, C. A., Smaczniak, C., Krajewski, P., et al. (2009). Target genes of the MADS transcription factor SEPALLATA3: intergration of developmental and hormonal pathways in the *Arabidopsis* flower. *PLoS Biology*, *7*, e1000090.
Kim, K.-J., Choi, K.-S., & Jansen, R. K. (2005). Two chloroplast DNA inversions originated simultaneously during the early evolution of the sunflower family (Asteraceae). *Molecular Biology and Evolution*, *22*, 1783–1792.
Kim, M., Cui, M., Cubas, P., Gillies, A., Lee, K., Chapman, M. A., et al. (2008). Regulatory genes control a key morphological and ecological trait transferred between species. *Science*, *322*, 1116–1119.
Kloos, W. E., George, C. G., & Sorge, L. K. (2004). Inheritance of the flower types of *Gerbera hybrida*. *Journal of American Society for Horticultural Science*, *129*, 802–810.
Kosugi, S., & Ohashi, Y. (1997). PCF1 and PCF2 specifically bind to *cis* elements in the rice proliferating cell nuclear antigen gene. *The Plant Cell*, *9*, 1607–1619.
Kotilainen, M., Elomaa, P., Uimari, A., Albert, V., Yu, D., & Teeri, T. H. (2000). *GRCD1*, an *AGL2* like MADS box gene, participates in the C function during stamen development in *Gerbera hybrida*. *The Plant Cell*, *12*, 1893–1902.
Koyama, T., Sato, F., & Ohme-Takagi, M. (2010). A role of *TCP1* in the longitudinal elongation of leaves in *Arabidopsis*. *Bioscience, Biotechnology, and Biochemistry*, *74*, 2145–2147.

Kramer, E. M., Dorit, R. L., & Irish, V. F. (1998). Molecular evolution of genes controlling petal and stamen development: duplication and divergence within the APETALA3 and PISTILLATA MADS-box gene lineages. *Genetics*, *149*, 765–783.

Kuang, Q., Li, L., Peng, J., Sun, S., & Wang, X. (2013). Transcriptome analysis of *Gerbera hybrida* ray florets: putative genes associated with gibberellin metabolism and signal transduction. *PLoS One*, *8*, e57715.

Laitinen, R. A. E., Broholm, S., Albert, V. A., Teeri, T. H., & Elomaa, P. (2006). Patterns of MADS-box gene expression mark flower-type development in *Gerbera hybrida (Asteraceae)*. *BMC Plant Biology*, *6*, 11.

Laitinen, R. A. E., Immanen, J., Auvinen, P., Rudd, S., Alatalo, E., Paulin, L., et al. (2005). Analysis of the floral transcriptome uncovers new regulators of organ determination and gene families related to flower organ differentiation in *Gerbera hybrida (Asteraceae)*. *Genome Research*, *15*, 475–486.

Leseberg, C. H., Eissler, C. L., Wang, X., Johns, M. A., Duvall, M. R., & Mao, L. (2008). Interaction study of MADS-domain proteins in tomato. *Journal of Experimental Botany*, *59*, 2253–2265.

Litt, A. (2007). An evaluation of A-function: evidence from the APETALA1 and APETALA2 gene lineages. *International Journal of Plant Science*, *168*, 73–91.

Lundberg, J., & Bremer, K. (2003). A phylogenetic study of the order Asterales using one morphological and three molecular data sets. *International Journal of Plant Sciences*, *164*, 553–578.

Luo, D., Carpenter, R., Copsey, L., Vincent, C., Clark, J., & Coen, E. (1999). Control of organ asymmetry in flowers of Antirrhinum. *Cell*, *99*, 367–376.

Luo, D., Carpenter, R., Vincent, C., Copsey, L., & Coen, E. (1996). Origin of floral asymmetry in Antirrhinum. *Nature*, *383*, 794–799.

Mandel, M. A., & Yanofsky, M. F. (1995). A gene triggering flower formation in *Arabidopsis*. *Nature*, *377*, 522–524.

Marshall, D. F., & Abbott, R. J. (1980). On the frequency of introgression of the radiate (T_r) allele from *Senecio squalidus* L. into *Senecio vulgaris* L. *Heredity*, *45*, 133–135.

Marshall, D., & Abbott, R. (1984). Polymorphism for outcrossing frequency at the ray floret locus in *Senecio vulgaris* L. II: Confirmation. *Heredity*, *52*, 331–336.

Martín-Trillo, M., & Cubas, P. (2010). TCP genes: a family snapshot ten years later. *Trends in Plant Science*, *15*, 31–39.

Melzer, R., Verelst, W., & Theißen, G. (2009). The class E floral homeotic protein SEPALLATA3 is sufficient to loop DNA in 'floral quartet'-like complexes in vitro. *Nucleic Acids Research*, *37*, 144–157.

Mitchison, G. J. (1977). Phyllotaxis and the Fibonacci series. *Science*, *196*, 270–275.

Navaud, O., Dabos, P., Carnus, E., Tremousaygue, D., & Herve, C. (2007). TCP transcription factors predate the emergence of land plants. *Journal of Molecular Evolution*, *65*, 23–33.

Palmer, J. H., & Marc, J. (1982). Wound-induced initiation of involucral bracts and florets in the developing sunflower inflorescence. *Plant and Cell Physiology*, *23*, 1401–1409.

Pelaz, S., Ditta, G. S., Baumann, E., Wisman, E., & Yanofsky, M. F. (2000). B and C floral organ identity functions require SEPALLATA MADS-box genes. *Nature*, *405*, 200–203.

Pickard, B. G. (2007). Delivering force and amplifying signals in plant mechanosensing. In O. P. Hamill (Ed.), *Current Topics in Membranes: Mechanosensitive Ion channels* (Vol. 58, pp. 361–392). New York: Elsevier/Academic Press.

Pozner, R., Zanotti, C., & Johnson, L. A. (2012). Evolutionary origin of the Asteraceae capitulum: Insights from Calyceraceae. *American Journal of Botany*, *99*, 1–13.

Preston, J. C., & Hileman, L. C. (2009). Developmental genetics of floral symmetry evolution. *Trends in Plant Science*, *14*, 147–154.

Preston, J. C., Hileman, L. C., & Cubas, P. (2011). Reduce, reuse, and recycle: developmental evolution of trait diversification. *American Journal of Botany*, *98*, 397–403.

Reinhardt, D., Mandel, T., & Kuhlemeier, C. (2000). Auxin regulates the initiation and radial position of plant lateral organs. *The Plant Cell, 12*, 507–518.

Riechmann, J. L., Krizek, B. A., & Meyerowitz, E. M. (1996). Dimerization specificity of Arabidopsis MADS domain homeotic proteins APETALA1, APETALA3, PISTILLATA, and AGAMOUS. *Proceedings of the National Academy of Sciences of the United States of America, 93*, 4793–4798.

Rijpkema, A. S., Zethof, J., Gerats, T., & Vandenbussche, M. (2009). The petunia AGL6 gene has a SEPALLATA-like function in floral patterning. *The Plant Journal, 60*, 1–9.

Rosin, F. M., & Kramer, E. M. (2009). Old dogs, new tricks: regulatory evolution in conserved genetic modules leads to novel morphologies in plants. *Developmental Biology, 332*, 25–35.

Ruokolainen, S., Ng, Y. P., Albert, V. A., Elomaa, P., & Teeri, T. H. (2010). Large scale interaction analysis predicts that the *Gerbera hybrida* floral E function is provided both by general and specialized proteins. *BMC Plant Biology, 10*, 129.

Ruokolainen, S., Ng, Y. P., Albert, V. A., Elomaa, P., & Teeri, T. H. (2011). Over-expression of the *Gerbera hybrida* At-SOC1-like1 gene Gh-SOC1 leads to floral organ identity deterioration. *Annals of Botany, 107*, 1491–1499.

Ruokolainen, S., Ng, Y. P., Broholm, S. K., Albert, V. A., Elomaa, P., & Teeri, T. H. (2010). Characterization of SQUAMOSA-like genes in *Gerbera hybrida*, including one involved in reproductive transition. *BMC Plant Biology, 10*, 128.

Schwarz-Sommer, Z., Hue, I., Huijser, P., Flor, P. J., Hansen, R., Tetens, F., et al. (1992). Characterization of the Antirrhinum floral homeotic MADS-box gene deficiens: evidence for DNA binding and autoregulation of its persistent expression throughout flower development. *The EMBO Journal, 11*, 251–263.

Schwarz-Sommer, Z., Huijser, P., Nacken, W., Saedler, H., & Sommer, H. (1990). Genetic control of flower development by homeotic genes in *Antirrhinum majus*. *Science, 250*, 931–936.

Shchennikova, A. V., Shulga, O. A., Immink, R., Skryabin, K. G., & Angenent, G. C. (2004). Identification and characterization of four chrysanthemum MADS-Box Genes, belonging to the APETALA1/FRUITFULL and SEPALLATA3 subfamilies. *Plant Physiology, 134*, 1632–1641.

Shulga, O. A., Mitiouchkina, T. Y., Schennikova, A. V., Skryabin, K. G., & Dolgov, S. V. (2011). Overexpression of AP1-like genes from Asteraceae induces early-flowering in transgenic Chrysanthemum plants. *In Vitro Cellular & Developmental Biology-plant, 47*, 553–560.

Shulga, O. A., Shchennikova, A. V., Angenent, G. C., & Skryabin, K. G. (2008). MADS-box genes controlling inflorescence morphogenesis in sunflower. *Russian Journal of Developmental Biology, 39*, 2–5.

Smith, R. S., Guyomarc'h, S., Mandel, T., Reinhardt, D., Kuhlemeier, C., & Prusinkiewicz, P. (2006). A plausible model of phyllotaxis. *Proceedings of the National Academy of Sciences of the United States of America, 103*, 1301–1306.

Soltis, D. E., Chanderbali, A. S., Kim, S., Buzgo, M., & Soltis, P. S. (2007). The ABC model and its applicability to basal angiosperms. *Annals of Botany, 100*, 155–163.

Sommer, H., Beltran, J. P., Huijser, P., Pape, H., Lonnig, W. E., Saedler, H., et al. (1990). Deficiens, a homeotic gene involved in the control of flower morphogenesis in *Antirrhinum majus*: the protein shows homology to transcription factors. *EMBO Journal, 9*, 605–613.

Stuessy, T. F., Spooner, D. M., & Evans, K. A. (1986). Adaptive significance of ray corollas in *Helianthus grosseserratus (Compositae)*. *American Midland Naturalist, 115*, 191–197.

Stuessy, T. F., & Urtubey, E. (2006). Phylogenetic implications of corolla morphology in subfamily Barnadesioideae (Asteraceae). *Flora-Morphology, Distribution, Functional Ecology of Plants, 201*, 340–352.

Sun, M., & Ganders, F. R. (1990). Outcrossing rates and allozyme variation in rayed and rayless morphs of *Bidens pilosa*. *Heredity, 64*, 139–143.

Tähtiharju, S., Rijpkema, A. S., Vetterli, A., Albert, V. A., Teeri, T. H., & Elomaa, P. (2012). Evolution and diversification of the *CYC/TB1* gene family in Asteraceae – a comparative study in gerbera (Mutisieae) and sunflower (Heliantheae). *Molecular Biology and Evolution, 29*, 1155–1166.
Tank, D. C., & Donoghue, M. J. (2010). Phylogeny and phylogenetic nomenclature of the Campanulidae based on an expanded sample of genes and taxa. *Systematic Botany, 35*, 425–441.
Theiβen, G., & Saedler, H. (2001). Floral quartets. *Nature, 409*, 469–471.
Thomas, M. M., Rudall, P. J., Ellis, A. G., Savolainen, V., & Glover, B. J. (2009). Development of a complex floral trait: the pollinator-attracting petal spots of the beetle daisy, *Gorteria diffusa (Asteraceae)*. *American Journal of Botany, 96*, 2184–2196.
Tröbner, W., Ramirez, L., Motte, P., Hue, I., Huijser, P., Lonnig, W. E., et al. (1992). *GLOBOSA*: a homeotic gene which interacts with *DEFICIENS* in the control of *Antirrhinum* floral organogenesis. *The EMBO Journal, 11*, 4693–4704.
Trow, A. H. (1912). On the inheritance of certain characters in the common groundsel - *Senecio vulgaris*, Linn. – and its segregates. *Journal of Genetics, 2*, 239–276.
Turing, A. M. (1952). The chemical basis of morphogenesis. *Philosophical Transactions of the Royal Society of London, Biological Sciences, 237*, 37–72.
Uimari, A., Kotilainen, M., Elomaa, P., Yu, D., Albert, V. A., & Teeri, T. H. (2004). Integration of reproductive meristem fates by a *SEPALLATA*-like MADS box gene. *Proceedings of the National Academy of Sciences of the United States of America, 101*, 15817–15822.
Vieten, A., Sauer, M., Brewer, P. B., & Friml, J. (2007). Molecular and cellular aspects of auxin-transport-mediated development. *Trends in Plant Science, 12*, 160–168.
Wang, H., Jiang, J., Chen, S., Qi, X., Peng, H., Li, P., et al. (2013). Next-generation sequencing of the *Chrysanthemum nankingense* (Asteraceae) transcriptome permits large-scale unigene assembly and SSR marker discovery. *PLoS One, 8*, e62293.
Wang, Z., Luo, Y., Li, X., Wang, L., Xu, S., Yang, J., et al. (2008). Genetic control of floral zygomorphy in pea (*Pisum sativum* L.). *Proceedings of the National Academy of Sciences of the United States of America, 105*, 10414–10419.
Wang, Y., Ma, Y., Fu, J., Qi, S., Ma, H., & Dai, S. (2013). Isolation and functional analysis of the *ClM8-FRUITFULL*-like MADS-box gene from *Chrysanthemum lavandulifolium*. *Scientia Horticulturae, 161*, 125–133.
Wellmer, F., Alves-Ferreira, M., Dubois, A., Riechmann, J. L., & Meyerowitz, E. M. (2006). Genome-wide analysis of gene expression during early *Arabidopsis* flower development. *PLoS Genetics, 2*, e117.
Wolpert, L. (1969). Positional information and the spatial pattern of cellular differentiation. *Journal of Theoretical Biology, 25*, 1–47.
Yu, D., Kotilainen, M., Pöllänen, E., Mehto, M., Elomaa, P., Helariutta, Y., et al. (1999). Organ identity genes and modified patterns of flower development in *Gerbera hybrid (Asteraceae)*. *The Plant Journal, 17*, 51–62.
Zahn, L. M., Kong, H., Leebens-Mack, J. H., Kim, S., Soltis, P. S., Landherr, L. L., et al. (2005). The evolution of the SEPALLATA subfamily of MADS-box genes: a preangiosperm origin with multiple duplications throughout angiosperm history. *Genetics, 169*, 2209–2223.

AUTHOR INDEX

Note: Page numbers with "*f*" denote figures; "*t*" tables.

C

D

G

H

K

L

S

T

U

V

W

SUBJECT INDEX

Note: Page numbers with "*f*" denote figures; "*t*" tables.

G

R

S

COLOUR PLATES

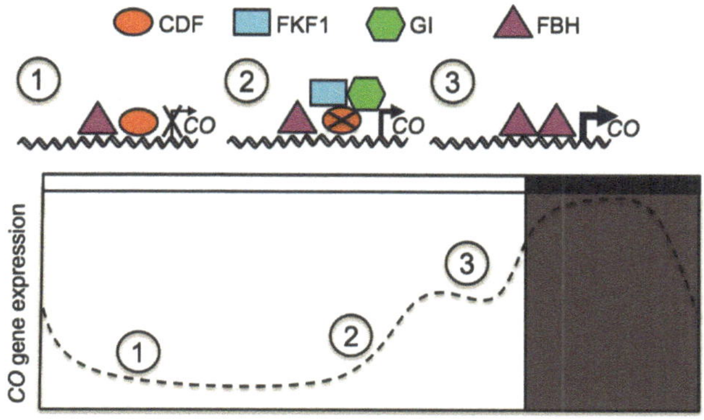

Takato Imaizumi *et al.*, Figure 1.2 *CONSTANS (CO)* oscillatory transcription is dependent on multiple factors throughout the day. Under inductive long-day conditions, the peak of *CO* expression is constrained to the afternoon before dusk. In the morning, CYCLING DOF FACTOR (CDF) family transcription factors bind to the *CO* promoter to repress its transcription. Beginning in the afternoon, FLAVIN-BINDING, KELCH REPEAT, F-BOX 1 (FKF1) and GIGANTEA (GI) form a protein complex that ubiquitinates CDFs through FKF1 and targets them for proteasomal degradation, releasing the *CO* promoter from repression. FLOWERING BHLH (FBH) transcriptional activators are then recruited to the *CO* genomic locus, resulting in increased transcription of *CO* before dusk. Constraining *CO* mRNA expression to the late afternoon and stabilisation of resultant CO protein result in *FLOWERING LOCUS T* expression at dusk and promotion of flowering in long days.

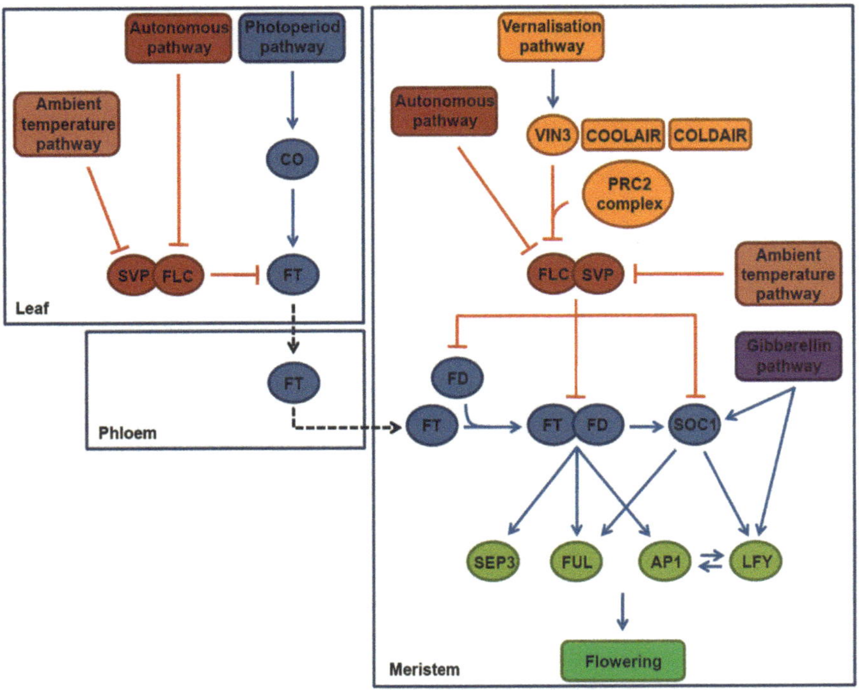

Sibum Sung *et al.*, Figure 2.1 Flowering time pathways in *Arabidopsis*. Timing of flowering is controlled by the integration of various flowering pathways that incorporate environmental and developmental cues. There are five major flowering pathways in *Arabidopsis*. In the photoperiod pathway, CO activates the transcription of *FLOWERING LOCUS T (FT)* in response to inductive long days in the leaf. FT protein moves to the shoot apical meristem (SAM) via the phloem. In the SAM, FT protein interacts physically with FD protein. The FT–FD complex promotes the expression of *SUPPRESSOR OF OVER-EXPRESSION OF CONSTANS 1 (SOC1)* and several other floral meristem identity genes, including *SEP3*, *FUL*, *AP1* and *LFY*. In the vernalisation pathway, *VERNALIZATION INSENSITIVE 3* and two long noncoding RNAs (COOLAIR and COLDAIR) are induced at various times during exposure to cold temperatures. *FLOWERING LOCUS C (FLC)* is negatively regulated by the autonomous pathway. Thus, both the vernalisation and autonomous pathways converge to repress *FLC*. FLC protein physically interacts with SVP protein and the FLC–SVP complex represses the expression of the floral integrator genes, such as FT, FD and SOC1 in the leaf and SAM. In the ambient temperature pathway, SVP protein is accumulated under cooler temperatures and represses the expression of floral integrator genes and delays flowering. The gibberellin (GA) pathway promotes flowering through the activation of *SOC1* and *LFY*. Arrows indicate the positive regulation and bars indicate the negative regulation.

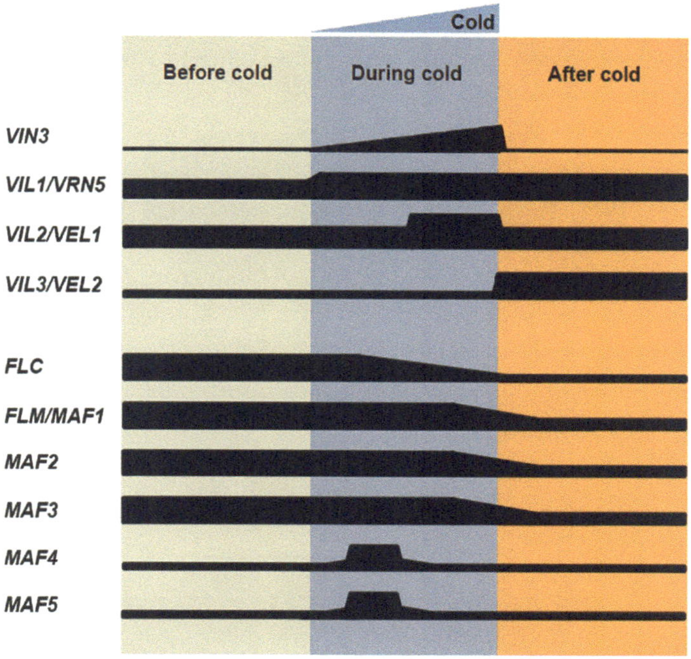

Sibum Sung *et al.*, Figure 2.2 Vernalisation-mediated changes in expressions of *VERNALIZATION INSENSITIVE 3 (VIN3)* and *FLOWERING LOCUS C (FLC)* gene families. All members of the VIN3 gene family are differentially expressed over the course of vernalisation. VIN3 is only expressed during cold. *VIN3-LIKE 1* (VIL1) and VIL2 are rather constitutively expressed with a small increase during cold. VIL3 is induced after cold. All members of the FLC gene family are also differentially expressed during the course of vernalisation. FLC, *FLOWERING LOCUS M, MADS AFFECTING FLOWERING 2 (*MAF2) and MAF3 are repressed by vernalisation. The expression of MAF4 and MAF5 is transiently increased during early periods of cold exposure but eventually reduced to the basal level when cold persists. The VIN3 gene family is required for the regulation of expression of the FLC gene family by vernalisation.

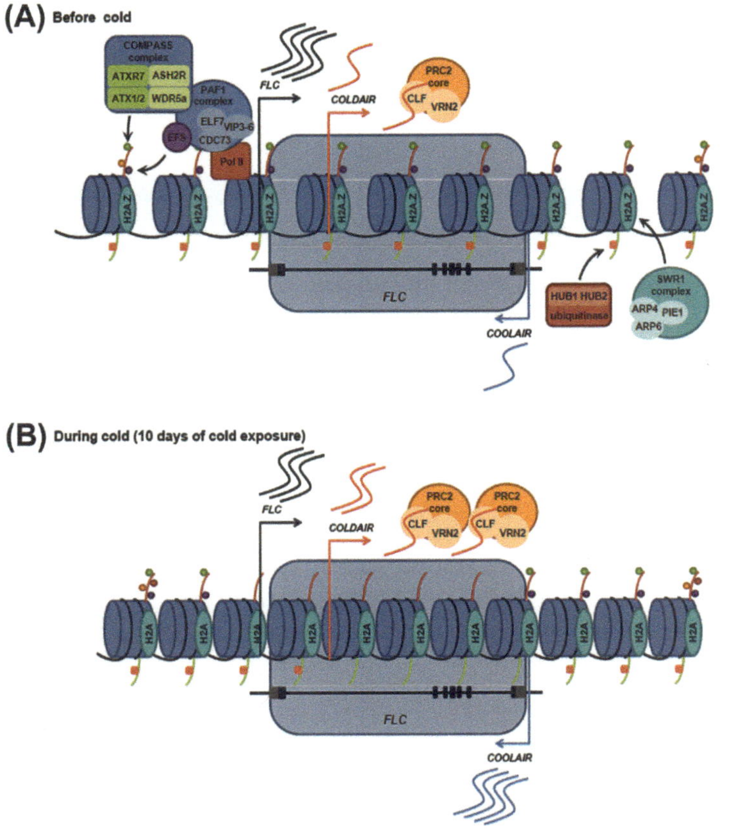

Sibum Sung *et al.,* Figure 2.3 The vernalisation-mediated changes at *FLC* chromatin. (A) Before cold, *FLOWERING LOCUS C (*FLC) is highly expressed in winter-annual strains of *Arabidopsis*. H3 Lys 4 trimethylation (H3K4me3) and H3K36me3 are enriched at FLC chromatin. In addition, H2B mono-ubiquitination (H2Bub1) and H2A.Z are enriched at FLC chromatin when FLC is actively transcribed. Low levels of H3K27me3 and Polycomb Repressive Complex 2 (PRC2) enrichment at FLC chromatin are detected. COOLAIR and COLDAIR are expressed at low levels. (B) During cold, COOLAIR expression is rapidly increased and reaches a peak within 14 days of cold exposure.

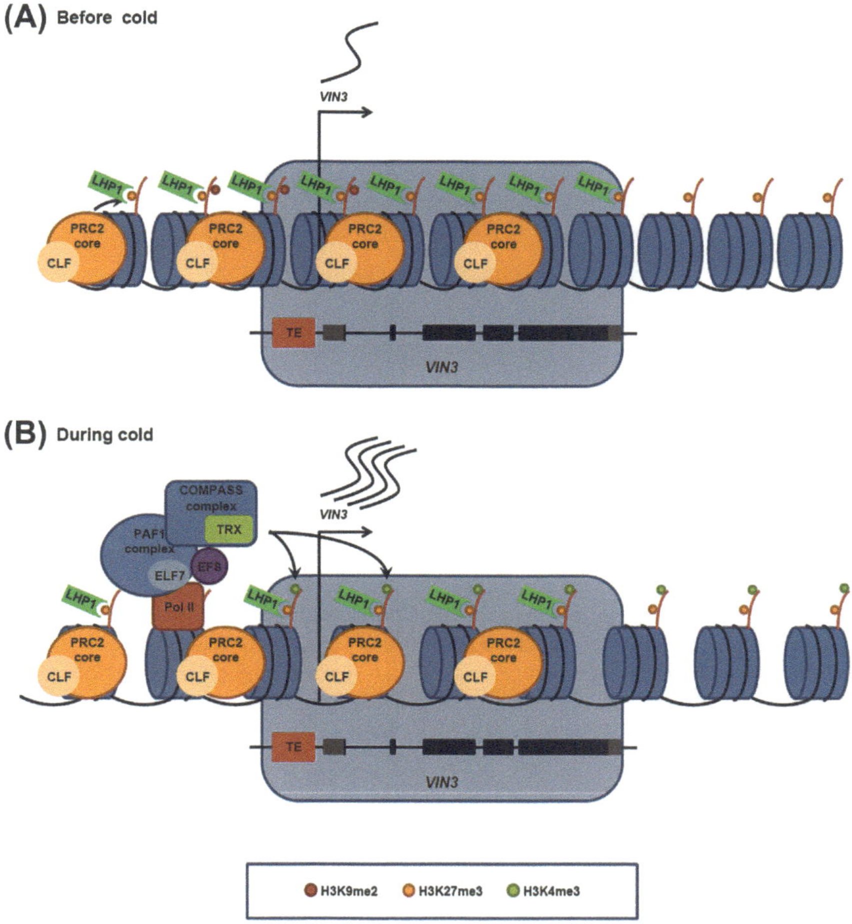

Sibum Sung *et al.*, Figure 2.4 Changes at *VIN3* chromatin during the course of vernalisation. (A) Before cold, VIN3 is expressed at a very low level. Polycomb Repressive Complex 2 (PRC2) and LIKE HETEROCHROMATIN PROTEIN 1 (LHP1) are associated with VIN3 chromatin. VIN3 chromatin is enriched with repressive histone marks, H3 Lys9 dimethylation (H3K9me2) and H3 Lys 27 trimethylation (H3K27me3). In addition, a transposable element (TE)-derived sequence is present at the VIN3 promoter region and H3K9me2 is enriched around the TE region. (B) During cold, H3K9me2 decreases whereas an active histone mark, H3K4me3, increases at the transcription start site of VIN3. Polymerase II-associated factor 1 (PAF1) complex and *EARLY FLOWERING IN SHORT DAYS (EFS)* are necessary for the fullest extent of VIN3 induction. PRC2 and LHP1 are still associated with VIN3 chromatin when VIN3 is induced. After cold, H3K4me3 is decreased while H3K9me2 is increased again at VIN3 chromatin.

Markus Schmid *et al.,* Figure 3.1 Effect of mutations in gibberellic acid (GA) biosynthesis and signalling genes on flowering in *Arabidopsis thaliana*. Wild type *Arabidopsis thaliana* (L*er*-1), GA signalling mutant lacking four DELLA proteins (*ga1-3 gai-t6 rga-t2 rgl1-1 rgl2-1*) and the GA biosynthesis mutant *ga1-3* grown under long day photoperiod at 23 °C. The *ga1-3* mutant flowers slightly late compared to L*er*-1 plants, while *ga1-3 gai-t6 rga-t2 rgl1-1 rgl2-1* flowers early.

Markus Schmid *et al.,* Figure 3.3 *Arabidopsis thaliana* plants impaired in trehalose-6-phosphate synthesis are late flowering. The *tps1* mutant carrying the chemically inducible rescue construct *pGVG::TPS1* flowers extremely late under inductive photoperiodic conditions compared to control plants. Depicted are 20 (Col-0) and 50 (*tps1 pGVG::TPS1*)-day-old plants grown under long day at 23 °C. *Picture credit: Jathish Ponnu.*

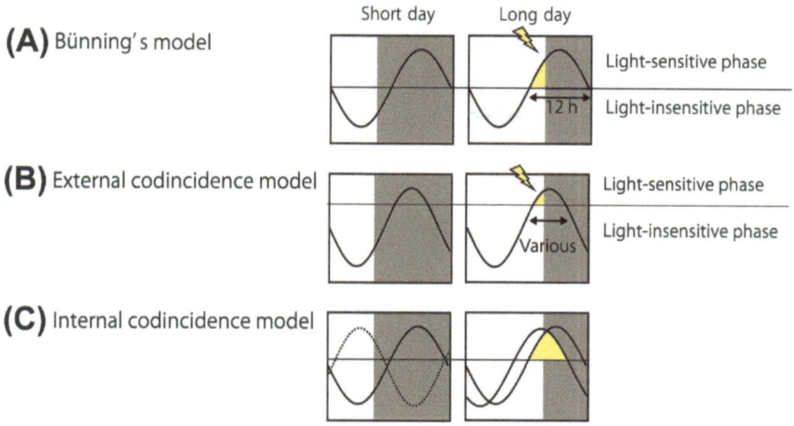

Takeshi Izawa *et al.*, Figure 4.1 The models for day-length recognition. (A) Bünning's model. The light-sensitive phase is set by the circadian clock during the second half of the day (evening and night). When the light signal is present during the light-sensitive phase, plants perceive it as long day. (B) The external coincidence model. As in Bünning's model, the light-sensitive phase is set by the circadian clock, and the light signal during the light-sensitive phase is perceived as long day, but the light-sensitive phase is considerably shorter than in Bünning's model. (C) The internal coincidence model. Two distinct diurnal rhythms are formed by the circadian clock, or circadian clock and other diurnal rhythms coexist. When the two rhythms overlap, the day length is recognised.

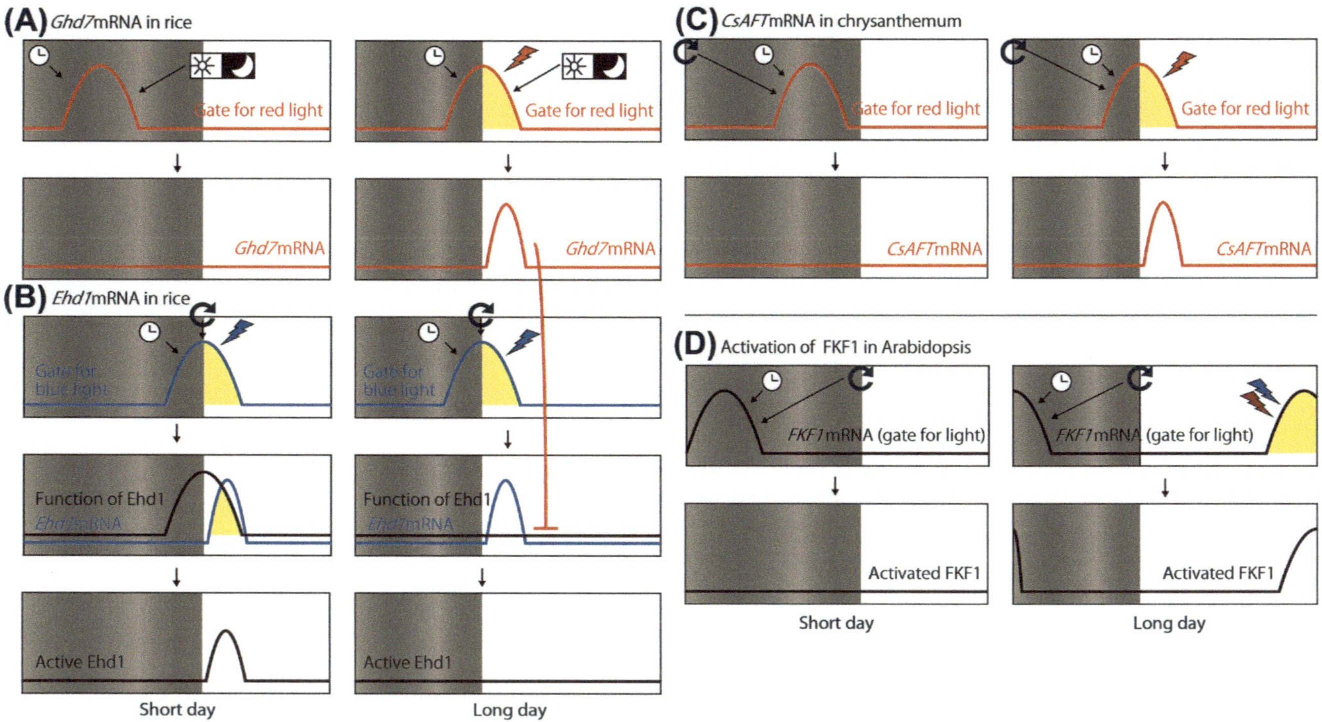

(A) Ghd7mRNA in rice
Gate for red light
Ghd7mRNA
(B) Ehd1mRNA in rice
Gate for blue light
Function of Ehd1
Ehd1mRNA
Active Ehd1
Short day
Long day
(C) CsAFTmRNA in chrysanthemum
Gate for red light
CsAFTmRNA
(D) Activation of FKF1 in Arabidopsis
FKF1mRNA (gate for light)
Activated FKF1
Short day
Long day

Takeshi Izawa *et al.*, Figure 4.2 Gating mechanisms for day-length recognition. (A) The gate for red light to induce *Ghd7* mRNA in rice (Itoh et al., 2010). The *Ghd7* gate is entrained by day length and is open around midnight under short-day conditions and around dawn under long-day conditions. Red light perception by phytochromes coincident with the gate opening time under long-day conditions results in *Ghd7* mRNA induction. (B) The gate for blue light to induce *Ehd1* mRNA in rice (Itoh et al., 2010). The *Ehd1* gate is entrained by the circadian clock, and is open around dawn regardless of day length. As a result, *Ehd1* mRNA is induced at dawn, but is repressed by Ghd7 only under long-day conditions. Moreover, Ehd1 activity promotes *Hd3a* expression only in the morning under short-day conditions. Thus, when *Ehd1* mRNA and the Ehd1 activity are coincident at dawn under short-day conditions, *Hd3a* mRNA is induced. (C) The gate for red light to induce *CsAFT* mRNA in *Chrysanthemum* (Higuchi et al., 2013). The *CsAFT* gate is entrained by the circadian clock or light-to-dark transition, and is open around midnight under short-day conditions and around dawn under long-day conditions. Under long-day conditions, red light signals are perceived when the gate for *CsAFT* expressions opens, and *CsAFT* mRNA is induced. (D) The gate for blue light to photoactivate *FKF1* in *Arabidopsis* (Imaizumi et al., 2003). The peak of the diurnal rhythm of *FKF1* mRNA is around dusk, and is similar under long- and short-day conditions. Under long-day conditions, the *FKF1* mRNA level starts to increase before dusk, whereas under short-day conditions it increases after dusk. Therefore, the FKF1 protein synthesis can be coincident with the blue light signal under long-day conditions, and is followed by FKF1 photoactivation. The clock symbols indicate entrainment by the circadian clock, the circular arrows indicate light transitions, the sun/moon symbol indicates day-length information and the red and blue lightning symbols indicate red or blue light signalling, respectively.

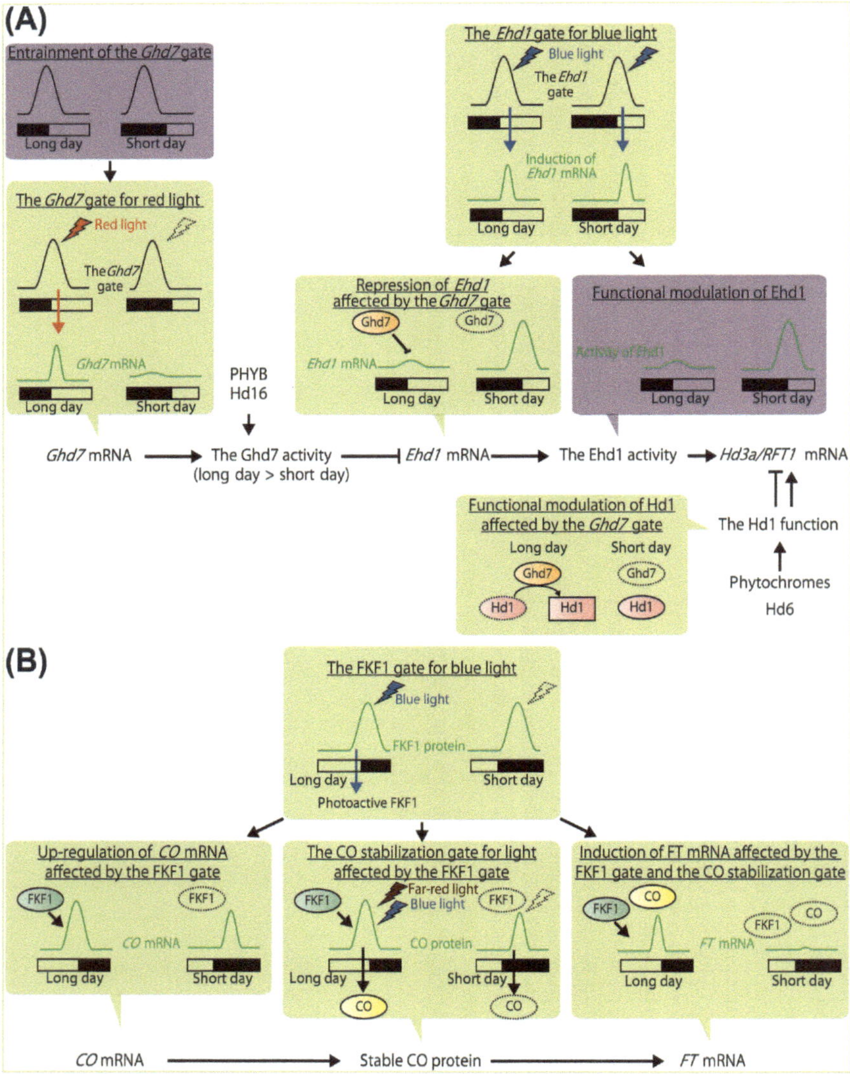

Takeshi Izawa *et al.*, Figure 4.3 Multiple molecular mechanisms for day-length recognition in (A) rice and (B) *Arabidopsis*. Green rectangles show responses to day length regulated by gating mechanisms or their direct effects. Blue rectangles show mechanisms whose regulation is currently unknown. Black waves represent diurnal rhythms of gates. Green waves represent diurnal rhythms of mRNA expression or protein activity. Circles show proteins. Dashed circles indicate proteins showing reduced activity or abundance. Circled and squared *Hd1* indicate that the protein promotes or represses *Hd3a* expression, respectively.

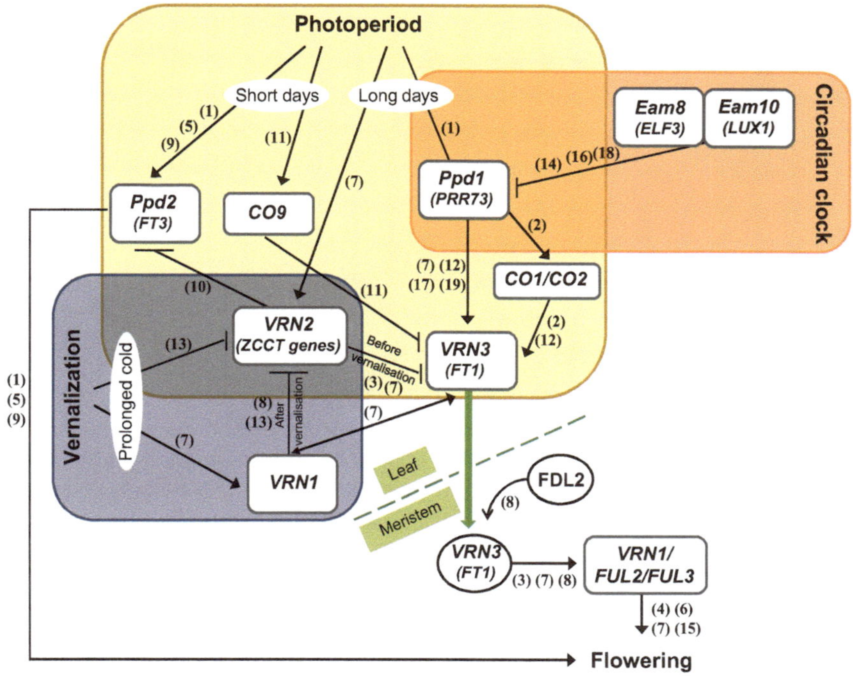

Maria von Korff and Chiara Campoli, Figure 5.1 Model of flowering time control pathways in wheat and barley. The different external and internal cues are highlighted in different colours. Positive and negative regulatory actions are indicated by arrows and lines with bars, respectively. Boxes indicate genes, while circles indicate proteins. The green arrow shows that the FT1 protein moves from the leaf to the meristem. The figure incorporates different aspects of previously published wheat and barley models. Numbers in brackets indicate literature in which experimental evidences support the model: (1) Laurie et al. (1995); (2) Turner et al. (2005); (3) Yan et al. (2006); (4) Adam et al. (2007); (5) Faure et al. (2007); (6) Shitsukawa et al. (2007); (7) Hemming et al. (2008); (8) Li and Dubcovski (2008); (9) Kikuchi et al. (2009); (10) Casao, Iguarta, et al. (2011); (11) Kikuchi et al. (2011); (12) Campoli, Drosse, et al. (2012); (13) Chen and Dubcovski (2012); (14) Faure et al. (2012); (15) Kinjo et al. (2012); (16) Mizuno et al. (2012); (17) Shaw et al. (2012); (18) Campoli et al. (2013); (19) Shaw et al. (2013).

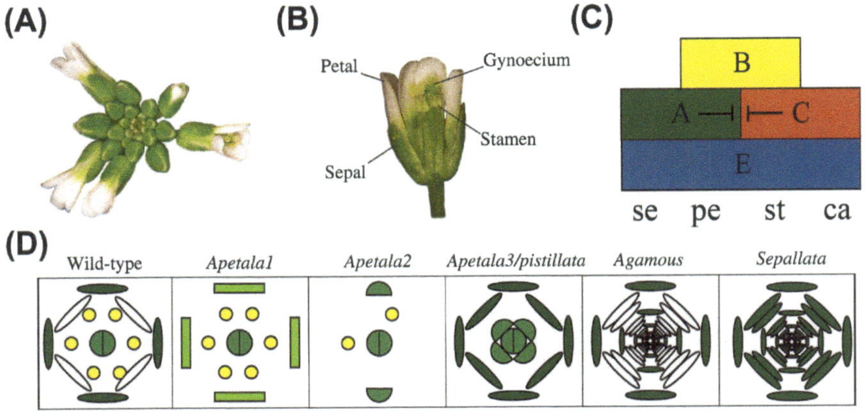

Frank Wellmer *et al.*, Figure 6.1 Overview of flower development and the ABCE model. (A) Top view on an *Arabidopsis* wild-type inflorescence. Because floral primordia are sequentially initiated from the inflorescence meristem in a spiral pattern, no two flowers in an inflorescence are at the exact same stage of development (Smyth et al., 1990). (B) Four distinct organ types are present on *Arabidopsis* flowers. These organ types are arranged in concentric circles termed whorls. Typically, four sepals are present in the outermost whorl (the first whorl), four petals are present in the second whorl, six stamens are present in the third whorl and two fused carpels, which form the gynoecium that houses the ovules, are present in the fourth whorl (Smyth et al., 1990). (C) The ABCE model states that the overlapping activities of four classes of homeotic genes specify the four types of floral organs. A and E class genes are required for sepal identity; A, B and E class genes are required for petal identity; B, C and E class genes specify stamens; and C and E class genes specify carpels (Coen & Meyerowitz, 1991; Krizek & Fletcher, 2005). (D) Schematic of the phenotypes of the floral homeotic mutants. In strong *ap1* alleles, sepals are transformed into bract-like organs while petals are mostly absent (Irish & Sussex, 1990). In strong *ap2* alleles, sepals are transformed to carpels, while petals are absent and stamen numbers are reduced (Bowman et al., 1991). Strong *ap3* and *pi* alleles have sepals in place of petals and carpels in place of stamens (Bowman et al., 1991; Jack et al., 1992). Strong mutant alleles of the C function gene *AG* have petals in place of stamens and sepals in place of carpels while the floral meristem fails to terminate resulting in the indefinite reiteration of sepals and petals (Bowman et al., 1989). Quadruple *sep1 sep2 sep3 sep4* flowers reiterate leaf-like organs indefinitely (Ditta et al., 2004; Pelaz et al., 2000).

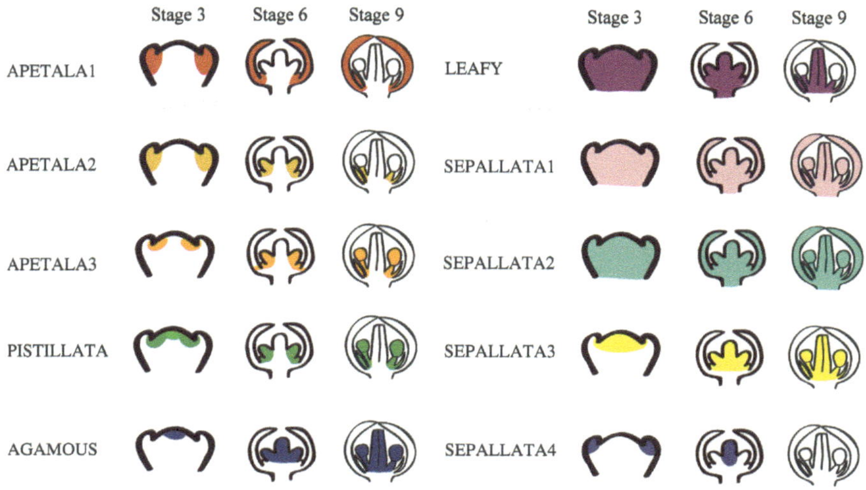

Frank Wellmer *et al.*, Figure 6.2 Expression patterns of key floral regulators at three stages of flower development. *AP1* is expressed in the outer two floral whorls at stages 3, 6 and 9 (Mandel et al., 1992). *AP2* is expressed in the outer two whorls at stage 3. By stage 6, *AP2* is expressed throughout the second and third whorls before being restricted from the anthers at stage 9 (Wollmann et al., 2010). *AP3* and *PI* are expressed throughout the second and third whorls from stage 3 to stage 9 (Goto & Meyerowitz, 1994; Jack et al., 1992). *AP3* is expressed at the base of sepals at stage 6 (Jack et al., 1992), while *PI* is expressed at low levels in the centre of stage 3 floral meristems (Goto & Meyerowitz, 1994). *AG* is expressed throughout the third and fourth whorls from stages 3–9 (Gustafson-Brown et al., 1994; Yanofsky et al., 1990). *SEP1* and *SEP2* are expressed throughout the flower from stage 3–9 (Flanagan & Ma, 1994; Savidge et al., 1995). *SEP3* is expressed in the inner three whorls from stage 3 to 9 (Mandel & Yanofsky, 1998). *SEP4* is expressed in the first whorl at stage 3, the fourth whorl at stage 6, and expression is not detected at stage 9 (Ditta et al., 2004). The floral meristem identity gene *LFY* is expressed in all whorls at stage 3 and 6 before it becomes restricted to petals and stamen filaments at stage 9 (Weigel et al., 1992).

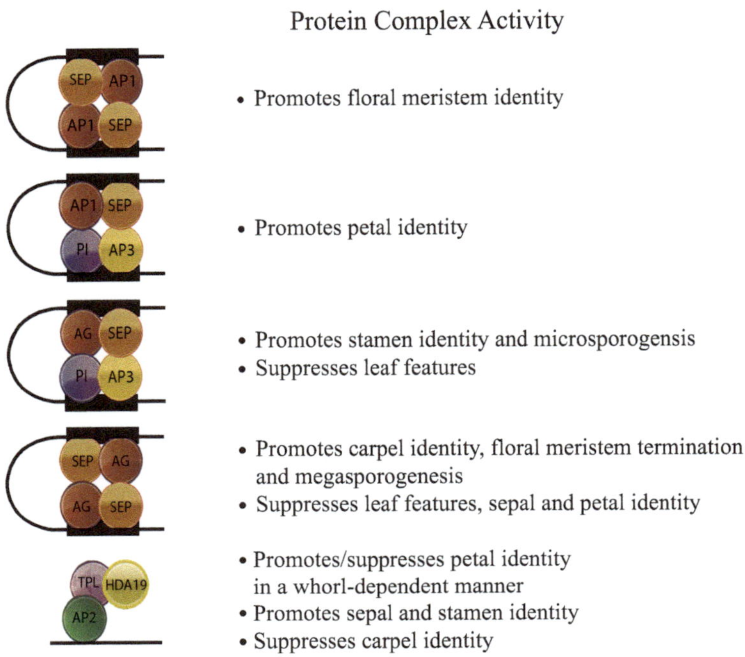

Frank Wellmer *et al.,* Figure 6.3 Functions of the complexes formed by the floral organ identity factors. The MADS domain-containing floral organ identity factors can form higher-order complexes, termed floral quartets, that are thought to bind two CArG boxes simultaneously via DNA looping to regulate their downstream targets (Honma & Goto, 2001; Riechmann, Krizek et al., 1996). Functions can be ascribed to these putative complexes based on molecular and genetic analysis (as reviewed in (Krizek & Fletcher, 2005; O'Maoileidigh et al., 2014)). The non-MADS domain-containing complex AP2–TPL–HDA19 transcriptionally regulates *SEP3*, *AP3* and *AG* to control floral patterning (Krogan et al., 2012).

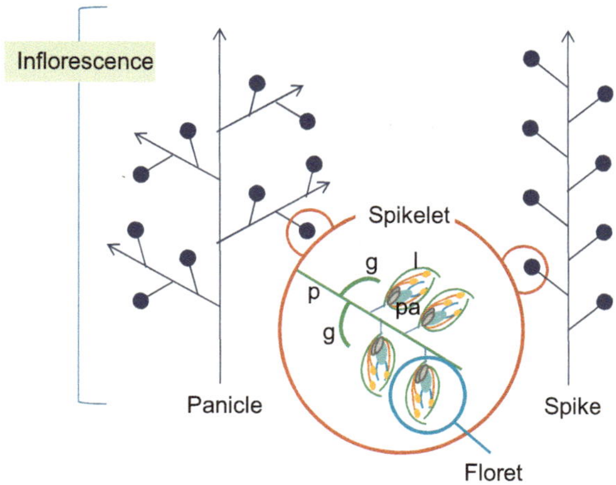

Junko Kyozuka, Figure 7.1 Structure of the grass inflorescence. A schematic representation of a grass inflorescence is shown. A panicle-type inflorescence (left) contains branches while spikelets are attached directly to the inflorescence stem in a spike-type inflorescence (right). A spikelet is subtended by a pair of suppressed leaves called glumes. A flower (floret) contains a lemma, palea, lodicules, stamens and a pistil. Closed circle, spikelet; p, pedicel; g, glume; l, lemma; pa, palea.

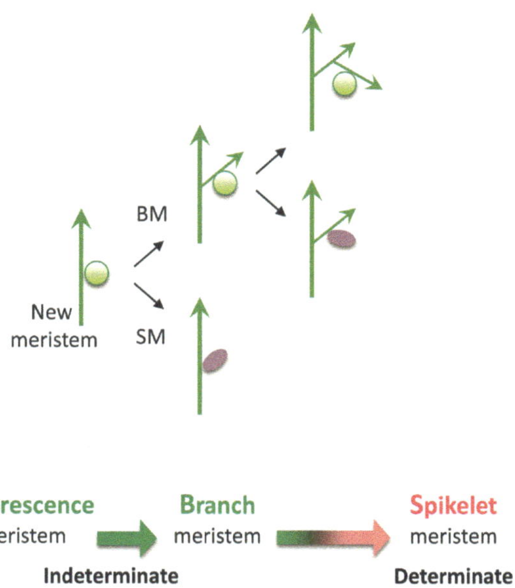

Junko Kyozuka, Figure 7.2 Concept of meristem identity specification. A new meristem (green circle) acquires the identity of a indeterminate branch meristem (BM, green arrow) or a determinate spikelet meristem (SM, purple oval). The BM further produces new meristems that acquire either SM or BM identities.

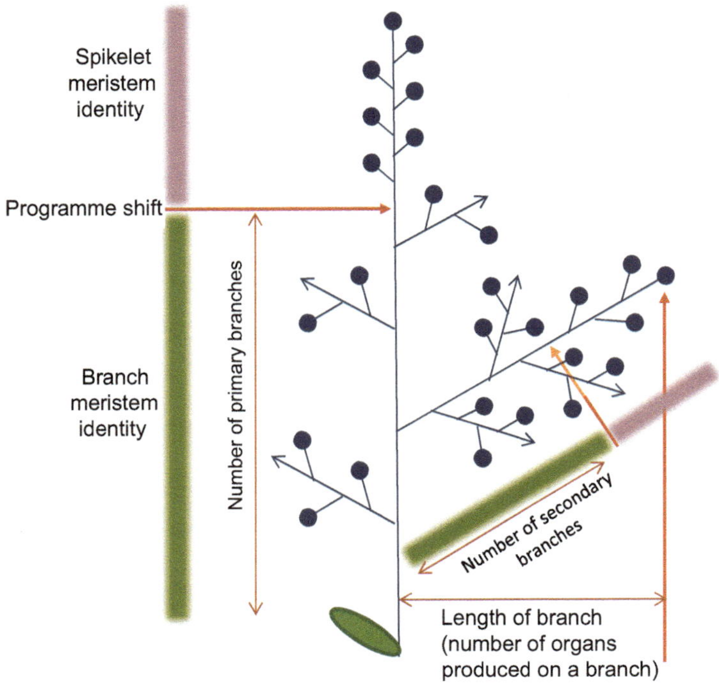

Junko Kyozuka, Figure 7.3 Basic concept of grass inflorescence architecture. The pattern of inflorescence branching is determined by the timing of the meristem phase shift from the BM to the SM. A new meristem acquires the identity of a branch meristem or a spikelet meristem. Among meristems produced by a branch meristem, early lateral meristems acquire a BM identity and grow as next order branches at proximal positions (green region), while later lateral meristems are specified as spikelet meristems (purple region). The timing of the programme shift determines the number of branches and the shape of the branch. Timing of the transition to spikelet identity at the top of each branch determines the total number of organs produced on the branch.

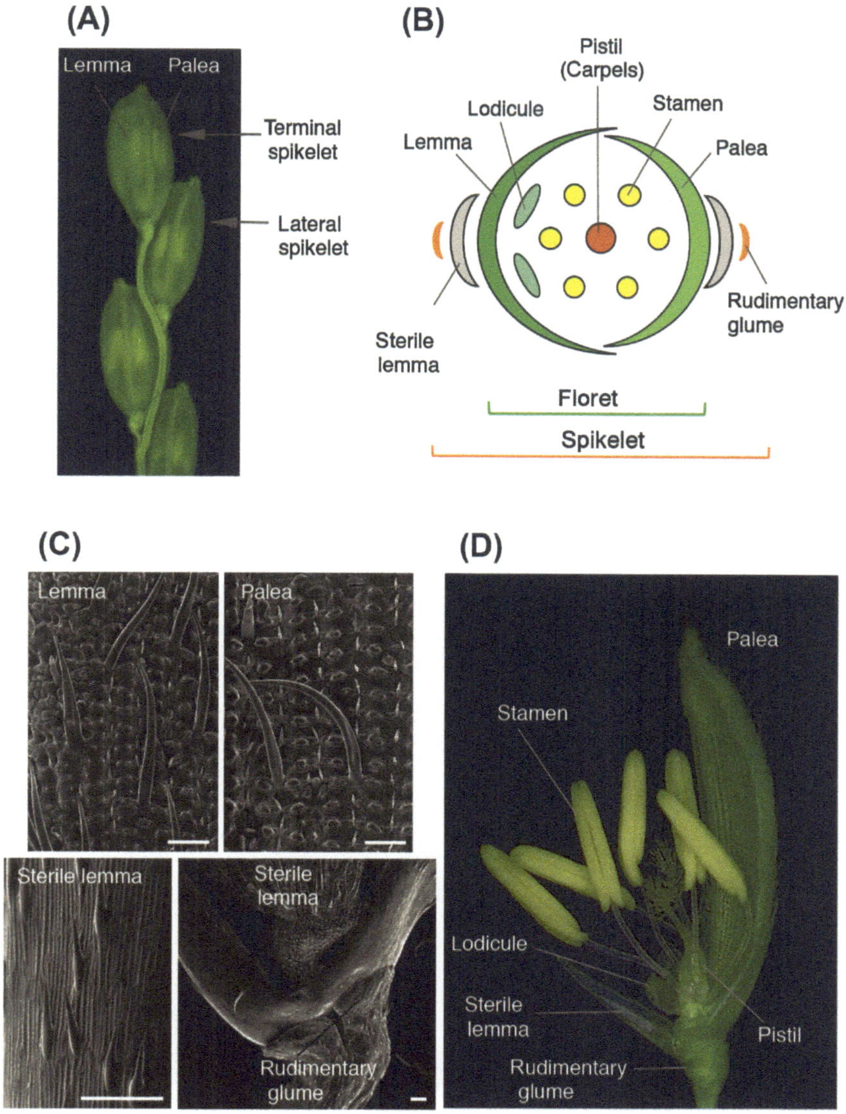

Hiro-Yuki Hirano *et al.,* Figure 8.1 Spikelet and flower structure of rice. (A) An apical region of the primary branch producing the terminal and lateral spikelets. (B) Schematic representation of the rice spikelet. (C) Epidermal abaxial surface of the lemma, palea and sterile lemma, and the basal region of the spikelet. The lemma and palea have a rough abaxial surface with many tubercles, whereas the abaxial surface of the sterile lemma is smooth. (D) Flower and spikelet organs. The lemma and a part of the palea are removed. Scale bars: 100 μm.

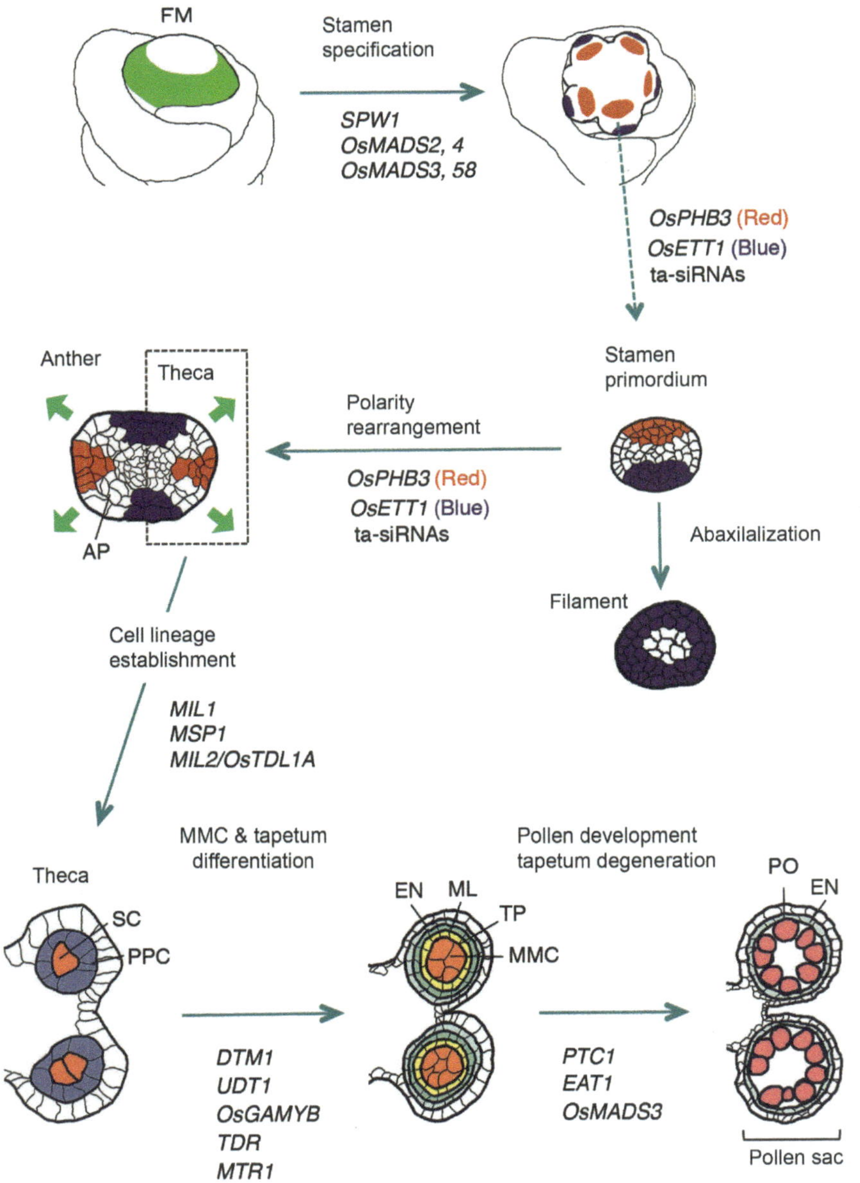

Hiro-Yuki Hirano *et al.*, Figure 8.4 Stamen development and anther differentiation. Stamen initiation is specified by the combination of class B genes (*SPW1, OsMADS2, OsMADS4*) and class C genes (*OsMADS3, OsMADS58*). The adaxial–abaxial polarity is initially established in the stamen primordia in a similar way as in the leaf primordia, and then the polarity is rearranged and re-established in the theca as a developmental unit. The cells in the area between the regions of adaxial and abaxial identity (marked by expression of *OsPHB3* and *OsETT1*, respectively) proliferate and differentiate into pollen sacs. AP, archesporial cells; EN, endothecium; FM, flower meristem; ML, middle layer; MMC, microspore mother cell; PO developing pollen; PPC, primary parietal cell; SC, sporogenous cell; TP, tapetum.

Beth Thompson, Figure 9.1 Mature maize inflorescences. (A) Normal maize plant. The male inflorescence, the tassel, is formed at the apex of the plant, and the female inflorescence, the ear, is formed in the axil of a leaf. (B) Mature tassel. (C) Mature ear. (D) Mature spikelet pair from a tassel. (E) Tassel spikelet opened to expose two male florets (glumes have been removed). (F) Close-up of ear spikelets that each contain a single female floret.

Paula Elomaa *et al.*, Figure 10.1 Flower heads in *Gerbera hybrida* (top) and *Helianthus annuus* (below) are compressed structures comprised of distinct flower types. The marginal ray flowers (RF) are large and bilaterally symmetrical. The gerbera ray flowers are female while those in sunflower are neutral in sex. The ray flowers have showy ventral ligule (vl) formed by three fused petals while the two dorsal petals are rudimentary (dp). Gerbera has intermediate trans flowers (TF) similar to ray flowers but just smaller in size. In both species, the central disc flowers (DF) are hermaphrodite and bear carpels and pollen-producing stamens. In gerbera the disc flowers are gradually changing from bilateral symmetry towards radial symmetry the more central their position in the capitulum is, while in sunflower all the disc flowers are fully radially symmetrical (tubular).

Paula Elomaa *et al.*, Figure 10.3 Transgenic line that ectopically expresses the gerbera CYC-like gene GhCYC2 (top) highly resembles a gerbera cultivar showing the 'crested' or 'double' phenotype with enlarged, male sterile trans and disc flowers (below).

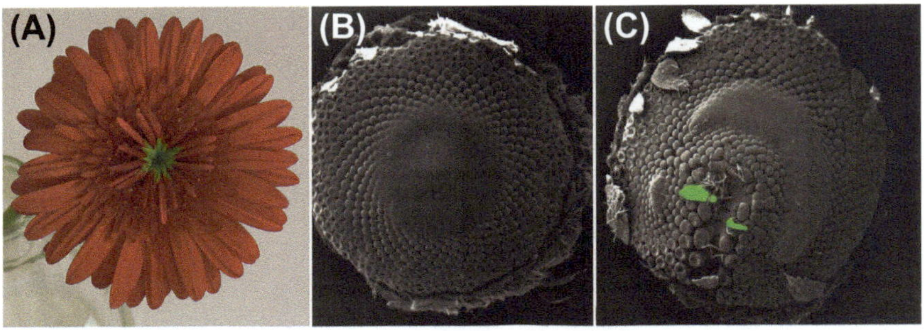

Paula Elomaa *et al.*, Figure 10.4 (A) A gerbera cultivar displaying ectopic involucral bracts and ray flowers in the centre of the inflorescence. (B) Scanning electron microscopy (SEM) image of a wild type undamaged capitulum of gerbera showing normal initiation of flower primordia. (C) SEM image of a wounded wild type capitulum shows that wounding creates new margins that re-initiate bracts (coloured in green) and induce flower primordia in an abnormal pattern. Scale bars 1 mm.

FSC
www.fsc.org
MIX
Papier | Fördert
gute Waldnutzung
FSC® C083411

Kolibri 360 GmbH
Ferdinand-Jühlke-Straße 7
99095 Erfurt, Deutschland
produktsicherheit@kolibri360.de